PROCEEDINGS OF THE 1ST INTERNATIONAL CONGRESS ON SIGNAL AND INFORMATION PROCESSING, NETWORKING AND COMPUTERS (ICSINC 2015), BEIJING, CHINA, 17–18 OCTOBER 2015

Signal and Information Processing, Networking and Computers

Editors

Na Chen & Tingting Huang
Beijing University of Posts and Telecommunications, China

CRC Press is an imprint of the
Taylor & Francis Group, an **informa** business

A BALKEMA BOOK

CRC Press/Balkema is an imprint of the Taylor & Francis Group, an informa business

© 2016 Taylor & Francis Group, London, UK

Typeset by V Publishing Solutions Pvt Ltd., Chennai, India

Published by: CRC Press/Balkema
P.O. Box 11320, 2301 EH Leiden, The Netherlands
e-mail: Pub.NL@taylorandfrancis.com
www.crcpress.com – www.taylorandfrancis.com

ISBN: 978-1-138-02881-4 (Hbk)
ISBN: 978-1-315-64880-4 (eBook PDF)

Table of contents

Preface   ix

The first international conference on signal and information processing,
networking and computers

Pilot decontamination in a massive MIMO system: Exploiting pilot sequence hopping   3
Y. Li, X. Jing, S. Sun & H. Huang

A novel 3D multivariate data visualization method   9
F. Hou, H. Huang & X. Jing

A fast CU depth decision algorithm based on learning how to use machines   17
L. Wang, X. Jing, S. Sun & H. Huang

Kernel-based learning technique for Cooperative Spectrum Sensing in Cognitive Radio   27
X. Chen, X. Jing, S. Sun, H. Huang, X. Wang & D. Chen

Future mobile network: A framework for enabling SDN in mobile operator networks   35
X. Xia, Z. Sun, H. Zhao, F. Shi, C. Mao, Q. Sun, L. Cao, B. Wang,
S. Zhang & A. de Boer

A novel Continuous Action-set Learning Automaton algorithm   43
Y. Guo, H. Ge, Y. Yan, Y. Huang & S. Li

Full-Duplex Relay based on GMD block diagonalization in MIMO relay systems   49
Y. Wang, X. Zhang & D. Yang

An effective and scalable algorithm for hybrid recommendation based on Learning To Rank   59
P. He, H. Yuan, J. Chen & C. Zhao

A Joint Source Channel Coding scheme based on environment vector clustering
over WiFi channels   69
T. Huang, S. Sun, Y. Guo, N. Chen & Z. Zhou

A robust topology control algorithm for channel allocation in the Cognitive
Radio Ad Hoc networks   77
Y. Chen, X. Jing & H. Huang

Adaptive spread spectrum audio watermarking based on data redundancy analysis   83
R. Li, S. Xu, B. Rong & H. Yang

A graph coloring based resource allocation in Heterogeneous Networks   91
Z. Ye, X. Jing & H. Huang

Time drift detection in process mining   99
H. Che, Q. Machu & Y. Zhou

Spectrum sensing for cognitive radio systems with unknown non-zero-mean noise   109
M. Sun, T. Tian, C. Zhao & B. Li

A secure fuzzy-based cluster head election algorithm for Wireless Sensor Networks   115
J. Shi & X.N. Han

Using game theory to optimize GOP level rate control in HEVC 123
J. Zhu, S. Sun & Y. Guo

Reinforcement learning based cooperative sensing policy in Cognitive Radio network 131
X. Ye, X.J. Jing, S. Sun, Y. Li, X. Wang & D. Cheng

A combined BD-SO precoding scheme for next generation HetNets 139
D. Lu, J. Sun & C. Qian

Secure transmission with Artificial Noise in heterogeneous massive MIMO network 147
C. Li, S. Sun, W. Liu & H. Huang

User scheduling algorithm based on Null-Space precoding scheme for
heterogeneous networks 157
Q. Zhou, C. Fu & M. Lu

Risk assessment method of multidimensional AHP based on SoS architecture 167
Y. Huang, H. Yan, Y. Zheng & C. Zhao

Line outage identification based on partial measurement of
Phasor Measurement Unit (PMU) 175
J. Guo, T. Yang, H. Feng & B. Hu

Mitigating Primary User Emulation attacks in Cognitive Radio networks
using advanced encryption standard 185
H. Jiang, X.J. Jing, S. Sun, H. Huang, Y. Li, X. Wang & D. Cheng

Research and realization of music recommender algorithm based on
hybrid collaborative filtering 195
H. Che & Z. Wang

Research of AdaBoost robustness based on Learning Automata 203
S. An, Y. Guo, X. Guo, Y. Yan & S. Li

On applying Confidence Interval Estimator to the pursuit learning schemes: Various
algorithms and their comparison 209
H. Ge, J. Li, S. Li, Y. Yan & Y. Huang

Modified HARQ and ARQ for LTE broadcast 215
X.P. Zhu, D.C. Yang, J. Wang, X.X. Zhang & J.Q. Zhang

Secure transmission with Artificial Noise in massive MIMO systems 223
Z. Ma, S. Sun & X. Zhang

Parameter estimation of 60-GHz millimeter wave communications using
the nonlinear Power Amplifier based on PSO algorithm 233
M. Zhang, X. Sun, Z. Zhou, Y. Xing & Z. Zhang

The ICSINC2015 workshop on telecom big data based research and application

Analytical method of gridding user perception and market development
based on big data 243
M. Li, J. Yuan & J. Xu

WCDMA data based LTE site selection scheme in LTE deployment 249
L. Xu, Y. Luan, X. Cheng, X. Cao, K. Chao, J. Gao, Y. Jia & S. Wang

A novel LTE network deployment scheme using telecom big data 261
T. Zhang, X. Cheng, L. Xu, X. Cao, M. Yuan & Y. Wang

Five-dimension labeled 4G user immigration model based on big data analysis 271
H. Zhang, L. Zhang, C. Song & X. Cheng

Telecom big data based investment strategy of value areas 281
M. Mu, Y. Wang & W. Chen

A Coverage of Self-Optimization Algorithm using big data analytics
in WCDMA cellular networks 289
J. Gao, X. Cheng, L. Xu, L. Cao & K. Chao

A user perception based value-added service strategy for 4G mobile networks 299
L. Cao, Y. Zhou, X. Cheng, K. Chao & M. Yuan

Big data assisted value areas of mobile internet 307
C. Song, X. Cheng, H. Zhang & Y. Jia

Spectrum allocation based on data mining in heterogeneous cognitive wireless networks 315
C. Cheng, X. Cheng, M. Yuan, L. Xu, S. Zhou, J. Guan & T. Zhang

Big data based mobile terminal performance evaluation 329
C. Song, Y. Jia, X. Cheng & X. Liu

Human visual system based interference management method for
video applications in LTE-A 337
S. Zhou, X. Cheng, M. Yuan & C. Cheng

A novel complaint calls handle scheme using big data analytics in mobile networks 347
Y. Wang, X. Cheng, L. Xu, J. Guan, T. Zhang & M. Mu

A novel big data based problematical sectors detection algorithm in WCDMA networks 357
P. Ren, X. Cheng, L. Xu, M. Yuan & K. Chao

Big data assisted human traffic forewarning in hot spot areas 367
Y. Jia, K. Chao, X. Cheng, T. Zhang & W. Chen

A novel big data based Telecom User Value Evaluation method 375
K. Chao, P. Wang, L. Xu, D. Wu, X. Cheng & M. Mu

A novel Big Data based Telecom Operation architecture 385
X. Cheng, L. Xu, T. Zhang, Y. Jia, M. Yuan & K. Chao

Mining characteristics of a 4G cellular network based on big data analysis 397
G. Li & Q. Fan

Author index 403

Preface

The First International Conference on Signal and Information Processing, Networking and Computers (ICSINC) focuses on the key technologies and challenges of signal and information processing schemes, network application, computer theory and application, and so on. The ICSINC2015 Workshop on Telecom Big Data based Research and Application (IWT-BDRA) promotes the in-depth exploration of the most recent research and development findings on the basis of telecom big data. This telecom big data workshop welcomes the participation of the industry to showcase and commercialize their relevant technology and application as well as products. This book embraces a variety of research studies presented at the International Conference on Signal and Information Processing, Networking and Computers (ICSINC), held in Beijing, China, October 17–18 2015. The research areas are a convergence of the similarity with both algorithms and applications. The current state and trend of signal processing were exchanged in this conference.

48 papers were selected for publication in the proceedings. Each of the papers submitted to the conference was peer reviewed by at least two specialists in the author's research field. A paper typically goes through at least one revision cycle before being accepted for publication. The authors of the papers published in this book have enabled future investigators to gain ready access to the results of their research. Some highlights in this book:

- Software Defined Network, e.g. analyzing the requirements of future mobile Internet and illustrate the restrictions of using SDN in mobile operator networks to bypass the restrictions and address the challenges.
- Internet Application, e.g. a formal model based on learning to rank for hybrid recommendation that integrates diversity and the representation of diversity features by using entropy based on attributes of users and items.
- Signal Processing in Networked Systems, e.g. an algorithm that focuses on assigning channels to bring a high connectivity of the whole network while maintaining a robust network at the same time.
- Spectrum Estimation and Modeling, e.g. a new algorithm based on deep Q-learning for the channel selection model of a Secondary User (SU) with reconfigurable RF front-ends. The algorithm could help SU to select the channel for the maximum cumulative reward.
- Spectrum Shaping and Filters, e.g. an adaptive spread spectrum method or effective embedding and detection of audio watermarking. The method introduces audio data redundancy analysis at the watermark receiver to achieve enhanced extraction performance.
- Information Theory and Statistics, e.g. a method extracts time-related characteristics from processes and then compares all of them together by using statistical hypothesis tests in different successive populations. Such a method could not only allow accurate detection when some parts of the processes started to have abnormal behavior: longer or shorter, but also enable identification of which parts are involved.
- WCDMA Data, e.g. a novel WCDMA data based LTE site selection (WDLSS) scheme. The WDLSS scheme analyses five factors of existing WCDMA sites, including the height, the density, the signal power, the coverage control and the interference.
- Big Data Assisted Telecom Operation, e.g. logistic regression and neural network respectively identify the 2G/3G users who are most likely to immigrate to a 4G network.
- Big Data Analysis of Network Quality and Performance, e.g. a coverage self-optimization algorithm (CSOA) based on big data analytics, by obtaining and analyzing the counters (records the performance indicators of networks) from existing network.

– Big Data Based Network Optimization, e.g. a Single-objective Multivariable Quantum-inspired Particle Swarm Optimization (MQPSO) algorithm and a multi-objective M-QPSO based on data mining for spectrum allocation to improve the network benefit in heterogeneous network.

We extend sincere appreciation to all authors, members of the committee, and reviewers for giving their generous contribution of time and effort, who ensure that the papers of high scientific quality are published in the proceedings. We also acknowledge the admirable work of all editors from CRC Press/Balkema, Taylor & Francis with thanks.

The first international conference on signal and information
processing, networking and computers

Signal and Information Processing, Networking and Computers – Chen & Huang (Eds)
© 2016 Taylor & Francis Group, London, ISBN 978-1-138-02881-4

Pilot decontamination in a massive MIMO system: Exploiting pilot sequence hopping

Yue Li, Xiaojun Jing, Songlin Sun & Hai Huang
School of Information and Communication Engineering, Beijing University of Posts and Telecommunications, Beijing, China

ABSTRACT: With the demand for high speed data transmission and rapid growth of end users, massive Multiple-Input Multiple-Output (MIMO) has shown to be promising owing to its high spectrum efficiency. Uplink pilot contamination is the main performance bottleneck of a large-scale MIMO multicell multi-user TDD system. Channel estimation suffers from pilot contamination because of the shortage of orthogonal sequences. Orthogonal pilot sequences are used in one cell but the same pilot sequences must be reused in neighboring cells, which causes pilot contamination. In this paper, pilot sequence hopping is performed at each transmission slot, which provides a randomization of pilot contamination. Applying the blind equalization technique, it is shown that such randomized contamination can be significantly suppressed. Numerical results show that the channel estimation accuracy improved.

Keywords: massive MIMO; pilot decontamination; TDD

1 INTRODUCTION

With the rapid growth of high data rates and the number of users, massive Multiple-Input Multiple-Output (MIMO) appears to be a promising technology to meet this challenge, and it provides significant increment in reliability and data rate for wireless communications (Larsson et al. 2014, Marzetta. 2010, Hoydis. 2013). Recently, massive MIMO has been proposed as one of the important technologies of the fifth generation of wireless communication networks (5G). It has aroused wide interest among researchers. The gains of different users come from the corresponding channel gains between each Mobile Terminal (MT) antenna and the Base Station (BS) antennas. Thus, by using a large number of antennas at the base station, a significant gain can be achieved. In theory, when the number of antennas at the base station tends to infinity, the small scale fading and thermal noise can be eliminated completely.

Pilot decontamination has been the focus of several works recently. As discussed in Marzetta (2010), in a single cell scenario, the only factor that limits the capacity of the network is the correlation time of the channel which needs to be much larger than the number of MTs to obtain accurate channel estimation and data transmission. These are divided into two categories; one with coordination among cells and one without. The first category includes Yin et al. (2013); the desired and interfering signals can be distinguished in the channel covariance matrices as long as interfering signals and the desired angle-of-arrival spreads do not overlap. A pilot coordination scheme is proposed to meet this condition. The work in Ashikhmin & Marzetta (2012) utilizes coordination among base stations to share downlink messages. This eliminates interference when the number of BS antennas goes to infinity. Meanwhile non-cooperative multicellular Time-Division Duplex (TDD) networks suffering from a pilot contamination problem was first reported by Jose et al. (2009). In TDD systems, the Channel State Information (CSI) can be obtained at the BS during the uplink transmission due to the channel reciprocity. Since the length of the channel correlation time is not long enough for orthogonal pilot sequences being used in the different communities, the non-orthogonal pilots of neighboring cells will contaminate the each other's pilots. Gao et al. (2011) uses a multi-cell precoding technique, not only to minimize the mean squared error of the signals of interest within the cell, but also the interference imposed on other cells. The work in Payami & Tufvesson (2012) shows that the channel estimates can be found as eigenvectors of the covariance matrix of the received signal when the system has "favorable propagation" and the number of BS antennas grows.

The major contribution of this paper is a pilot decontamination, which does not require inter-cell coordination, and it can exploit past pilot signals. It is based on pilot sequence hopping performed within each cell. Pilot sequence hopping means that in each transmission slot, every user is able to choose a new pilot sequence. When the channel is time-variant and correlated across time slots, it remains possible to exploit the information about the channel across time slots by an appropriate equalizer and benefits from contamination randomization. In this paper, channel estimation across multiple time slots is performed using an equalizer with blind equalization property which is capable of tracking the channel and the channel correlation.

The rest of the paper is organized as follows. The applied system model is introduced in section II. Section III elaborates the proposed solution and numerical results are described in section IV. Finally, the conclusions are drawn in Section V.

2 SYSTEM MODEL

This paper treats a cellular system consisting of L cells with K users in each cell. A massive MIMO scenario is considered, where the BS has M antennas and the UE has a single antenna. We restrict our attention to the channel estimation performed in a single cell, which we term "the cell of interest" and assign the index "0". Simple schematic diagram is shown in Figure 1. The channel between the BS in the cell of interest and the k'th user in the l'th cell is denoted $h^{kl} = [h^{kl}(1)h^{kl}(2)...h^{kl}(M)]$, where the individual channel coefficients are complex scalars. Note that for $l > 0$, h^{kl} refers to a channel between the BS of interest and a UE connected to a different base station. We furthermore restrict our attention to the estimation of a single channel coefficient. Hence, a channel is denoted as the complex scalar h^{kl}. The work easily extends to vector estimations, in which case spatial correlation can be exploited for improved performance. A rich scattering environment is assumed, such that h^{kl} can be modeled using Clarke's model (Farhang–Boroujeny. 2003); hence,

$$h^{kl} = \frac{1}{\sqrt{N_s}} \sum_{m=1}^{N_s} e^{j2\pi f_d t \cos \alpha_m + \phi_m}$$ (1)

where N_s is the number of scatterers, f_d is the maximum Doppler shift, α_m and ϕ_m is the angle of arrival and initial phase, respectively, of the wave from the m'th scatterer. Both α_m and ϕ_m are i.i.d. in the interval $(-\pi, \pi)$ and $f_d = \frac{v}{c} f_c$, where v is the speed of the UE, c is the speed of light and f_c is the carrier frequency.

In a massive MIMO system, collection of Channel State Information (CSI) is performed using uplink pilot training. The CSI achieved this way is utilized in both downlink and uplink transmissions based on the channel reciprocity assumption. We define a pilot training period followed by an uplink and a downlink transmission period as a time slot. See Figure 2 for an example of a transmission schedule with two time slots. During the n'th pilot training period, the k'th user in the l'th cell transmits a pilot sequence $x_n^{kj} = [x_n^{kl}(1)x_n^{kl}(2)...x_n^{kl}(\tau)]^T$, where τ is the pilot sequence length. Ideally, all pilot sequences in the entire system are orthogonal, in order to avoid interference. However, this would require pilot sequences of at least length $L \cdot K$ which in most practical systems is not feasible. Instead, orthogonality within each cell only is ensured, i.e., $\tau = K$, thereby dealing with the potentially strongest sources of interference. As a result, all cells use the same set of pilots, potentially causing interference from neighboring cells. This is referred to as pilot contamination. We define the contaminating set C_n^{kl} as the set of all pairs i, j, which identify all UEs applying the same pilot sequence in the n'th time slot

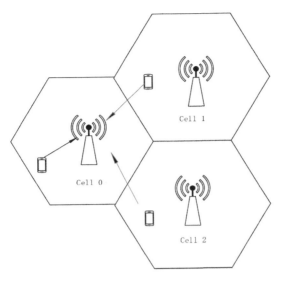

Figure 1. A cellular system with three cells. Cell 0 is of interest and the neighboring cells will poten-tially cause interference.

	Time slot 1			Time slot 2		
User1	Pilot	Uplink	Downlink	Pilot	Uplink	Downlink
User2	Pilot	Uplink	Downlink	Pilot	Uplink	Downlink

Figure 2. Scheduling example.

as the k'th user in the l'th cell. Hence, $x_n^{ij} = x_n^{kl} \forall i, j \in C_n^{kl}$. The pilot signal received by the BS of interest, concerning the k'th user in the n'th time slot can be expressed as

$$y_n^{k0} = h_n^{k0} x_n^{k0} + \sum_{i, j \in C_n^{k0}} h_n^{ij} x_n^{ij} + z_n^{k0}, \tag{2}$$

where $z_n^{k0} = \left[z_n^{k0}(1) z_n^{k0}(2) \ldots z_n^{k0}(\tau) \right]^T$ and $z_n^{k0}(j)$ are circularly symmetric Gaussian random variables with zero mean and unit variance for all j. Here, only signals leading to contamina-tion are included in the sum term, since any $h_n^{ij} x_n^{ij} \forall i, j \notin C_n^{kl}$ are removed when correlating with the applied pilot sequence. Hence, all contributions from the sum term are undesirable and will contaminate the CSI. Without loss of generality, we focus on the channel estimation for a single user in a single cell. Hence, in the remainder of the paper, we omit the superscript k for ease of notation.

3 PROPOSED PILOT DECONTAMINATION

Pilot sequence hopping is a technique where the UEs randomly switch to a new pilot sequence in between time slots. This must be coordinated with the BS, which in practice can be realized by letting the BS send a seed for a pseudorandom number generator to each UE. Random pilot sequence hopping is illustrated in Figure 3 in case of $\tau = K = 4$. Note how the identity of the contaminator changes between time slots, as opposed to a fixed pilot sequence sched-ule, where the contaminator remains the same UE. Consequently, the undesirable part of the

Cell 0	Cell 1			
UE of interest	UE 1	UE 2	UE 3	UE 4
X3	X2	X1	X4	X3
X2	X3	X2	X1	X4
X4	X4	X3	X2	X1
X1	X2	X1	X3	X4
X4	X1	X2	X4	X3
X2	X2	X3	X1	X4
X2	X4	X1	X3	X2
X1	X3	X2	X1	X4

Time slot (vertical label, left side). t_c (right side bracket).

Figure 3.   An example of a random pilot schedule for the UE of interest and potential contaminators in a neighboring cell.

pilot signal, i.e., the sum term in (2), varies rapidly between time slots compared to the variation caused by the mobility of a single contaminator in a fixed schedule. In fact, the impact of pilot sequence hopping, from a contamination perspective, can be viewed as a dramatic increase of the mobility of the contaminator. This in turn leads to a lowered autocorrelation, or decorrelation, in the contaminating signal, which is the motivation behind performing pilot sequence hopping.

The level of decorrelation is related to the time between two instances, where the same user acts as a contaminator. We refer to this as the collision distance, and we denote it t_c, see Figure 3. Note that in the case of a fixed pilot schedule, $t_c = 1$. The goal of pilot sequence hopping is to maximize t_c, either in an expected sense or maximum sense, i.e., maximization of the minimum value. The latter can be pursued through a minimal level of coordination of pilot sequence schedules among neighboring cells. However, this work is strictly restricted to a framework with no inter-cell coordination; hence, we focus on the expected value of t_c. If pilot sequence hopping is performed at random and $\tau = K$ then t_c follows a geometric distribution, such that

$$P(t_c = d) = (1-p)^{d-1} p, \; d = 1,2... \; p = \frac{1}{K} \tag{3}$$

To help the understanding of the benefit from pilot sequence hopping, consider the ideal case of a constant channel between BS and UE of interest and a single contaminating neighboring cell. Noise is disregarded in this example, since attention is on decontamination. Moreover, we assume an infinite amount of orthogonal pilot sequences and an infinite amount of users per cell, such that $\tau = K = \infty$ which means contaminating signals in all time slots are independent. For simplicity, we assume $x_n^H x_n = 1$. We consider an estimator whose $\bar{h}$ is the average of all estimates until time slot n. Hence,

$$\bar{h} = h + \frac{1}{n} \sum_{i=1}^{n} h_i' \tag{4}$$

In this case, the error in the estimate is solely composed of the average of the contaminating signals, which are independent and have variance σ_c^2. Hence, pilot sequence hopping had been performed so that the variance of the estimation error is $\frac{\sigma_c^2}{n}$. In a more practical example with a time-varying channel, the amount of information carried in a pilot signal decays over time. For this reason, we have chosen an appropriate blind equalization technique.

We propose to extend the blind equalization to massive MIMO networks [10]. The algorithm is exploited to adaptively correct the imperfect channel estimates. Massive MIMO application can be directly derived as:

$$w_{n+1} = w_n - 2\mu sign\left(\hat{s}_n\right)\left(\left|\hat{s}_n\right| - R\right) \bullet x_n, \tag{5}$$

where $\hat{s}_n = w_n^H x_n$, x_n is the n'th symbol of the received data and μ is a step-size parameter. The initialization is derived as:

$$w_0 = \frac{\hat{h}_n}{\hat{h}_n^H \hat{h}_n} \tag{6}$$

4 NUMERICAL RESULTS

In our simulations, we consider a massive MIMO network comprising three cells, where the pilot signals of the cell of interest are interfered with by the users of its adjacent cells (Figure 2). We assume one interferer in each neighboring cell and one user in the interested cell using the same pilot sequence as the users in its neighboring cells. One transmit antenna is assumed for each user and the number of antennas at the BS are N = 128.

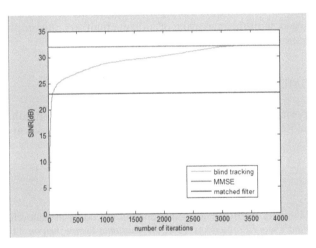

Figure 4.   SINR comparison of our proposed blind tracking technique with respect to the MF and MMSE detectors having the perfect CSI knowledge.

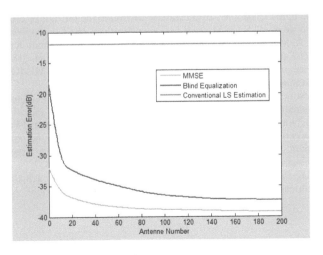

Figure 5.   Estimation MSE vs. antenna number.

Figure 4 shows the SINR performance of the proposed blind tracking technique in dB with respect to the number of iterations. As the figure depicts, there exists an abrupt SINR improvement during the first 50 iterations where the output SINR of the blind combiner reaches that of the MF combiner with the perfect CSI knowledge. Running larger numbers of iterations has shown that the output SINR of our blind channel tracking technique can suppress the pilot contamination effect and converge toward that of the MMSE combiner.

As for Figure 5, it's obviously that the pilot contamination is quickly eliminated when the number of antennas increase. Our proposed blind equalization technique is much better than the conventional LS estimation with the increasing of antenna numbers while it is close to the case which adopts MMSE with the perfect CSI knowledge.

5 CONCLUSION

In this paper, we addressed the pilot contamination problem in a TDD multi-cellular massive MIMO network. The pilot contamination problem can adversely affect the performance of the massive MIMO networks and as a result create a great amount of multi-cell interference in both uplink and downlink transmissions. In the numerical results, it shows that after running a small number of iterations our algorithm performs better than the matched filter with perfect CSI. Also, with the increasing of antenna numbers, the proposed technique will decrease the estimation error evidently.

ACKNOWLEDGMENTS

This work is supported by project NSFC 61471066 and the open funding project of State Key Lab of Virtual Reality Technology and Systems at Beihang University under Grant No. BUAA-VR-15 KF-19.

REFERENCES

Ashikhmin, A. & Marzetta, T. 2012. Pilot contamination precoding in multicell large scale antenna systems. *Information Theory Proceedings. IEEE International Symposium on.* 1137–1141.
Farhang-Boroujeny, B. 2003. Multicarrier modulation with blind detection capability using cosine modulated filter banks. *Communications, IEEE Transactions on.* 51(12): 2057–2070.
Gao, X., Edfors, O., Rusek, F. & Tufvesson, F. 2011. Linear pre-coding performance in measured very-large MIMO channels. *IEEE Vehicular Technology Conference.* 1–5.
Hoydis, J., ten Brink, S. & Debbah, M. 2013. Massive MIMO in the UL/DL of cellular networks: How many antennas do we need? *Selected Areas in Communications, IEEE Journal on.* 31(2): 160–171.
Jose, J., Ashikhmin, A. Marzetta, T. & Vishwanath, S. 2009. "Pilot contamination problem in multi-cell TDD systems. *IEEE International Symposium on Information Theory.* 2184–2188.
Larsson, E.G., Tufvesson, F., Edfors, O. & Marzetta, T.L. 2014. Massive MIMO for next generation wireless systems. *Communications Magazine, IEEE.* 52(2): 186–195.
Marzetta, T. 2010. Noncooperative cellular wireless with unlimited numbers of base station antennas. *Wireless Communications, IEEE Transactions on.* 9(11): pp. 3590–3600.
Marzetta, T.L. 2010. Noncooperative cellular wireless with unlimited numbers of base station antennas. *Wireless Communications, IEEE Transactions on.* 9(11): 3590–3600.
Payami, S. & Tufvesson, F. 2012. Channel measurements and analysis for very large array systems at 2.6 Ghz. Proc. *6th European Conference on Antennas and Propagation.* 433–437.
Yin, H., Gesbert, D., Filippou, M. & Liu, Y. 2013. A coordinated approach to channel estimation in large-scale multiple-antenna systems. *Selected Areas in Communications, IEEE Journal on.* 31: 264–273.

Signal and Information Processing, Networking and Computers – Chen & Huang (Eds)
© 2016 Taylor & Francis Group, London, ISBN 978-1-138-02881-4

A novel 3D multivariate data visualization method

Fukang Hou, Hai Huang & Xiaojun Jing
Key Laboratory of Trustworthy Distributed Computing and Service (BUPT), Ministry of Education, Beijing University of Posts and Telecommunications, Beijing, China

ABSTRACT: 3D visualization of multivariate data is of great importance in many scientific disciplines especially while analyzing information from a macro point of view. However, visualization of a large number of special multivariate volume data is a very challenging task. In this paper, we propose a method to present several kinds of data at the same time. Texture generated by the Gabor noise function is mapped onto transparency to make internal data visible. We show the practicability of our approach through visualizing hurricane Isabel data, and its effectiveness is approved while comparing with normal direct volume rendering method and some other visualization techniques. Furthermore, we evaluate our approach through user study. Overall, the result of our study point that users get more accurate judgment while using our novel volume rendering technique. High subjective ranking implies that our method is an important step in the challenge of visualization of multivariate data.

Keywords: multivariate data visualization; volume rendering; 3D visualization

1 INTRODUCTION

The visualization of multiple variables is often necessary while we analyzing data in many scientific disciplines. Applications such as medicine and meteorology are often in need of viewing the data on the whole or in the detail. However, visualization of multiple volumetric data is still an unsolved challenging problem.

Showing multivariate data in a single visualization brings several challenges. First, visual information that a person can perceive is very limited. Color and texture are used to display various information that is able to be seen directly or not. However, when the information data goes out of the three-dimensional space, such as multivariate 3d data used in this paper, it becomes hard for a person to understand all the information in one visualization. Secondly, when user try to understand the internal data, the influence of external data on his/her judgment is inevitable.

The most straightforward approach to display data is color mapping. However, it is not suitable when multiple co-located values are to be visualized. This problem can be solved by color blending or color weaving in T. Urness et al. (2003), and a study by H. Shenas et al. (2007) indicates that users perform better while using color weaving method.

However, color rendering is inherent in 3D volume rendering even for single variable data set, as transparency has to be used to reveal the internal structures. Noticing that a user always zoom out to gain an overview of the data and zoom in to see more details, in this paper, we propose a novel 3D visualization method to display the trends of the data from the overall perspective and get accurate numerical judgment when zoom in.

To reduce the effect of color blending (Gama et al. 2014) on value judgment, an algorithm using noise to map opacity is used in our work. We choose Gabor noise function to redistribute the transparency in a voxel. We also presented an improved volume ray casting method to enhance Information that we are more concerned about to show the trends of data set more clearly.

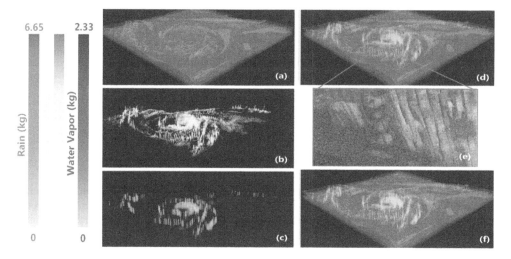

Figure 1. Visualization of hurricane Isabel data: rain, cloud water, water vapor. The first three images (a, b, c) are visualization of single data attribute. The image (d) shows the results of multivariate data visualization using our method, and (e) give us the details when we zoom in. However, if some data (e.g. rain) is more important for us, it could be enhanced as shown in picture (f).

We applied our method for the visualization of several 3d data sets, and hurricane Isabel data is mainly discussed in this paper (Fig. 1). Furthermore, we have evaluated and compared our method with other visualization approaches by user study. The results show that our novel 3d visualization method performs better, i.e. users get significantly more accurate numerical judgement while their overall trends analysis is still perfect.

2 METHOD

Our work is designed to achieve two major objectives: (1) Displaying multiple co-located variables accurately as many as possible in one visualization while avoiding the impact between each other. (2) Showing the trend of each variable, i.e. maintaining the see-through properties for each kind of data.

Unfortunately, there are conflicts between the two objectives: to gain accurate judging values of data, higher opacity is recommended while it is against the see-through property. However, we notice that users usually zoom out to gain an overview over the data or zoom in to see more details. Inspired by this, we solve the problem by texture rending and information enhancement.

2.1 *Gabor noise based texture rendering*

Gabor noise function is used to redistribute the transparency in a voxel (Khlebnikov et al. 2012, 2013). In this way, color blending is avoided when displaying multiple co-located data variables, and users could see the original color which represents data type and value. In this section, our work has three parts: setting Gabor noise function, mapping noise function to opacity, filtering.

Random-phase Gabor noise is introduced with the phase-augmented Gabor kernel, which is designed to overcome the problem of phase shift (Fig. 2) in A. Lagae & G. Drettakis (2011). The phase shift ϕ is introduced by slicing (solid green line).

The Gabor kernel g of Lagae et al. is defined as

$$g(x; a, \omega, \phi) = e^{-\pi a^2 |x|^2} \cos(2\pi x \cdot \omega + \phi), \tag{1}$$

10

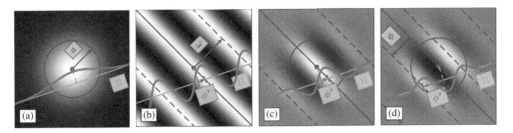

Figure 2. (a) 2D Gaussian (b) 2D harmonic (c) 2D Gabor kernel (d) 2D phase-augmented Gabor kernel.

where a is the bandwidth, ω is the frequency, and $\phi \in [0, 2\pi)$ is the phase of the harmonic. Thus, n-dimensional random-phase Gabor noise n using Gabor kernel is

$$n(\mathbf{x}; a, \omega) = \sum_i g(x - x_i; a, \omega, \phi_i),$$ (2)

where the random positions $\{x_i\}$ are distributed according to a n-dimensional Poisson process, and $\{\phi_i\}$ are the random phases distributed according to uniform distribution:

$$\{\phi_i\} \sim U[0, 2\pi].$$ (3)

Once the random-phase Gabor noise function is set up, we need to define the mapping function of its values to opacity, which will be used in final volume rending step. To maintain the average opacity value that is defined by the regular transfer function, we impose the following constraint on the noise transfer function:

$$\forall \alpha : \int_{-\infty}^{\infty} p(t) \cdot M_\alpha(t) \, dt = \alpha,$$ (4)

where $p(t)$ is the probability density function of the intensity distribution of the noise and $M_\alpha(t)$ is the mapping function that corresponds to a particular opacity value α.

In theory, the opacity mapping function could be chosen arbitrarily as long as it satisfies condition (4). Gaussian mapping function is used in our work.

2.2 Color mapping

Observer perceives an unique hue as a pure which has no admixture of the other colors. Statistical data are shown by mapping from a numerical data space to an appropriate color space. HSV color space is adopted in our work (Fig. 3(a)). The HSV model describes colors in terms of hue, saturation and value. Perceptual variables are intuitively aligned within the resulting space while using empirically derived color space such as HSV.

Thus, we use this attribution to discriminate different kinds of variables. However, not the whole domain of saturation or value is available (Fig. 3(b)). The colors in blue square is used in our work (both saturation and value range from 0.2 to 1), as hue is distinguishable in perception.

Analyzing Figure 3(b) we can get three conclusions: (1) Zero saturation produces white, a shade of gray, or black, while full saturation produces no tint; (2) Specifying a hue with full saturation and value produces pure colors; (3) Specifying zero value produces black, regardless of hue or saturation.

Thus our mapping function designed which map data values onto saturation and value is shown as Figure 4.

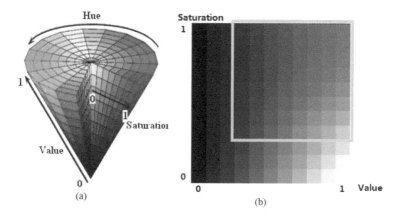

(a)   (b)

Figure 3.   HSV color space.

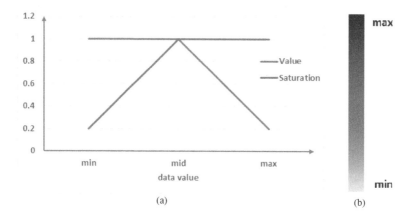

(a)   (b)

Figure 4.   (a) Mapping function (b) Mapping result (hue = 60).

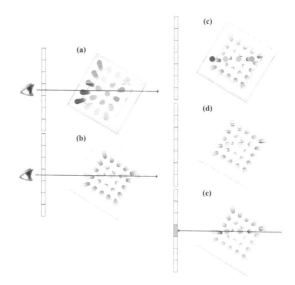

Figure 5.   Our ray casting steps: (a) Ray casting (b) Sampling (c) Shading (d) Weighting (e) Compositing.

2.3 *Ray casting based information enhancement*

In our work, the volume ray casting algorithm(Livingston et al. 2012) used to enhance information we need comprises five steps:

(1) Ray casting. As shown in Figure 5 (a), the volume for each pixel of the final image on the screen is cast through by a ray of sight. The volume is considered to be enclosed within a bounding cuboid, which is used to intersect the volume and the ray of sight. (2) Sampling. We select equidistant sampling points along the ray of sight. However, generally, the volume is not aligned with the ray of sight and sampling points are usually located in between voxels. Because of that, we use linear interpolation to get sample values from its surrounding voxels. (3) Shading. At this stage, the sample points are shaded (i.e. textured and colored) according to their values. (4) Weighting. In order to highlight the information we care about, sampling points are assigned different weights according to attention, and then, transparency values of the samples are multiplied by related weights. (5) Compositing. All sampling are composited along the ray of sight from back to front points after being shaded and weighted, resulting in the final color value for the processing pixel.

3 APPLICATIONS

In this section, two different data sets are visualized by our method. The first example is our simple small data set which is designed to show the details of the results obtained by our method. The second application deals with large climate data set analysis. We implemented our method based on OPENGL and tested the performance on Inter Core i5-2400, 4G RAM, AMD Radeon HD 6450 PC. In total, the frame rate for climate data set with resolution of $500 \times 500 \times 100$ is between 5 and 10 frames per second.

The distribution of multivariate data in space has three kinds of cases: subset, intersection, deviation. To show all kinds of situation, our simple data set is shown as Figure 6. It can be seen that the internal data have good readability. The observer can accurately understand the data, because there is no confusion introduced like color blending.

Climate data set obtained through simulation or measurement is always a mixture of various attributes such as water vapor, temperature, pressure etc. Analyzing and understanding the relationship between each attributes is often needed for observers. We applied our method to this kind of application by visualizing hurricane Isabel. The resolution of the data sets is $500 \times 500 \times 100$ voxels. Detailed data description and WRF model data of hurricane Isabel can be obtained at its official website.

Our method is shown (Fig. 7) on the left comparing itself with Direct Volume Rendering (DVR) on the right. Five images of our method are respectively indicated single water vapor data, multivariate data (water vapor, cloud water and rain) without enhancement, multivariate data with enhanced rain data, multivariate data with enhanced cloud water data and detail of

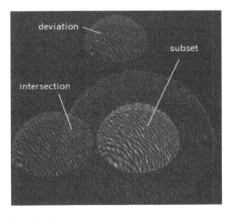

Figure 6. Simple data set visualization.

13

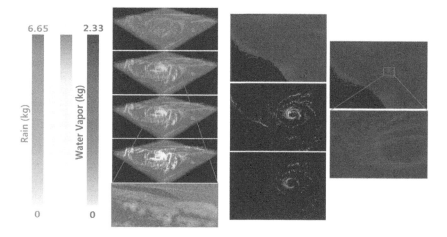

Figure 7.   Three attributes of Hurricane Isabel data set visualization: Water Vapor (red, $0 \sim 2.33 \times 10^{-2}$ kg), Cloud Water (yellow, $0 \sim 2.11 \times 10^{-3}$ kg) and Rain (green, $0 \sim 6.65 \times 10^{-3}$ kg).

our method. And five pictures of DVR method separately represent single water vapor data, single cloud water data, single rain data, multivariate data and detail of the method.

When an observer tries to get an overview of data, it could be seen that our method shows the trend of data well while it is hard for DVR method to distinguish rain data or cloud data from water vapor data. And our enhancement method allow us to analyze a certain data more clearly. While analyzing the details of data, our method shows all attributes (we can see three original colors: red, yellow and green) while this information is lost in DVR because of color blending.

4   USER STUDY

To evaluate the practicality and effectiveness of our method, we designed two comparative experiments. Twelve professionals (aged 21 to 30, 7 males and 5 females) and thirty regular users (aged 18 to 35, 15 males and 15 females) from a local university are invited to take part in the evaluation. Professional participants were graduate students and doctoral students in computer graphics while regular users were ordinary college students.

We displayed and compared five techniques for multivariate data analysis in our user study: DVR switching method (the data sets were visualized separately using conventional volume rendering), noise-based (our method without enhancement), noise-based switching method (the data sets were visualized separately using noise base method) and noise-based enhanced method.

The two contrast experiments were detail contrast and overview contrast. Detail contrast is to evaluate the accuracy of the method where participants were asked to determine the value of the specified position by observing the color value when zoom in, all three kinds of cases (subset, intersection and deviation) were shown to the participant. And for overview contrast, participants were asked to evaluate the understanding of the data trends by observing the color trends when zoom out.

The participants were provided with a laptop with a 15.6" monitor with 1366×768 resolution. To enter the judging values, the participants were provided with a slider dialog box (see Fig. 8). Task completion time was also measured by our software.

Once the participants completed the experiment, they are asked to fill out a questionnaire, where they assessed their agreement with following three statements on a five-point Likert scale.

- Overview – It is easy to understand and analyze the trends of the data set.
- Detail – The value of the specified position is shown clearly using this method.
- Confidence – I am confident that my answers are correct using this method.

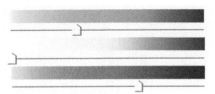

Figure 8.   Slider dialog box.

Finally, the participants ranked all five methods by their own impression on their overall personal preference.

5   RESULTS AND DISCUSSION

We noticed that the participants were not familiar with our system at the beginning and produced unreliable data. We neglected this part of the data as a warm-up.

Our hypothesis was that three noise-based method performs better than other techniques in accuracy and noise-based enhanced method have an advantage in terms of overall visualization.

Since a Shapiro-Wilk test rejected normality in all cases, error, consuming time, subjective assessment and ranking were all analyzed using Friedman non-parametric tests ($\alpha = 0.05$) for main effects, then, post-hoc analysis with Wilcoxon Signed-Rank Tests was conducted with a Bonferroni correction applied. Four evaluated measurement are defined as follow (see Fig. 9):

- Error – The difference between the judging value and the real value.
- Time – The time from the image appearing on screen until the participants determine values for each kind of data.
- Assessment – Subjective assessment value of five-point Likert scale.
- Rank – The rank given by participant from one (the highest evaluation) to five (the lowest evaluation).

There was a statistically significant difference in error depending on which type of method was adopted ($\chi^2(4) = 147.8$, $P < 0.01$). Post-hoc comparison showed that has the error for DVR was highest and DVR switching method was second, while noise-based switching method was lower than the others. There were no significant differences between noise-based method and noise-base enhanced method. And for time ($\chi^2(4) = 173.9$, $P < 0.01$), the conclusion of post-hoc comparison was similar with error.

We also found significant effect for overview ($\chi^2(4) = 149.3$, $P < 0.01$), detail ($\chi^2(4) = 186.3$, $P < 0.01$) and confidence ($\chi^2(4) = 178.3$, $P < 0.01$). Post-hoc comparison revealed that overview for noise-based enhanced method was higher than other methods, while there was no difference between noise-based switching method and noise-based enhanced method for detail and confidence. DVR was lowest and DVR switching method was second lowest for all three questions.

For rank ($\chi^2(4) = 186.5$, $P < 0.01$), noise-based switching method was No.1, DVR switching method and DVR ranked fourth and fifth respectively, while noise-based method and noise-based switching method had little difference.

Our study validates our hypothesis that three noise-based method performs better than other techniques in accuracy and our noise-based enhanced method is superior to other methods in terms of overview visualization. The other performance of our noise-based method is equivalent to noise-based method. Moreover, the study shows that three noise-based method are significantly better than two DVR method in all aspects. It is difficult and time-consuming to distinguish between different kinds of data for participant when using two DVR methods. Noise-based switching method performs better in detail because of no confusion.

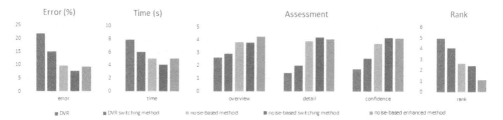

Figure 9. The result of the user study.

6 CONCLUSIONS AND FUTURE WORK

We have presented and evaluated a novel method for the 3D display of multivariate data visualization. Noise-based method gives a possibility to read data values accurately, and ray-casting based information enhancement method makes it more clear when analyzing the trends of the data set from the overall perspective. Our method shows significantly lower error for reading data values and higher subjective ranking than DVR techniques.

There are several directions for extending and improving our method that we will work on in the near future. Firstly, how well our visualization method scales with the number of visualized variable will be tested. Secondly, we noticed that background color will affect our results since opacity mapping is used, thus, new solutions are needed to avoid the error that occurs in this case. Thirdly, we will also attempt to combine our method with other multivariate visualization approaches. And fourthly, we will broaden the scope of applications for our method. Currently, we use our method to analyze climate data, other fields such as medicine will be tried later.

ACKNOWLEDGEMENTS

This work is supported by project NSFC 61471066 and the open funding project of State Key Lab of Virtual Reality Technology and Systems at Beihang University under Grant No. BUAA-VR-15 KF-19.

REFERENCES

Gama, Sandra, Gonçalves & Daniel. 2014. Guidelines for using color blending in data visualization. *Proceedings of the Workshop on Advanced Visual Interfaces AVI*: 363–364.

Khlebnikov, R., Kainz, B., Steinberger, Streit, M. & Schmalstieg, D. 2012. Procedural texture synthesis for zoom-independent visualization of multivariate data. *Eurographics/IEEE Symposium on Visualization, EuroVis*, v 31, n 3: 1355–1364.

Khlebnikov, Rostislav, Kainz, Bernhard, Steinberger, Markus, Schmalstieg & Dieter. 2013. Noise-based volume rendering for the visualization of multivariate volumetric data. *IEEE Transactions on Visualization and Computer Graphics*, v 19, n 12: 2926–2935.

Lagae, A. & Drettakis, G. 2011. Filtering solid gabor noise. *ACM Transactions on Graphics*, SIGGRAPH 2011, v 30, n 4.

Livingston, Mark, A., Decker & Jonathan, W. 2012. Evaluation of multivariate visualization on a multivariate task. *IEEE Transactions on Visualization and Computer Graphics*, v 18, n 12: 2114–2121.

Shenas, H., Kim, S., Interrante, V. & Healey, C1.G. 2007. Weaving versus blending: a quantitative assessment of the information carrying capacities of two alternative methods for conveying multivariate data with color. *IEEE Transactions on Visualization and Computer Graphics*: 1270–1277.

Urness, T., Interrante, V., Marusic, I., Longmire, E. & Ganapathisubramani, B. 2003. Effectively visualizing multi-valued flow data using color and texture. *In Proceedings of the 14th IEEE Visualization 2003, VIS '03, Washington, DC, USA*: 115–121.

Signal and Information Processing, Networking and Computers – Chen & Huang (Eds)
© *2016 Taylor & Francis Group, London, ISBN 978-1-138-02881-4*

A fast CU depth decision algorithm based on learning how to use machines

Li Wang, Xiaojun Jing, Songlin Sun & Hai Huang
*Key Laboratory of Trustworthy Distributed Computing and Service (BUPT), Ministry of Education,
Beijing University of Posts and Telecommunications, Beijing, China*

ABSTRACT: High Efficiency Video Coding (HEVC) is the latest video coding standard, which has been proposed and developed by the Joint Collaborative Team on Video Coding (JVT-VC). With the evolution of the HEVC test model (HM), plenty of efficient coding tools are integrated into HM, and the HEVC outperforms H.264/AVC by almost 50% bitrate reduction for an equal perceptual video performance. However, it imposes a great deal of computational complexity on the encoder because of the optimization, especially with regard to the Rate Distortion Optimization (RDO) process. In this paper, we propose a fast Coding Unit (CU) depth decision algorithm to alleviate the heavy computational burden of the encoder, which attains a better trade-off between the Rate-Distortion (RD) performance and coding complexity. Then, a Support Vector Machine (SVM) is applied to training the data points, which has a better prediction of the CU depth decision. Experimental results indicate that under the "low delay" configuration, our proposed algorithm can achieve about 49.15% complexity reduction on an average with only a 2.35% BDBR increase when compared to the original HM12.0.

Keywords: HEVC; RDO; coding depth decision; machine learning

1   INTRODUCTION

The High Efficiency Video Coding (HEVC) standard (Sullivan et al. 2012) is the joint video project of the Joint Collaborative Team on Video Coding (JCT-VC), which is an achievement of the two organizations namely the ITU-T Video Coding Experts Group (VCEG) and the ISO/IEC Moving Picture Experts Group (MPEG) standardization organizations. The new HEVC video compression standard was finalized in January 2013. With the technological leap of the display and the higher pursuit of a better human experience, High Definition (HD) and Ultra-High Definition (UHD) video contents have become more and more popular all over the world. In the near future, we can expect that there will be a high demand of video compression technologies.

As we all know, the video coding layer of HEVC adopts the well-known hybrid-approach coding scheme which is used in previous coding standards since H.261. The core of the coding layer in HEVC is the Coding Tree Unit (CTU) that is conceptually analogous to the macroblock (MB) in H.264/AVC. One CTU consists of a luma Coding Tree Block (CTB), the corresponding Chroma CTBs, and syntax elements. And HEVC also adopts the highly flexible quad-tree coding block partitioning structure. The most significant features of quad-tree coding block partitioning are the basic units of the Coding Unit (CU), Prediction Unit (PU), and Transform Unit (TU). Based on the hierarchical structure, CU, PU, and TU have different block sizes, and the following Rate Distortion Optimization (RDO) process imposes enormous computational complexity on the HEVC encoder.

CU block may take size from 64×64 to 8×8 pixels (Sullivan et al. 2012). A PU structure has its root at the CU level. So whether to code a frame block using inter or intra prediction is decided at the CU level. HEVC supports variable Prediction Block (PB) size from 64×64 to 4×4 pixels (Sullivan et al. 2012). A TU tree structure also has its root at the CU level, and the square PB sizes contain 4×4, 8×8, 16×16 and 32×32.

Many researchers and experts are dedicated to HEVC for decades. Due to their efforts, plenty of efficient coding algorithms have been integrated into HEVC, and the performance of HEVC outperforms H.264/AVC by almost 50% bitrate reduction for equal perceptual video performance. Xiong et al. (2014) proposed a CU selection algorithm based on Pyramid Motion Divergence (PMD), then, employed the neighboring K-nearest like algorithm to decide whether to split the CU or not. Basically, it exploits motion properties and spatial neighboring correlations of the video content while doing the CU depth decision optimization. However, on one hand the PMD comparison among hundreds of samples in the First-In-First-Out (FIFO) stack causes a large complexity overhead, on the other hand the algorithm doesn't change the "try all and find the best" optimization strategy of the RDO process. So the effect of optimization is limited. Shen et al. (2013) authors exploited the mode correlations among different depth levels and the corresponding RD cost to obtain the rarely used CU and PU modes for intra coding.

Machine learning is a hotspot of current research, which has evolved from computational learning theory in artificial intelligence and the study of pattern recognition. It learns from the training set and makes predictions on data. In machine learning, Support Vector Machines (SVMs) are widely used. SVMs are supervised learning models with associated learning algorithms which have been applied to analyze data and recognize patterns and so on. SVMs already have shown the excellent performances in high dimensional spaces, which is the most important advantage over other algorithms. Inspired by the excellent property, many researchers have attempted to apply different algorithms to HEVC for better coding performance. In Shen & Yu (2013), a CU splitting early termination algorithm based on weighted SVM was proposed. The CU splitting was model as a binary classification problem, and RD loss due to misclassification was introduced as weights in SVM training. Meanwhile, the main idea of Shen & Yu (2013) is to select features and obtain the most suitable separating hyperplane between the two classes. Zhang et al. (2015) proposed an efficient machine learning program based on the CU depth decision method. First, CU depth decision process was modeled as a three-level hierarchical decision problem, and a novel CU depth decision structure was presented. Then, misclassification was considered in the SVM training and a multiple classifier is designed to control the risk of the false prediction. Finally, the optimal training parameters are derived by a sophisticated RD-complexity model.

In this paper, we concentrate on the trade-off between complexity reduction for equal perceptual video quality and the time saving, so we proposed a CU depth decision for HEVC based machine learning algorithm. The main contributions are: 1) We model the HEVC CU depth decision as a binary classification problem; 2) We deeply exploit the motion properties and spatial correlations of the video content in the same CU depth level among different frames; 3) A suitable machine learning algorithm is proposed for the CU depth allocation, which is proved to have a good trade-off between the complexity and the RD-cost. The rest of the paper is organized as follows. Section 2 presents the analysis of the CU depth decision in the whole HEVC coding process. The analysis of the proposed machine learning algorithm is illustrated in Section 3. Section 4 gives the experimental environments and corresponding results. Finally, conclusions are drawn in Section 5.

2 A BRIEF REVIEW OF HEVC CU MODE DECISION PROCESS

2.1 *CU modes in HEVC*

In HEVC, each frame is divided into plenty of Coding Tree Units (CTUs). Also, generally, the CTUs in the margin of the picture are not the same as the CTUs in the middle, as the pixels of the picture may not be as integral or multiple as that of the CTU size. For HEVC encoder, a CTU can be recursively divided into four equally sized CUs owing to the quad-tree structure. The CU size of luma samples varies from 64×64 to 8×8, so a CTU may contain

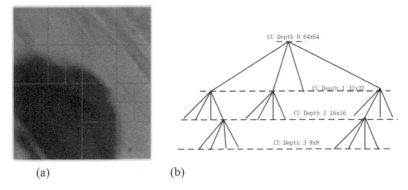

Figure 1. Subdivision of a CTB into CBs. (a) CTB with its partition. (b) Corresponding quad-tree.

only one CU or may be split into four identical CUs, and each CU can be partitioned into Prediction Units (PUs) and a tree Transform Units (TUs). For the HEVC encoder, it checks all the possible CU partitions in the depth level, $i \in [0,3]$.

The CU depth selection in HEVC is a recursive process, as shown in Figure 1. All the number of the possible CU partitions is $1 + 4 + 16 + 64 = 85$. However, in the whole coding process only a small partition of them will be eventually selected as the optimal CU partitions, and at least one half CU modes are not necessarily to be checked.

2.2 Overview of the RDO process

As we mentioned above, each CU may contain 1, 2, or 4 PUs depending on the partition mode, and there are up to 8 partition modes for the inter-coded CUs. In the RDO process, the optimal modes including CU, PU, and TU are chosen as the one with the minimal RD cost, and the following equation is the corresponding decision formula:

$$
\begin{aligned}
&\min\{J = D + \lambda \cdot R\} \\
&J_{mode} = \omega_{chroma} \cdot SSD_{chroma} + SSD_{luma} + \lambda \cdot R_{bit}
\end{aligned}
\tag{1}
$$

where SSD_{luma} denotes the sum of the squared differences between the luma original blocks and the reconstructed luma blocks. Similarly, SSD_{chroma} equals to the sum of the squared differences between the two chroma blocks. Rbit specifies the bit cost of the mode decision. The parameter ω_{chroma} is the regulation factor in order to weigh the chroma part of the SSD, and λ denotes the Lagrangian multiple.

In HM, the encoder will recursively choose the optimal size combinations of the CU, PU, and TU, according to the "try all and find the best" strategy. In the ith CU level, there are just two results of the division, namely split and non-split. And the RD cost is correspondingly calculated in the split and non-split manner, respectively. The RD cost of CU_i is calculated as follows:

$$
J_{min\_CU_i} = \min(J_{CU_i\_unsplit}, J_{CU_i\_split})
\tag{2}
$$

$$
J_{CU_i\_split} = \sum_{i=0}^{3} J_{min\_CU_i}^{j} \quad \text{and} \quad J_{CU_3\_split} = \sum_{k=0}^{3} J_{4 \times 4}^{k}
\tag{3}
$$

According to the equation (4), we can judge whether to split the CU_i or not.

$$
splitflag_{CU_i} =
\begin{cases}
unsplit & J_{CU_i\_unsplit} \leq J_{CU_i\_split} \\
split & J_{CU_i\_unsplit} > J_{CU_i\_split}
\end{cases}
\tag{4}
$$

19

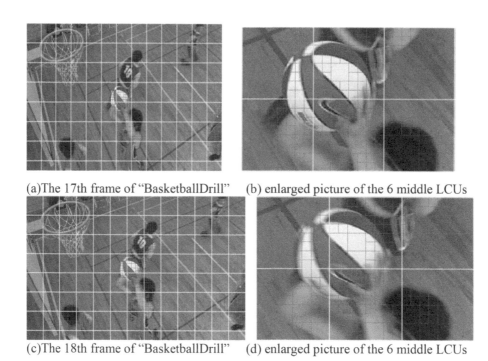

(a)The 17th frame of "BasketballDrill"   (b) enlarged picture of the 6 middle LCUs

(c)The 18th frame of "BasketballDrill"   (d) enlarged picture of the 6 middle LCUs

Figure 2.   CU modes between the current picture and the neighboring frames.

where $splitflag_{CU_i}$ denotes the optimal CU split. $J_{CU_{i\_split}}$ and $J_{CU_{i\_split}}$ denotes the RD costs of the CU encoded in the *unsplit* and *split* manner, respectively.

2.3   *CU correlations in HEVC*

In HEVC, when compared with H.264/AVC, many coding technologies are extended significantly, especially the intra and inter prediction. The intra predication of HEVC operates in the spatial domain, to eliminate redundant information in space. Similarly, interpicture predication operates in the temporal domain. This mechanism of prediction tries to achieve higher compression rates. In Figure 2, we can clearly see that in the "BasketballDrill" video, the final CU modes between the current picture and the co-located area of the neighboring frame are similar. So we can fully exploit the correlations of the CU modes in the neighboring frames to predict the following frames' CU modes, and we can skip the global searching which is the most complex and time consuming process.

We can clearly see that in Figure 2(a) and Figure 2(c), CU modes in the co-located neighboring frame have a majority of similarities, especially in Figure 2(b) and Figure 2(d), which are the enlarged pictures of the 6 middle LCUs of the left pictures, respectively. In the right two pictures, the CU modes are almost the same in the upper-middle LCU, and more than 80% of the area of the enlarged picture has the same CU modes. Therefore, it is an efficient method to predict the following frames using machine learning.

3   THE PROPOSED METHOD BASED ON MACHINE LEARNING

3.1   *The proposed CU depth decision structure*

The CU depth decision is a recursive process, Figure 3(a) shows the original computational process in the HM. Where D_n represents a process of checking the CU at the depth level n. $P_n(i)$ is the CU depth recursive process, where $n \in \{0, 1, 2, 3\}$ is the depth level, $i \in \{0, 1, 2, 3\}$ is the index of the sub-CUs. This is the typical "try all and find the best" strategy (Sullivan et al. 2012), which has the best RD performance but with the highest complexity. In Figure 3(b), D_n is firstly checked and then a classification is applied to predict whether the CU coding shall be early terminated or not (Shen & Yu. 2013). In this structure, D_n is the essential and necessary component. For example, if the $P_3(i)$ is the optimal, then D_0, D_1, and D_2 checking is unnecessary. In order to skip the unnecessary checking and minimize the coding complexity, a new structure was proposed in Zhang et al (2015), as shown in Figure 3(c). In the new structure, classification is performed firstly to predict the CU depth, and the D_n checking will be skipped.

However, the disadvantage is obvious, as the misclassification is not of any concern in the structure and the prediction accuracy of the classifier cannot guarantee a sufficient high level. To solve the problem in Figure 3(c), we propose a new structure, which is shown in Figure 3(d).

Next, the most important but also challenging work is to design a suitable classifier which has a great relationship with the features of the video and the parameters which are the input part of the SVM.

3.2 The analysis of the fast CU depth decision structure

In order to reduce the complexity, we introduce several representative features which are related to CU splitting. Good features have many advantages, such as reducing training time, improving the prediction accuracy, and reducing storage requirements. As we all know, the video content itself has abundant texture features, so making full use of these texture features will make a big difference. We learn from experience, by Shen & Yu (2013), who adopt an F-score method. We treat these features as the input elements of the learning algorithm, then we get the appropriate CU partition level of a picture. However, the predicted accuracy may not always be kept at the high level, there certainly exist some CUs which have not attained the best performance. The following is the design of the verification.

We put the MV feature of the CU block in the first place. If the value of the MV feature is less than a certain threshold, we can draw the conclusion that the motion of the vertical direction and horizontal direction are relatively small, so the pictures can be treated as homogeneous pictures. We define a MV set $\{MV_0, MV_1, MV_2, MV_3, MV_4\}$, which represents the value of the CU in the current layer and its four sub-blocks, respectively. If all the five parameters are less than a certain value, we can obtain that this CU block is relatively flat at the current level, and the recursive split process should be early terminated. Otherwise, it means that the current CU block should be split, and then skip the CU encoding process of the current layer directly to do the division of the next layer.

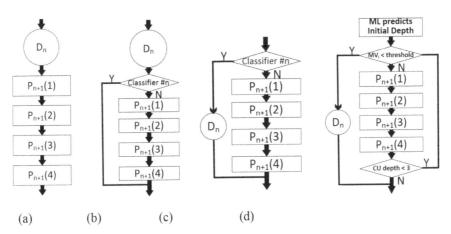

Figure 3.   Different structures. (a) Structure in HM. (b) Early termination structure (Shen's scheme) (Shen & Yu. 2013). (c) Fast scheme structure. (d) The proposed structure.

$$splitflag_{CU_i} = \begin{cases} unsplit & MV_j < \text{threshold} \quad j \in \{0, 1, 2, 3, 4\} \\ split & \text{otherwise} \end{cases} \tag{5}$$

Through the above analysis, the algorithm is as follows:

1. First, according to the video content and the situation of the CU modes, texture complexity and the correlation in temporal domain, put these features as the input of machine learning algorithm, and then predict the appropriate CU partition level, which can be called the initial split numbers, the abbreviation is ISN.
2. Second, verification process is going on. Use formula (5) to decide whether the current CU split is right. If the parameter set MV_j is less than the threshold, we judge the current CU partition to be acceptable. Otherwise, go to step 3.
3. Finally, continue the recursive CU split process for those CUs whose MV_j is bigger than the threshold. Once the split process is completed, if the parameter set MV_j satisfies the condition that all the members are less than the threshold, then the whole process is over. Otherwise, repeat step 3 until the CU depth equals to the maximum CU depth.

3.3 Weighted SVM

We model the CU splitting as a binary classification problem (a CU which is split into four sub-parts is labeled as +1, the rest is labeled as −1), and we employ the SVM to tackle the classification problem. As the classification conditions are unable to handle all the cases, so the misclassification has a small probability of occurrence. On the one hand, in order to reduce the impact of unexpected cases, on the other hand in order to maintain the RD performance, we introduce the SVMs and set RD loss caused by the misclassification as weights in SVM training. As we all know, the most challenging but vital principle is to derive the most suitable separating hyperplane that can effectively distinguish and separate the two classes. Then the following is a brief derivation process of the weighted SVM.

Given m training examples, we set the training samples as:

$$\{x_i, y_i, W_i\}_{i=1}^m \quad x_i \in R^N, y_i \in \{-1, 1\}, W_i \in R \tag{6}$$

$$f(x) = w^T \phi(x) + b, \tag{7}$$

where $\{x_i, y_i, W_i\}$ is the ith training example, x_i is the input feature and vector of the hypothesis, and y_i is the corresponding class label that when it equals to 1 indicates the current CU has been split. On the contrary, −1 means the CU is not split. The discrimination function of the hyperplane is represented by $f(x)$, and $f(x)$ equals to 0 when the most suitable parameters are found. When the input x_i is mapped into a high dimensional space that is when we see $\phi(x)$ as a nonlinear operator.

In the model, the weights are defined as the percentage of RD loss caused by misclassification, and the formula is as follows:

$$\begin{cases} W_i = \dfrac{RD_i(s) - RD_i(u)}{RD_i(u)}, & \text{when CU is finally encoded in one CU} \\ W_i = \dfrac{RD_i(u) - RD_i(s)}{RD_i(u)}, & \text{otherwise} \end{cases} \tag{8}$$

where $RD_i(s)$ and $RD_i(u)$ are the RD cost of the splitting mode and the un-splitting mode, respectively.

And then the cost function becomes

$$J(w) = \frac{1}{2} w^T w + C \sum_{i=1}^m \varepsilon_i W_i \tag{9}$$

22

Combined with the corresponding constraints, we performed the Lagrange method on formula (9), and then we get the formula (Chung & Jen. 2011) as follows:

$$L(w,b,\alpha,\gamma,\varepsilon) = \frac{1}{2}w^2 + C\sum_{i=1}^{m}\varepsilon_i W_i - \sum_{i=1}^{m}\alpha_i(y_i(w^T\phi(x_i))+b)-1+\varepsilon_i) - \sum_{i=1}^{m}\varepsilon_i\gamma_i \qquad (10)$$

then the parameter is as follows:

$$\alpha^* = \arg\max_{\alpha}\sum_{i=1}^{m}\alpha_i - \frac{1}{2}\sum_{i=1}^{m}\alpha_i\alpha_j y_i y_j\phi^T(x_i)\phi(x_j) \qquad (11)$$

subject to

$$\sum_{i=1}^{m}\alpha_i y_i = 0, 0 \le \alpha_i \le CW_i, i = 1, 2, ..., m \qquad (12)$$

With the above parameters, we adopt the offline training mode. First, the video sequences in the different classes are encoded with the original HM encoder to generate training data. Then, we use the offline learning algorithm to train a model. Finally, the trained model is used as a predictor to encode the video sequences in turn.

4 EXPERIMENTAL RESULTS AND ANALYSIS

In this section, to verify the accuracy of the proposed algorithm, three video sequences, Cactus, BQMall, and FourPeople were selected as the input training data source. Thirty frames in each sequence were encoded with four QPs, namely 22,27,32, and 37. Considering all the samples in the different CU levels, seventy percent of them were used as the training examples and the rest were used as testing examples, which means that the number of the split mode samples are generally the same amount of the un-split mode samples.

To evaluate the performance of the proposed fast CU splitting algorithm, a serial of comprehensive experiments were conducted by comparing the proposed algorithm with the benchmarks on the HM12.0. We perform the coding experiments in the Low delay B main profile. To measure the performance of the proposed algorithm the average BD-rate and the time reduction ratio were mentioned in the Table 1. The time reduction ratio is defined as:

$$\Delta T = \frac{T_{proposed} - T_{HM}}{T_{HM}} \times 100\% \qquad (13)$$

We compared the performance of the proposed algorithm with the three benchmarks, including Shen's scheme (Shen & Yu. 2013) (denoted by Shen), Xiong's scheme (Xiong et al. 2014) (denoted by Xiong), and Zhang's scheme (Zhang et al. 2015) (denoted by Zhang). Different test sequences from class A to class E were treated as input to the trained encoder, and the video sequences were encoded with four QPs, which are 22, 27, 32, and 37. The BDBR and ΔT comparisons (Bjøntegaard. 2001) between the proposed algorithm and the three benchmarks are shown in the Table 1.

In Table 1, we can observe that the complexity reduction of Shen is from 22.42% to 70.32%, and 48.56% on an average. Regarding BDBR, Shen's scheme achieves a maximum of 5.62% increase with respect to the HEVC reference software HM 6.0 with an average of 4.74%. At the same time, we can obtain the Xiong's scheme that can reduce the complexity from 61.15% to 26.70%, 44.47% on an average when compared to the original HM. The BDBR is 6.95% on an average. Xiong's scheme achieves a significant reduction in time; however, the RD degradation is relatively large. As for the Zhang's scheme, it achieves a complex reduction from 29.01% to 70.91%, 50.29% on an average, which is better than the other two

23

Table 1. BDBR and the time reduction ratio comparisons between proposed algorithm and the benchmarks.

Class	Sequence	Shen vs HM		Xiong vs HM		Zhang vs HM		The Proposed vs HM	
		ΔT %	BDBR %	ΔT %	BDBR %	ΔT %	BDBR %	ΔT %	BDBR %
A	PeopleOnStreet	-26.48	4.93	-32.04	10.95	-44.44	1.24	-42.49	1.97
	Traffic	-52.37	4.88	-46.62	6.36	-56.10	2.05	-57.32	2.56
	NebutaFestival	—	—	—	—	—	—	—	—
	SteamLocomotive Train	—	—	—	—	—	—	—	—
B	BasketballDrive	-48.40	3.05	-52.10	10.18	-50.76	2.25	-49.84	2.06
	BQTerrace	-50.62	2.80	-40.08	5.95	-45.01	2.16	-47.04	2.51
	Cactus	-49.00	5.24	-52.01	11.94	-48.42	1.97	-47.85	1.89
	Kimono	-52.32	4.83	-46.60	6.38	-56.13	2.08	-57.32	2.12
	ParkScene	-55.41	4.49	-42.32	5.30	-47.82	1.46	-46.28	2.04
C	BasketballDrill	-39.42	4.71	-38.90	7.12	-47.53	1.86	-49.30	1.96
	BQMall	-41.53	5.38	-37.01	5.62	-49.56	1.78	-49.49	1.77
	PartyScene	-39.94	5.41	-32.50	5.52	-36.63	1.30	-32.97	1.43
	RaceHorses	-38.28	5.62	-31.46	5.40	-35.49	1.69	-31.33	1.85
D	BasketballPass	-28.35	5.25	-40.45	2.76	-36.43	1.29	-34.68	1.26
	BlowingBubbles	-34.42	5.08	-26.70	4.60	-29.01	0.89	-32.81	0.92
	BQSquare	-40.38	4.68	-29.45	3.20	-30.67	0.84	-34.35	0.89
	RaceHorses	-22.42	5.49	-30.33	4.11	-33.37	1.23	-28.73	2.51
E	FourPeople	-64.80	5.42	-58.32	11.75	-66.28	2.83	-64.09	3.53
	Johnny	-70.32	4.35	-61.15	8.89	-70.91	2.48	-66.74	3.29
	KristenAndSara	-67.41	4.16	-57.70	7.20	-68.23	2.28	-65.32	3.23
Others	Vidyo1	-65.60	4.52	-60.16	8.57	-68.17	2.66	-66.05	2.85
	Vidyo3	-65.62	4.60	-58.90	7.63	-67.47	3.42	-65.07	3.68
	Vidyo4	-66.73	4.55	-59.15	6.45	-67.81	2.90	-63.20	3.01
Average		-48.56	4.74	-44.47	6.95	-50.29	1.93	-49.15	2.35

benchmarks. Meanwhile, the BDBR is within 3.42% to 0.84%, 1.93% on an average, which is also the smallest among the three benchmarks.

As for the proposed method, it achieves the average time saving about 49.15% with the time saving that goes up to 66.74% in the sequence "Johnny." Concerning the RD performance, it gains 2.35% in terms of the BDBR on an average, and the worst case appears in the sequence "Vidyo3". As we all know, the more there is an increase in the BDBR, the more degradation there is in the video quality. Based on this perception, we can find that the proposed method with a simpler algorithm structure than Zhang's scheme has a better performance than the Shen's scheme and the Xiong's scheme, and roughly the same level of performance is compared with the Zhang's scheme.

5 CONCLUSION

In this paper, we propose a fast CU depth decision algorithm based on machine learning. First, we review the CU partition process in the original HEVC, and find the shortcomings of the current standards. Second, we propose our CU depth selection scheme, then we take advantage of the features of the video content as the validation. Finally, we present the training process and the final experiment results. Through the analysis, a conclusion can be drawn: our proposed method is effective. In the future, we will fully exploit and select the features of the video content, which will in turn improve the performance of the whole algorithm.

ACKNOWLEDGMENTS

This work is supported by project NSFC 61471066 and the open funding project of State Key Lab of Virtual Reality Technology and Systems at Beihang University under Grant No. BUAA-VR-15 KF-19.

REFERENCES

Bjøntegaard, G. Apr. 2001. Calculation of average PSNR differences between RD-Curves. *ITU-T VCEG-M33, Austin, Texas, USA.*

Bossen, F. Jan. 2013. Common test conditions and software reference configurations. *Joint Collaborative Team on Video Coding*, document Receive JCTVC-L1100.

Chung, C. & L.C. Jen. 2011. LIBSVM: A library for support vector machines. *ACM Transactions on Intelligent System and Technology,* vol. 2, no. 27, pp. 1–27.

Shen, L., Z. Zhang & P. An. Feb. 2013. Fast CU size decision and mode decision algorithm for HEVC intra coding. *IEEE Transactions on Consumer Electronics.* vol. 59, no. 1, pp. 207–213.

Shen, X. & L. Yu. Jan. 2013. CU splitting early termination based on weighted SVM. *EURASIP Journal on Image and Video Processing.* vol. 2013, Art. ID 4.

Sullivan, G.J., J. Ohm, W.J. Han & T. Wiegand. Dec. 2012. Overview of the high efficiency video coding (HEVC) standard. *IEEE Transactions on Circuits and System for Video Technology*, vol. 22, no. 12, pp. 1649–1668.

Xiong, J., H. Li, Q. Wu & F. Meng. Feb. 2014. A fast HEVC inter CU selection method based on Pyramid Motion Divergence. *IEEE Transactions on Multimedia,* vol. 16, no. 2, pp. 559–564.

Zhang, Y., S. Kwong, X. Wang, H. Yuan, Z. Pan & L. Xu. Jul. 2015. Machine learning-based coding unit depth decisions for flexible complexity allocation in High Efficiency Video Coding. *IEEE Transactions on Image Processing,* vol. 24, no. 7, pp. 2225–2238.

Signal and Information Processing, Networking and Computers – Chen & Huang (Eds)
© 2016 Taylor & Francis Group, London, ISBN 978-1-138-02881-4

Kernel-based learning technique for Cooperative Spectrum Sensing in Cognitive Radio

Xin Chen, Xiaojun Jing, Songlin Sun & Hai Huang
Key Laboratory of Trustworthy Distribute Computing and Service (BUPT), Ministry of Education, Beijing University of Post and Telecommunication, Beijing, China

Xiaohan Wang & Dongmei Chen
305 Hospital of PLA, Beijing, China

ABSTRACT: In this paper, a novel Cooperative Spectrum Sensing (CSS) algorithm for Cognitive Radio (CR) networks based on the machine learning technique is proposed. As a very powerful tool in machine learning, the kernel method is applied for pattern classification in CSS. The received signals from different spectrum sensors are fed into the classifier for channel availability decision. In this regard, the kernel-based Principle Component Analysis (PCA) technique, with a leading eigenvector is implemented for feature extraction from the received signal. The classification is according to the feature's similarity between the received signal and the primary signal. In order to improve the performance of the CR system, especially when the RF environment changes dramatically, an adaption mechanism is applied to allow efficient adaption towards the CR environment rather than a fixed model. A low SNR value after the spectrum access indicates that an adaption step (i.e. a new training process) is required for the latest RF environment. The performance of the proposed scheme is quantified in terms of the detection probability, the false alarm probability and the analysis of the Receiver Operating Characteristic (ROC) curve. The simulation results clearly reveal that the proposed scheme can be very appropriate for spectrum sensing, especially when the RF environment changes with time.

Keywords: cognitive radio; cooperative spectrum sensing; principle component analysis; kernel; primary user classification

1 INTRODUCTION

With the fast development of wireless communication technology, the need for a higher data transmission rate is increasing, which results in the scarcity of spectrum resources. However, according to the report from the Federal Communications Commission (FCC), the spectrum utilization ranges from 15% to 80%, which indicates that large parts of licensed bands are underutilized. In order to improve the efficiency of spectrum utilization, Cognitive Radio (CR) technology is proposed to provide a more intelligent manner of spectrum management. As for the utilization of these portions of unoccupied spectrum resources, CR allows dynamic spectrum access to those licensed bands that remain to be occupied. The appropriate DSA scheme is a kind of learning engine, which can make decisions according to the observation result. Among the various challenges in cognitive radio, spectrum sensing is the major problem that constraints the development of cognitive radio. The precondition of vacating channel utilization is that SU should sense its surrounding environment and identify all types of RF activities. G.Q. Ning et al. (2011) says a spectrum sharing model based on channel classification is built for a centralized network. The channel assignment is according to the chart eristic of CR users including the channel busy duration, the location of SUs and the fairness of channel access opportunity. The system model in J.Q. Duan et al. (2011) is similar to that in G.Q. Ning et al. (2011). SUs at the edge of the cell can acquire more channel resources by allowing multi-hop access to the CR-BS. In S. Sarkar et al. (2008), CR users are divided into two categories. The system will reserve some idle channel for SUs in the prior group.

Cooperative Spectrum Sensing (CSS) can be exploited especially when the CR devices are distributed in different locations. In I.F. Akyildiz et al. (2011), it is possible for the CR devices to cooperate in order to achieve higher sensing reliability than individual sensing does by yielding a better solution to the hidden PU problem that arises because of shadowing and multi-path fading (D. Cabric et al. 2004, A. Sahai 2006).

Since the degree of freedom in the cognitive radio network is increasing, it is difficult to decide the channel state with one fixed or static model. In this case, a dynamic scheme of spectrum sensing is required, which can adjust several parameters and policies simultaneously (e.g. transmit power, sensing time duration, access policy, hand-off mechanism, threshold evaluation, coding scheme, sensing algorithm, etc.), when the RF environment changes dramatically. Thus, an adaption mechanism can be applied to improve the performance of spectrum sensing.

In this paper, a CSS scheme under the framework of kernel (G. Lanckriet et al. 2004) PCA with adaption mechanism is proposed. As a kind of fast machine learning method, the training process can be finished without external support, which is appropriate for an automatical training process in the adaption process. To begin with, the kernel PCA scheme is exploited for feature extraction from the collection of received signal form SUs. In the feature space, the inner-product is taken as a measure of similarity between eigenvectors of the received signal and the original primary signal. If they are similar to each other under a certain threshold, the channel is treated to vacate the channel and SUs can access the licensed band. In the meantime, an adaption mechanism towards the previous kernel PCA scheme is exploited whenever the SNR value for spectrum access is unacceptable. If that happens, the CR system will restart the training process based on the current received signal with the label "channel available" and maintain a new eigenvector from the training sample, which is more appropriate for the current RF environment. With the help of the kernel PCA method, the CR system can automatically restart the training process that treated the latest received signal as training sample within a short period of time. Compared to the traditional way of spectrum sensing, the proposed scheme allows efficient adaption towards the RF environment.

The rest of the paper is organized as follows. In Section 2, the system model of a CR network is presented. The CSS scheme under the framework of kernel PCA and its adaption mechanism is described in Section 3. The experiment scenario and simulation result will be presented in Section 4. Finally, the conclusion is shown in Section 5.

2 SYSTEM MODEL

2.1 *Network model*

The CR network consists of N secondary users with the indexed number n = 1, ..., N and one primary transmitter (or base station) located in the center of the cell. In the SU network, CR is in the central part of the cell, while SUs are randomly distributed around the base station. The SU shall do the spectrum sensing work through the entire bandwidth within the PU's coverage area, which is defined according to the distance between the primary transmitter and SUs. Only the licensed channel within the coverage area of the primary transmitter will be sensed by SU. The coverage area in the CR network is shown in Figure 1. SU located in the overlap area of two PUs shall participate in the CSS of the licensed band towards both the two PUs. SU located outside of the coverage area shall not participate in the CSS. In this way, the accuracy of the detection can be improved and avoid inter-PU interference.

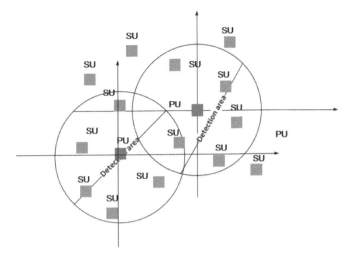

Figure 1.   Coverage area of the CR network.

2.2   *Detection model*

The availability judgment of the licensed band is based on the analysis of the received signal $y(n)$ from SUs, which are in the coverage area of the primary transmitter. According to the feature of the received signal, the licensed channel condition can be divided into two scenarios: one is that the licensed channel is occupied H_1 and the other is that the licensed channel is vacant H_0.

In practice, $y(n)$ is implemented for the judgment of primary user and the result of spectrum sensing can be classified into two scenes:

$$y(n) = \begin{cases} w(n) & H_0 \\ x(n) + w(n) & H_1 \end{cases} \tag{1}$$

where $y(n)$ is the received signal from the n-th SU and consists of $x(n)$ and $w(n)$, which represent the primary user's signal and the noise, respectively.

In order to improve the accuracy of the spectrum sensing algorithm and maintain appropriate training data that can represent the real primary signal, SUs in the coverage area of the primary transmitter will work together. The final training data is the collection of the received primary signal from all SUs, represented by N-dimensional vector $y = (y(1), y(2), ..., y(N))^T$ with a label of channel availability. $y(i)$ means the received primary signal form the i-th SU. The first training sample signal is known as priority and the training sample signal in the adaption module is the received signal, directly with the default label of H_1.

3   SPECTRUM SENSING ALGORITHM

3.1   *Kernel-based PCA detection algorithm*

Compared to the classical PCA approach, a kernel based PCA is exploited in the spectrum sensing algorithm. As a nonlinear method, the problem can be discussed in a higher dimensional feature space, which is assumed to be more suitable for problem discussion. In this way, a better performance of spectrum sensing can be achieved, compared to that in the original space. With the help of kernel function, the mapping φ between the original feature space and the high dimensional feature space is not necessary. In this way, the channel detection algorithm with kernels can achieve a better performance than the linear way of problem discussion.

In the proposed spectrum sensing algorithm, $y(i)$ indicates the labeled training sample and $y(j)$ represents the received signal samples from several SUs without a label (i.e. the test sample). The kernel-based PCA method is implemented in the training sample and the test samples for eigenvector extraction. The similarity between the two eigenvectors is exploited for channel availability decision.

The kernel function is exploited to map the original data $y(n)$ towards higher dimensional feature space represented by $\varphi(y_n) = \varphi(y_1), \varphi(y_2), \dots, \varphi(y_n)$. The mean value of both the training sample and the testing sample in the feature space are zero, e.g. $\frac{1}{N}\sum_{n=1}^{N}\varphi(y_n) = 0$. Thus, the sample covariance matrix of $\varphi(y_n)$ is

$$R_{\varphi(y)} = \frac{1}{N}\sum_{n=1}^{N}\varphi(y_n)\varphi(y_n)^T \tag{2}$$

The received signal $Y(n), n = 1, 2, \dots, N$ is collected from different SUs within the coverage area, and the leading eigenvector $\hat{v}_1^f$ of the sample covariance matrix $R_{\varphi(y)}$ is the linear combination of the feature space data represented as:

$$\hat{v}_1^f = \sum_{n=1}^{N}\tilde{\beta}_n\varphi(Y_n), \tag{3}$$

where $\tilde{\beta}_n = (\tilde{\beta}_1, \tilde{\beta}_2, \dots, \tilde{\beta}_N)^T$ is the leading eigenvector of the kernel matrix.

$$\tilde{K} = (k(y_i, Y_j))_{ij} = (<\varphi(y_i), \varphi(Y_j)>)_{ij} \tag{4}$$

The similarity between the training sample v_1^f and the testing sample $\hat{v}_1^f$ is measured by the inner-product.

$$\begin{aligned}
\rho = <v_1^f, \tilde{v}_1^f> &= <\sum_{i=1}^{N}\beta_i\varphi(y_i), \sum_{j=1}^{N}\beta_j\varphi(Y_j)> \\
&= \{\beta_1(\varphi(y_1), \varphi(y_2), \dots, \varphi(y_N))\}^T \cdot \{\tilde{\beta}_1(\varphi(Y_1), \varphi(Y_2), \dots, \varphi(Y_N))\} \\
&\quad k(y_1, Y_1), k(y_1, Y_2), \dots, k(y_1, Y_N), \\
&= \beta_1^T \left(k(y_2, Y_1), k(y_2, Y_2), \dots, k(y_2, Y_N) \right)\tilde{\beta}_1 \\
&\quad k(y_N, Y_1), k(y_N, Y_2), \dots, k(y_N, Y_N), \\
&= \beta_1^T K \tilde{\beta}_1
\end{aligned} \tag{5}$$

The equation above indicates a measurement of similarity between v_1^f and $\hat{v}_1^f$ without knowing v_1^f and $\hat{v}_1^f$. K is the kernel matrix between $\varphi(y_i)$ and $\varphi(Y_j)$ represented by:

$$K = (k(y_i, Y_j))_{ij} = (<\varphi(y_i), \varphi(Y_j)>)_{ij} \tag{6}$$

The flow chart of the for spectrum sensing mentioned above is shown in Figure 2 and Figure 3. The proposed detection algorithm with leading eigenvector under the frame work of kernel PCA is summarized here as follows:

1. The received signal of primary user $y_1, y_2, \dots, y_M$ is treated as training data and the kernel function k is exploited for feature extraction, the kernel matrix is $K = (k(y_i, y_j))_{ij}$. The result of feature extraction is the leading eigenvector β_1 obtained by eigen-decomposition of K.
2. The received vectors are $y_1, y_2, \dots, y_M$. According to the same kernel function mentioned in the first step, the kernel matrix $\tilde{K} = (k(Y_i, Y_j))_{ij}$ is obtained. By eigen-decomposition of $\tilde{K}$, the feature of kernel matrix (leading eigenvector) is obtained represented by $\tilde{\beta}_1$.
3. The expression of leading eigenvectors for both the training signal and received signal is:

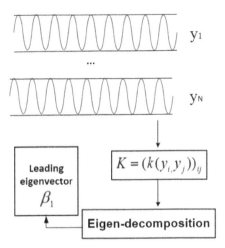

Figure 2.   Feature extraction from the training sample.

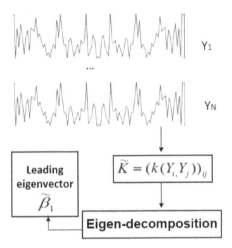

Figure 3.   Feature extraction from the testing sample.

$$v_1^f = (\phi(y_1), \phi(y_2), ..., \phi(y_M))\beta_1$$
$$\tilde{v}_1^f = (\phi(Y_1), \phi(Y_2), ..., \phi(Y_M))\tilde{\beta}_1 \qquad [7]$$

4. The similarity between the leading eigenvectors of v_1^f and $\tilde{v}_1^f$ is calculated by:

$$\rho = \beta_1 K' \tilde{\beta}_1 \qquad [8]$$

5. The presence or absence of primary signal x(n) in y(n) is determined by threshold evalua-
    tion (i.e. whether $\rho > \eta$ or not) where η is the value of threshold.

3.2   *Adaption mechanism*

In order to achieve an adaptive spectrum sensing process under uncertainty and mutable RF
environment, we proposed an adaption (or feedback) mechanism for the spectrum sensing
algorithm. When the previous model or classifier is not appropriate to measure the feature

space, the SUs will stop their spectrum access towards the licensed band and restart the training process to maintain a new model following the same kernel PCA method mentioned above.

Whether a model is appropriate or not depends on the SNR (Signal Noise Ratio) value that SU can achieve when accessing the licensed channel. The SNR value is a measurement of performance for the CR system, which will be calculated through the whole spectrum sensing process. In order to make an appropriate evaluation towards the existing spectrum sensing model or classifier, the mean value of the latest 30 SNR result is exploited, which is defined as:

$$A_i = \sum_{i=i-30}^{i} A_i \qquad [9]$$

where A_i is the SNR value in the i-th moment. If the latest average SNR value is equal to or higher than the average value within the previous 30 time slots, it means that the existing CR classifier has a better understanding of the CR environment and can make a correct pattern recognition of the primary signal. However, the classifier with lower SNR value indicates that the RF environment has changed greatly. The primary signal may be in a different form and contain a different feature, which is unknown in the existing model. Thus, a new training process based on the latest type of primary signal is required.

The training sample is the latest primary signal received from the spectrum sensors. The SUs will stop spectrum access as long as the adaption (or feedback) mechanism begins. The received primary signal results from all N spectrum sensors that will be gathered to form the available training vectors $y = (y(1), y(2), ..., y(N))^T$. Exploiting the latest received primary signal as the training data can get a suitable model towards the existing environment.

After the training process is completed, SUs can get ready for the dynamic spectrum access until the next adaption process begins.

4 SIMULATION RESULT AND DISCUSSION

The experiment results will be compared with the results of AND Rule, linear FDA and kernel FDA. The primary user's signal assumes to be the sum of three sinusoidal functions with unit amplitude of each.

In this study, 5 PUs (i.e. 5 licensed channel) and 50 SUs uniformly located in a circle of 10,000 πm^2 area are shown in Figure 2. The probability that a PU is in the active state

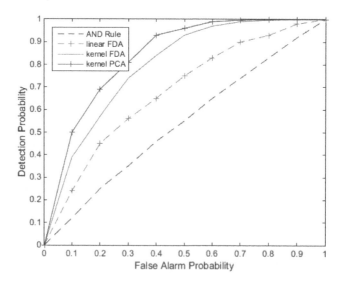

Figure 4. ROC curve for CSS.

follows a Poisson distribution with mean value λ. The time slot for SNR calculation is 1 ms.

The important simulation parameters are shown as follows: channel bandwidth (W) is 10 MHz, sensing duration T is 100 us, spectral density of noise N is −174d Bm, the path-loss exponent is 3, and the standard deviation of shadowing is 8dB. The transmit power of each PU is 200 mW. The location of both PUs and SUs are considered to be fixed.

From the spectrum sensing performance comparison in the ROC curve in Figure 4, it is shown that the detection performance of the proposed kernel PCA method has a better performance than the rest of the classification scheme. Besides, the performance of the kernel method can achieve a much higher detection probability with a low false alarm probability. The reason for this phenomenon is that the kernel method can describe the model with non-linear function, which is more appropriate for the CR network.

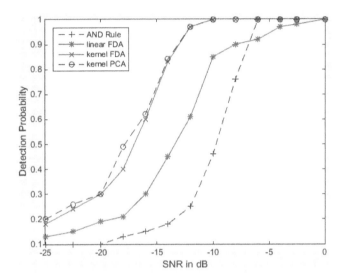

Figure 5.   Detection probability with SNR.

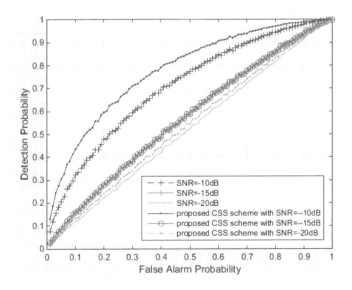

Figure 6.   Comparison of the CSS scheme with and without adaption module.

33

It is shown in Figure 5 that the detection probability of the spectrum sensing model increased with the SNR value increasing, which indicates that an SNR value of the RF environment has a great influence on the performance of spectrum sensing. In order to maintain a stable detection probability, we need a stable environment with a static SNR value or an adaptive CSS scheme towards the RF environment.

The ROC curves for the comparison of the kernel PCA method and the adaptive way of the kernel PCA are shown in Figure 6. The training sample of the two methods is received primary signal with channel SNR value of -15 in db. In order to evaluate the performance of the adoption module, we test our scheme in different RF environments. The simulation result is a comparison of the CR system, between a scheme with and without the adaption module under the channel SNR value of −15 db, −20 db and −10 db. It is observed in Figure 6 that the performance of the proposed scheme can be improved with the help of the adaption module when the RF environment changes.

5 CONCLUSION

In this paper, we have designed a Cooperative Spectrum Sensing (CSS) algorithm in CR networks based on pattern recognition techniques. The received signal of every SU within the coverage area is exploited for PU detection under the framework of kernel PCA. The inner product between leading eigenvectors is taken as the similarity measurement between the received signal and the original primary signal. As a dynamic scheme for classification, an adaption mechanism is proposed. Compared to the static way of classification, the proposed scheme can automatically adjust sensing policy and parameters of the classifier, based on the latest primary signal. Experiments are conducted to make a comparison between the proposed spectrum sensing scheme and the traditional way of vacant channel detection. The simulation result shows that the proposed scheme can improve the performance of the CR system, especially when the RF environment changes with time.

ACKNOWLEDGMENTS

This work is supported by project NSFC 61471066 and the open funding project of the State Key Lab of Virtual Reality Technology and Systems at Beihang University under Grant No. BUAA-VR-15 KF-19.

REFERENCES

Sarkar, S., C. Singh, A. Kumar, "A coalitional game model for spectrum pooling in wireless data access networks," in Proc. IEEE Conf. on Information Theory and Applications Workshop, 2008, pp. 310–319.

Akyildiz, I.F., Lo, B.F. and Balakrishnan, R. "Cooperative spectrum sensing in cognitive radio networks: A survey," Physical Communications, vol. 4, no. 1, pp. 40–62, March 2011.

Cabric, D., Mishra, S.M. and Brodersen, R. "Implementation issues in spectrum sensing for cognitive radios," in Proc. of 38rd Asilomar Conf. on Signals, Systems, and Computers, Pacific Grove, CA, Nov. 2004.

Lanckriet, G., Cristianini, N., Bartlett, P., Ghaoui, L. and Jordan, M. "Learning the kernel matrix with semidefinite programming," The Journal of Machine Learning Research, vol. 5, pp. 27–72, 2004.

Ning G.Q., Cao X.G. and Duan J.Q., et al. A Spectrum Sharing Algorithm Based on Spectrum Heterogeneity for Centralized Cognitive Radio Networks [C]. 2011 IEEE 73rd Vehicular Technology Conference (VTC Spring), 2011: 1–5.

Ning G.Q., Duan J.Q. and Su J, et al. Spectrum Sharing Based on Spectrum Heterogeneity and Multi-hop Handoff in Centralized Cognitive Radio Networks [C]. 2011 20th Annual Wireless and Optical Communications Conference, 2011: 1–6.

Sahai, A., Tandra, R., Mishra, S.M. and Hoven, N. "Fundamental design tradeoffs in cognitive radio systems," in Proc. of TAPAS'06, Boston, MA, Aug. 2006.

Signal and Information Processing, Networking and Computers – Chen & Huang (Eds)
© *2016 Taylor & Francis Group, London, ISBN 978-1-138-02881-4*

Future mobile network: A framework for enabling SDN in mobile operator networks

Xu Xia, Zhenqiang Sun, Huiling Zhao, Fan Shi, Congjie Mao & Qiong Sun
Beijing Research Institute, China Telecom, Beijing, China

Lei Cao & Bo Wang
China Telecom, Beijing, China

Shuang Zhang & Alexander de Boer
TNO, The Hague, The Netherlands

ABSTRACT: SDN (Software Defined Networking) makes the mobile network more flexible and efficient, and implies the potential to support new services that are not easily supported by legacy networks. In this paper, we analyze the requirements of future mobile Internet and illustrate the restrictions of using SDN in mobile operator networks to bypass the restrictions and address the challenges.

Keywords: SDN; WLAN; LTE

1 INTRODUCTION

Along with the fast development of the electronic industry and wireless communication technology, the number of mobile terminals has exceeded that of the fixed number of terminals, which implies that the mobile network is one of the most important infrastructures to connect people with each other, providing services to the people around the world. Just like food, air, and water, the mobile network is ubiquitous around us, making our life more comfortable and convenient. But with the expanding scale of the user group and the richer service types, the current mobile network that was originally built upon the legacy of cellular networks, is facing too many challenges, such as ossified network architecture, inflexibility of service deployment, unawareness of user traffic, high CAPEX and OPEX, lack of efficient controllability on the network and so forth, which hinder the further development of the mobile network itself.

As is known that the Internet was originally designed to share the fixed and expensive hardware resources between different users, instead of serving today's Internet service scenarios, and thus the legacy Internet does not support mobility naturally. IETF (Internet Engineering Task Force) tried to introduce MIP (Mobile IP) the Internet to support the mobility at layer 3, but the result is not satisfying. Cellular network was primarily built to provide a voice communication between mobile users, but it has been evolved to support high speed data transmission for Internet services in the past decades. So, the current mobile Internet leans against the cellular network, and the mobility is supported at layer 2 in the individual cellular networks.

SDN (Software Defined Networking), which was proposed to provide an innovative networking research platform for the research community through separating the control plane from the data plane of network, has attracted much attention in both the industry the and academia during the past few years. SDN makes the network more flexible and efficient, and implies the potentiality to support new services that are not easily supported by legacy networks. SDN is more and more frequently deployed in cloud data centers, e.g. Google, NTT DoCoMo, but it has not been widely used in mobile Internet. In this paper we will talk about SDN-enabled future mobile network. The paper is thus organized as follows. Section 2 gives an introduction to SDN, and section 3 illustrates the requirements for the future mobile network. In section 4 restrictions of the current mobile network are discussed, and section 5 presents an SDN-enabled future mobile network. Section 6 concludes this paper.

2 SOFTWARE DEFINED NETWORKING

Legacy network mixes the control plane and data plane together into a single element such as switch and router, but this integration causes many problems, e.g. deep coupling of control plane and data plane which hinders independent development of the two planes, high complexity of designing and manufacturing a network element involving both the control plane and the data plane, inflexibility to deploy new network functions for data plane functions according to service situations, closeness of the network element, high CAPEX, and OPEX. These problems make legacy network ossified and hard to develop further, so SDN is consequently proposed. The basic idea of SDN is separating the control plane from the data plane of the network, and providing a series of APIs (Application Programming Interface) on the control plane to make the network potentially programmable.

2.1 *History of SDN*

The emergence of SDN is not abrupt. Being the precursor of SDN, programmable network was proposed more than ten years ago to make network resilient and facilitate network evolution. Efforts regarding programmable network are shown as follows.

Active Network: Active network was proposed in the 1990s to be a programmable network infrastructure for customized services. Two approaches were proposed for active network, i.e. user-programmable switch and capsule. The user-programmable switch receives messages from users through out-of-band interface and executes the command included in these messages. The data is transmitted through an in-bound interface. Capsule is a program fragment that can be carried in user messages to be understood and executed by switches. Users can connect several capsules into a chain to provide a series of network functions. But because of practical security and performance concerns, active network technology has never been widely deployed in commercial networks.

OPENSIG: The Open Signaling Working Group was established in 1995, and its main task was to make ATM network, Internet and mobile network more open, extensible, and programmable. The main consideration of OPENSIG was the separation between the communication hardware and the control software, which was very difficult to implement at that time because of the vertically integrated network elements. The core idea of OPENSIG was to provide access to the network hardware via open and programmable network interfaces allowing the deployment of new services through a distributed programming environment.

DCAN: DCAN (Devolved Control of ATM Network) is aimed to design and develop the necessary infrastructure for scalable control and management of ATM networks. A principle of DCAN was that the control and management functions of ATM switches should be moved to external controlling entities through which the network can be potentially programmable.

4D Project: The 4D Project was started in 2004, and its primary goal was to investigate a clean slate design that emphasized the separation between the routing decision logic and the protocols governing the interaction between network elements. It used a decision plane to collect the global view of the network, and a dissemination and discovery plane to control the forwarding behavior of the data plane.

NETCONF: NETCONF was proposed to be a management protocol for modifying the configuration of network devices, allowing the network devices to expose an API through which extensible configuration data could be sent and retrieved. NETCONF should not be regarded as fully programmable, because it was primarily designed to aid automated configuration, rather than for enabling direct control of the state or enabling quick deployment of innovative services and applications.

Ethane: Ethane, the immediate predecessor of OpenFlow, defined a new architecture for enterprise network, using a centralized controller to manage the policy and security in the network. Three components were used in Ethane, i.e. controller, switch, and secure channel. The controller is responsible for deciding if a packet should be forwarded, while the switch is responsible for forwarding a packet according to the flow table, and the secure channel connects the controller and the switch in a secure way.

2.2 Current SDN architectures

The most popular SDN architectures that are frequently talked about should be ForCES and OpenFlow. Both of the two architectures separate the control plane and the data plane of the network, and standardize the interfaces between the two planes, respectively. But they are technically different in architecture design, forwarding model, and protocol interfaces.

ForCES: ForCES (Forwarding and Control Element Separation) is proposed by IETF to redefine a new network device internal architecture through splitting the control plane and the forwarding plane. As we know legacy network device integrates control plane and forwarding plane into a single box, the tight coupling of which hinders their independent and parallel development. ForCES defines the two logic entities, i.e. the Forwarding Element and the Control Element. The Forwarding Element is responsible for packet processing resorting to the underlying hardware. In the meantime, the Control Element is in charge of executing the control and signaling functions, through which the Forwarding Element is instructed by means of the ForCES protocol to accordingly handle the packets. Another important component in ForCES is LFB (Logical Function Block), which is a well-defined functional block residing on the Forwarding Element that is controlled by the Control Element via the ForCES protocol. LFB enables the Control Element to control the Forwarding Element configuration and how the Forwarding Element processes the packets.

OpenFlow: OpenFlow is the most popular SDN architecture in both the research community and the industry. Like ForCES architecture, OpenFlow also separates the control plane and the data plane of the network, and turns the control plane into a logically centralized entity outside of the data plane devices. In OpenFlow the controller acts as the control plane and the switches act as the data plane. The data plane forwarding behavior is decided by the controller through setting the flow tables into the switches. The flow table consists of flow entries, and each entry contains three fields, i.e. match field, action field, and counter field. Match field is filled with a matching rule that is used to match incoming packets by comparing the corresponding fields of the packet header with the matching rule. Action field implies that if the incoming packet matches the rule that is specified in the match field then this action will be executed consequently. Counter field is responsible for collecting statistical information, e.g. number of matched packets of a flow, number of received bytes of a flow, and duration of a flow entry.

Though OpenFlow is similar to ForCES in terms of separating the control plane and data plane, they do adopt different approaches. ForCES modifies the internal architecture of network element to make the device programmable, whilst OpenFlow modifies the architecture of the network to makes the whole network programmable. Therefore, ForCES and Open-Flow are SDN solutions on different levels.

3 REQUIREMENTS OF FUTURE MOBILE NETWORK

As one of the most important communication infrastructure in the world, mobile network is gradually evolving in order to meet the new requirements of the unlimitedly growing Internet services. For future mobile network, many challenges should be addressed and many requirements should be met.

3.1 Ubiquitous mobile access

Current mobile network is built upon cellular networks, so mobile access is often limited to a single network operator, regardless of better coverage of other wireless networks such as WLAN coverage in a local area. Although, smart phones often provide several air interfaces, only one is allowed to use at a time and the upper layer services cannot be smoothly moved from one air interface to another. So future mobile Internet should be able to support ubiquitous access for mobile users and enable the users to access Internet wherever, whenever, and however needed.

3.2 Network convergence

Wireless communication technologies have been developed so fast that in the near future multiple wireless networks should be probably overlapped to cover the same area. In such a case, how to manage these ambient wireless networks under the same framework will be an issue. To support the ubiquitous mobile access in the same coverage as that of multiple wireless networks, network convergence technologies should be adopted, to better serve mobile users.

3.3 Capability to support new services

The number of Internet services is growing at a super linear rate, which requires future mobile Internet to be flexible enough to support the new services without negatively impacting the quality of legacy services.

3.4 Service oriented subscription

For current mobile networks, users often subscribe by data traffic or by on-line time. However, this is not a good way of user subscription, because users are often more interested in his favorite services and they are more likely to pay for their interested services. However, current mobile Internet is acting as just a pipe, unaware of user traffic, let alone mentioning the support of service oriented subscription. This requires future mobile Internet to become a smarter pipe.

3.5 Energy efficient communication

Green communication is very important for the sustainable development of the world. Future mobile Internet should be adaptive to optimize the network resource configuration according to the running network states. If the traffic is low then only a small number of network elements are needed and the others can be shut down to save energy. Energy efficiency should be a great concern for future mobile Internet.

4 CHALLENGES OF THE CURRENT MOBILE NETWORK

The current mobile network is facing many challenges that restrict its further development that is resulting from the existing architecture and formation of cellular network infrastructure. The restrictions of the current mobile network hinders the deployment of the new Internet services and makes the mobile Internet unsuitable to meet the expectedly possible requirements of future services. Future mobile networks should break all these restrictions by re-architecting the network infrastructure to bear new network functions to support new services.

4.1 Ossified network architecture

Cellular network is a dedicated network to provide voice communication and data transmission through the connections between the mobile terminal and the base station. The network is built according to 3GPP or 3GPP2 standardizations, which specify the details of the network elements and the corresponding signaling procedures. The devices of a cellular network are often provided by several vendors who use different hardware and software to implement their products, which makes the network ossified and hard to support new services. If a new network function is required to be supported in the cellular network, a series of network elements from different vendors should be simultaneously updated to support this function. This is a very huge task, and to some degree makes the network inflexible and ossified.

4.2 *Inability of moving services to heterogeneous network smoothly*

In the current mobile Internet, mobility management is only supported at layer 2 by the cellular network, but the IP mobility at layer 3 has not been widely deployed. This hurdles ubiquitous mobile access between several overlapped coverage of wireless networks. The service running on the mobile terminal is consequently not able to migrate from one air interface to another without service interruption.

4.3 *Scalability of centralized data plane element*

In current cellular networks, some centralized data plane functions impose a heavy load to the network element. Take LTE as an example, monitoring, access control, and quality-of-service functionality at the packet gateway that introduces scalability challenges to the centralized S-GW and P-GW. This also makes the equipment very expensive, e.g. more than 6 million dollars for a packet gateway of a renowned supplier.

4.4 *High CAPEX and OPEX*

The device in the cellular network is a dedicated equipment, thus, the price is so often very high, which implies a high CAPEX. As is mentioned above, a cellular network contains many types of network elements manufactured by different vendors. This makes the cellular network very complicated and hard to manage, implying a high OPEX consequently.

5 SDN-ENABLED FUTURE MOBILE NETWORKS

Since the above sections analyze the requirements of the future mobile Internet and the limitations of the current mobile Internet, this section presents an SDN-enabled future mobile Internet architecture to break these limitations and to meet these requirements.

5.1 *SDN-enabled future mobile network architecture*

According to the results from the discussion of the above sections, we propose an SDN-enabled framework for future mobile networks, as shown as in Figure 1. The future mobile Internet should not be restricted to only cellular networks but as a matter of fact including various wireless accesses, e.g. WLAN network. This framework enables SDN in all the mobile networks, so that these networks can be potentially involved with each other in this uniform framework, and thus to provide a standard interface to connect all these networks toward convergence.

5.2 *Software defined cellular network*

Legacy cellular network couples the control plane and data plane, the low resilience as a result of which makes the network unable to support new service paradigms.

SDN-enabled cellular network firstly splits the control plane and the data plane, and then provides programmable interfaces to the network operator on the control plane, as shown as in Figure 2. This enables the incremental deployment of new networking functions on the data plane by programming the control plane.

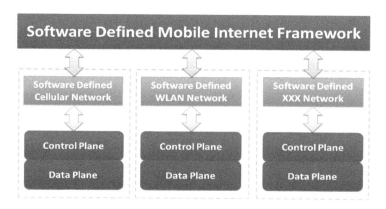

Figure 1.   SDN-enabled logical framework for future mobile network.

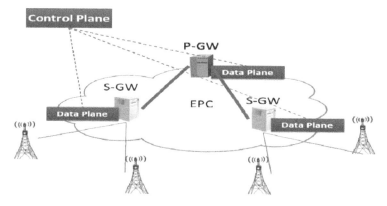

Figure 2.   Software defined cellular network.

To achieve service agility promised for service operators, the so-called north-bound and south-bound APIs centered on the control layer ought to be exploited. This archetype mutually reinforced with the NFV paradigm is most likely to shape a sustainable growth of the future mobile Internet. The Figure 3 illustrates the various aspects involved.

5.3   Software defined WLAN network

Legacy WLAN network can be implemented in two ways, i.e. FAT AP (Access Point), FIT AP + AC (AP Controller). FAT AP based implementation is suitable for a small scale network, e.g. home network and small office network. The carrier grade or enterprise grade WLAN are often implemented with FIT AP plus AC, and in this paper we focus only on this kind of WLAN networks. For this FIT AP plus AC based implementation, all the traffic from the APs should first be tunneled to the AC for centralized processing and then forwarded by the AC according to the forwarding table. Therefore, AC should bear high processing capability with superior performance, which brings about high CAPEX and OPEX.

SDN-enabled WLAN network makes the network flexible to deploy; the CAPEX and OPEX can be reduced by using general servers to execute the AC functions, shown as Figure 4. In SDN-enabled WLAN network, AP is only used to be the wireless antenna to transmit and receive data between the network and the mobile terminal. SDN controller is responsible for scheduling the traffic between a specific AP and AC according to pre-defined or intelligent policies. The AC can be dynamically deployed or removed based on the network traffic load. When the traffic load becomes very low, some ACs can be shut down to save energy, and the

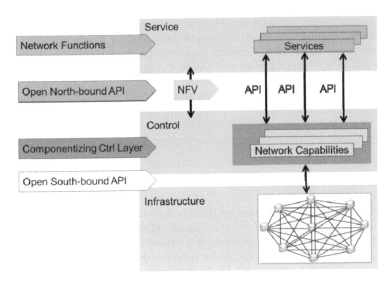

Figure 3.   High level sketch of areas for development of SDN-enabled future mobile Internet.

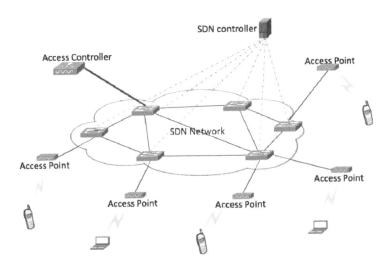

Figure 4.   Software defined WLAN network.

APs connected to these ACs can be redirected to other ones by the SDN controller automatically. The AC can be selectively implemented with virtual machine leveraging NFV (Network Function Virtualization) technologies.

5.4   *Vision of SDN-based future mobile network*

Future mobile networks should involve versatile mobile networks to support ubiquitous access for mobile terminals. Enabling SDN in future mobile network is an innovative consideration, which changes existing mobile networks fundamentally. We take the cellular network and WLAN network as the specimen to analyze the proposed software defined by future mobile network framework. By separating the control plane and the data plane, the cellular network and the WLAN network can both be potentially programmable to support new network functions, e.g. seamless mobility between heterogeneous mobile networks, service-oriented subscription and so on, which are not easily supported by legacy cellular networks.

This makes future mobile network flexible and adaptable. Additionally, SDN makes future mobile network potentially a green communication system by dynamically starting and stopping corresponding network elements based on the network running states. Many network optimization algorithms can be realized on the network by programming on the SDN controllers. Finally, the CAPEX and OPEX can both be reduced under such SDN-enabled framework. This is because the implementation of the data plane network element is greatly simplified and the complicated control plane is implemented with the software running on the general server with a price considerably lower than that of a dedicated network device, thereby, the network is simplified and easy to manage.

6 CONCLUSION

Mobile Internet is facing many challenges that hinder its further development, and it requires some breakthrough to nurture its further evolution. SDN as an innovative technology separating the control plane and the data plane of a network, providing programmable interface on the control plane, is an ideal candidate for future mobile Internet. We analyze the requirements of future mobile Internet and the limitations of the current mobile Internet, and present an SDN-enabled framework involving multiple mobile networks to meet the requirements of future mobile Internet. SDN-enabled cellular network and WLAN network is discussed as the specimen to illustrate the specific implementation instance of the framework.

REFERENCES

Kokku, R., Mahindra, R., Zhang, H. & Rangarajan, S. NVS: *A virtualization substrate for WiMAX networks, Proc MOBICOM, pp. 233–244, ACM, 2010.*

McKeown, N., Anderson, T., Balakrishnan, H., Parulkar, G., Peterson, L., Rexford, J., Shenker, S. & Turner, J. *Openflow: Enabling innovation in campus networks, Proc SIGCOMM CCR.*, vol. 38, no. 2, pp. 69–74, 2008.

Pentikousis, K., Wang Y. & Hu, W. 2013 Mobileflow: Toward software-defined mobile networks, Communications Magazine, IEEE 51(7).

Raghavan, B., Casado, M., Koponen, T., Ratnasamy, S., Ghodsi, A. & Shenker, S. *Software-de fined Internet architecture: Decoupling architecture from infrastructure, Proc. ACM SIGCOMM HotNets Workshop.*, October 2012.

Signal and Information Processing, Networking and Computers – Chen & Huang (Eds)
© 2016 Taylor & Francis Group, London, ISBN 978-1-138-02881-4

A novel Continuous Action-set Learning Automaton algorithm

Ying Guo, Hao Ge & Yan Yan
Department of Electronic Engineering, Shanghai Jiao Tong University, Shanghai, China

Yuyang Huang
Shanghai Starriver Bilingual School, Shanghai, China

Shenghong Li
Department of Electronic Engineering, Shanghai Jiao Tong University, Shanghai, China

ABSTRACT: A novel Continuous Action-set Learning Automaton algorithm named Adaptive Sampling CALA (ASCALA) is presented in this paper. In the proposed model, learning process is a combination of the sampling phase and the iteration phase. The new philosophy lies in the acquisition of a priori knowledge after sampling sufficiently. The learning step and the variance of action selection probability are adaptively adjusted during the two phases, varying from a small-step to a large-step learning and a large invariance to a gradually smaller one. The experiments with regard to function optimization evidently showed that our proposed method outshone the original algorithm regardless of the initial value of parameters. Besides, our algorithm is robust to the noise that added to the function.

Keywords: artificial intelligence; learning automaton; continuous action-set; adaptive; sampling

1 INTRODUCTION

1.1 *Learning automata*

Learning Automaton (LA), a promising field of artificial intelligence, is a self-adaptive machine that can learn the optimal action from a set of actions, by interacting with a random environment. During a cycle, the LA selects an action from a predefined action set according to a probability distribution of choosing each action. Then, this selected action will be sent to the environment to produce a response according to the reward or penalty probability of each action. After that, the LA takes this response and the aforementioned selected action into consideration to update its state, i.e. the probability of choosing actions. This cycle repeats until the LA selects the optimal action, which is defined as the one that achieves higher reward compared with the other actions. The learning process endows the automaton with the adaption to the environment. A broad range of LA applications are reported in areas, among which the latest such as mobile video surveillance system (Misra et al., 2015), event patterns tracking (Zhong et al., 2014), complex networks (Kumar et al., 2014), tutorial-like system (Oommen and Hashem, 2013) and so on.

Generally speaking, based on whether the action set is finite or continuous, Learning Automata can be divided into two types, i.e. Finite Action-set Learning Automaton (FALA) and Continuous Action-set Learning Automaton (CALA) (Thathachar and Sastry, 2002). FALA has finite number of actions, which has been studied extensively in the past decades. Since speed and accuracy are two conventional evaluation criteria to the performance of FALA, many scholars are devoted to accelerating the learning speed without compromising the accuracy. The concepts of discretization (Thathachar and Oommen, 1979) and estimator (L. Thathachar and Sastry, 1985) are two milestones in the evolution of FALA, furthermore Stochastic estimator algorithm (Papadimitriou et al., 2004) has been regarded as the most authoritative algorithm of FALA in the academic community. In recent years, some new approaches are published successively such as Discrete Bayesian Pursuit algorithm (DBPA) (Zhang et al., 2013), Discrete Generalized Confidence Pursuit algorithm (DGCPA) (Ge et al., 2015) and Last-position Elimination-based LA (LELA) (Zhang et al., 2014). Overall, the majority of research on the Learning Automata (LA) is referred to FALA, tending to be mature.

However, a number of limitations are exposed by FALA. For applications where actions evaluate continuously, usually FALA discretes the actions into finite ones to solve such issues. In order to improve the precision of discretization, we have no choice but increasing the number of actions. However, if the action size becomes large enough, FALA will converge too slowly with the initial probability of selection for any action being very small. On the other hand, when there is small number of actions, the finite action set destroys the thoroughness of searching over an entire action space, as the optima may be missed if they locate between action points. In such area CALA, whose actions are chosen from real line instead of finite number sets, displays an obvious superiority. Compared with FALA, there are relatively fewer studies on CALA in past literatures. Therefore, we are dedicated to putting forward a novel algorithm of CALA in this paper.

1.2 *Related work*

The action probability distribution of CALA, which is represented by r-dimensional vector in FALA, is generalized to a continuous function. The core of the problem is how this function updates at any stage. Based on this, a CALA algorithm is given in (Santharam et al., 1994), where the action probability distribution at instant n is a normal distribution with mean μ_n and standard deviation σ_n. Thus by updating these two parameters, the CALA updates its action probability distribution. The objective for CALA is to learn the value of α when $f = E[\beta(n)\,|\,\alpha(n) = \alpha]$ attains an optimum. That is, $N(\mu_n, \sigma_n)$ is supposed to converge to $N(\alpha^*, \sigma_l)$ where α^* is the optimum of function f and σ_l is a sufficiently small number. A theoretical analysis as well as the experiment results reveals its rationality. However, whether the performance is good or bad closely depends on its initial parameters.

In (Beigy and Meybodi, 2006) a new continuous action-set learning automaton is presented. Some modifications are conducted on the foundation of the first CALA algorithm. It is mainly the update strategy of σ_n that reflects the difference compared with the primitive one. In each iteration, σ_n does not rely upon two function values obtained on α_n and μ_n to alter its value. In contrast, it updates following a specific mathematical formula. However, we have met some problems when programming to verify this idea using the penalized Shubert function, which is one of the benchmark problems to test global optimization algorithms and we will explicitly introduce it in the latter pages. Indeed, we assert that the new algorithm is unable to converge as expected.

As a consequence of this flaw, in this paper we aim at identifying a gap in the aforementioned existing algorithm and presenting a more refined analysis to fill the gap. An adaptive sampling CALA (ASCALA) algorithm is proposed. Experimental results demonstrate that the revised method outperforms the original CALA algorithm. And the evaluation criteria is the number of results that converge to global optimum in repeated experiments, no matter what the initial parameter values are. Meanwhile, the proposed algorithm is also robust to the noise added to the function.

2 THE PROPOSED ALGORITHM

2.1 *Mathematic representation of CALA*

Generally, a learning automaton (LA) is represented as a quadruple $< A, B, P, T >$ and the environment interact with LA is defined as $< A, B, C >$, where in the CALA framework:

$A = \{\alpha \,|\, \alpha \in (-\infty, +\infty)\}$ is a continuous set of actions. Actions are chosen from real line.

$B = \{\beta \mid \beta \in [0,1]\}$ is the response set of the environment or the reinforcement signal. Throughout the paper, $\beta = 1$ means the environment penalizes the selected action to maximum extent.

$P(t) = \{p_\alpha(n) \mid \alpha \in (-\infty, +\infty)\}$ is the probability distribution function of the action set, α is selected on the basis of $p_\alpha(n)$ at the time instant n. In the proposed CALA, we use the Gaussian distribution $N(\mu, \sigma)$ for selection of actions.

T is the learning algorithms, $p_\alpha(n)$ is updated by learning algorithm at any stage n. Since in this paper the probability distribution function is specified by the mean and the variance of the Gaussian distribution at any instant, μ or σ is updated by T using the following equation:

$$\mu(or\ \sigma) = T[\mu(or\ \sigma), \alpha, \beta] \tag{1}$$

$C = \{C_\alpha \mid \alpha \in (-\infty, +\infty)\}$ is the penalty probability distribution over actions and the average penalty function is defined by

$$f(\alpha) = E[\beta(\alpha) \mid \alpha] \tag{2}$$

$f(.)$ is not known by the learning automaton and the only information is the function evaluation at the selected action. The objective for CALA is to learn the value of α when $f(\alpha)$ attains a minimum. Thus, it is desirable that with interaction of learning automaton and the environment, μ and σ could converge to their optimal values which results in the minimal value of $f(.)$.

2.2 Learning algorithm

In this paper, we propose a novel adaptive sampling CALA (ASCALA) algorithm as the learning algorithm. Learning process in our algorithm is composed of two phases: sampling procedure and iteration process. Compared to previous learning algorithms, the main characteristic of our method is that we add a behavior of sampling in order to ensure that the iteration process moves in the desired direction. To some extent, we obtain a priori knowledge after sufficiently sampling, making the learning process more targeted as well as purposeful.

In the early exploration, the environment is just like a "black box", about which we do not have any relevant information. So learning strategy is supposed to be cautious. That is why we set a small learning step in the sampling process. Besides, all of the actions should be fairly paid attention to since we have no prior knowledge, requiring a larger variance of action probability distribution function. This method allows that the under-sampled actions still have the chance of being further explored. On the contrary, exploration is gradually turned into exploitation after sampling, thus it is supposed to make a larger step since the learning turns increasingly accurate. Similarly, the variance of action selection probability is desired to be monotonically decreasing. Overall, the learning parameter is adaptively varied from sampling process to iteration process. That is why we call it adaptive sampling CALA algorithm.

The updating rule of parameter and in algorithm ASCALA is formulated below.

$$\mu_{n+1} = \begin{cases} \mu_n - a_s \beta(\alpha_n)(\alpha_n - \mu_n) & \text{if } n \le n_0 \\ \mu_n - a_l \beta(\alpha_n)(\alpha_n - \mu_n) & \text{if } n > n_0 \end{cases} \tag{3}$$

$$\sigma_{n+1} = \begin{cases} \sigma_0 & \text{if } n \le n_0 \\ \dfrac{1}{\lfloor n/10 \rfloor^{1/3}} & \text{if } n > n_0 \end{cases} \tag{4}$$

where a_s is a small learning step in sampling period while a_l is a larger one in iteration process. σ_0 is a predefined initial parameter of σ. n_0 is the sampling times, i.e. the turning point from the first phase to the latter one.

Now we make an intuitive explanation for the updating equations above.

If $\alpha_n > \mu_n$, which implies the selected action is on the right of the mean. Since $a_{s(l)}$, $\beta(\alpha_n)$, $\alpha_n - \mu_n$ is positive, the mean will move to left, which is the opposite direction to the selected action.

If $\alpha_n < \mu_n$, which implies the selected action is on the left of the mean. Since $a_{s(l)}$, $\beta(\alpha_n)$ is positive and $\alpha_n - \mu_n$ is negative, the mean will move to right, which is still the opposite direction to the selected action.

No matter what the situation is, the larger value for $\beta(\alpha_n)$, the greater degree of deviation for the mean. Only $\beta(\alpha_n) = 0$ leads to the convergence of the parameter μ, which is consistent with our learning objective in Eq. (2). That is, $E[\beta(\alpha)|\alpha]$ is close to unity when α is far away from its optimal value and it is close to zero when is near the optimal value. As the learning proceeds, we can learn the minimum of $f(\alpha)$.

3   EXPERIMENTAL RESULTS

In this section, experimental results are presented to test the implementation of ASCALA. We consider the optimization of the penalized Shubert function, which is one of the benchmark problems to test global optimization algorithms. It is formulated below.

$$F(x) = \sum_{i=1}^{5} icos((i+1)x+1) + u(x,10,100,2) \tag{5}$$

where the penalizing function is given by

$$u(x,b,k,m) = \begin{cases} k(x-b)^m, & x > b \\ 0, & |x| \le b \\ k(-x-b)^m, & x < -b \end{cases} \tag{6}$$

This function has 19 minima with interval $[-10,10]$, three of whom are global minima. The global minimum value of it is approximately equal to -12.87 attained at the point -5.9, 0.4, 6.8. And the local minimum value is close to -2.63, -2.72, -3.58, -3.74, -8.51 respectively.

In our experiment, we implement the proposed algorithm employing the following values of parameters: $\beta(\alpha) = [f(\alpha) - \lambda_2]/\lambda_1$ where $\lambda_1 = max\{f(x)\} - min\{f(x)\}$ and $\lambda_2 = min\{f(x)\}$ so that $\beta(\alpha) \in [0,1]$. The step size $a_s = 0.0001$ and $a_l = 0.01$. The sampling times n_0 is an adjustable parameter according to the initial μ_0 and σ_0. We carried out 1000 experiments, each consisting of 8000 iterations for the purpose of reducing randomness. Eventually, to evaluate the performance of our algorithm, we vote the 1000 results to one of the four optimal points named "global", "local1", "local2", "local3" respectively. In the process we allow a noise margin of 0.2. The mapping rule of vote is illustrated in Figure 1. For example, we cast a vote for "global" if one of the results is located in the range of $[-12.88, -12.68]$. Particularly, the noise-allowed regions of -3.74 and -3.58 have some areas in common thus we take their union set. That is why all the points between -3.75 to -3.39 are mapped to "local2". Other cases are analogous.

We conducted two groups of experiments, including one with no noise and the other with uniform noise $U(-0.5,0.5)$ added to the penalized Shubert function. In other words, for the former scenario, $f(x)$ is exactly equal to the evaluation of the penalized Shubert function

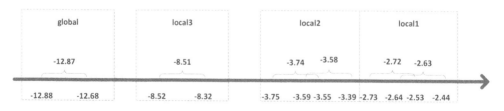

Figure 1.   The mapping rule of vote.

$F(x)$ while in the latter scenario $f(x) = F(x) + U(-0.5, 0.5)$. For CALA, we adopt absolutely the same initial setting and parameters as (Santharam et al., 1994), and for ASCALA the initial parameters are listed in Tables 1–2.

Given a set of initial values of $< \mu_0, \sigma_0, n_0 >$, we got 1000 results. Then we vote them to one of the four optima named "local1", "local2", "local3" or "global" utilizing the mapping rule in Figure 1. The number of results that are mapped to a certain optimum is its final votes. Experiments on CALA were carried out in a similar way as a contrast. Final simulation results where there is no noise are presented in Figure 2, and the noise-added results are presented in Figure 3. Both of the figures indicate that under any case of various initial parameters, the global optimum always won the highest number of votes when adopting the ASCALA algorithm. But for CALA, it can not achieve such a good effect. For instance, on the circumstance of no noise, "local1" got 883 votes but only 3 votes for "global" in case 1.

Table 1. The initial parameters of ASCALA with no noise.

Case	1	2	3	4	5	6	7
μ_0	3	3	-10	-10	10	10	7
σ_0	5	6	5	7	5	7	3
n_0	2000	2000	4000	4000	3000	3000	10

Table 2. The initial parameters of ASCALA uniform noise $U(-0.5, 0.5)$.

Case	1	2	3	4	5	6
μ_0	4	8	8	12	-10	-10
σ_0	6	3	5	6	5	6
n_0	1000	500	3000	1000	4000	4000

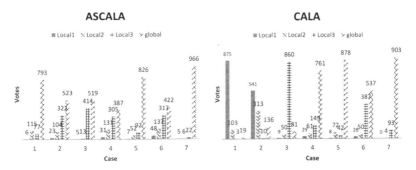

Figure 2. The simulation results with no noise.

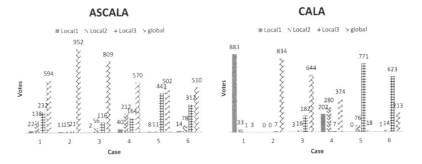

Figure 3. The simulation results with uniform noise $U(-0.5, 0.5)$.

47

A homologous outcome is reflected in case 2 as well as case 3, in which the results falls within the scope of the global optimum is extremely few. Similarly, when the evaluation is a noise-corrupted value, "local3", approximately equals to –8.51 rather than the global minimum won the majority of votes both in case 5 and case 6.

From the data analysis above, ASCALA surpasses the original algorithm. It offers a new vision in the research filed to the continuous action-set learning automaton.

4 CONCLUSION AND FUTURE WORK

In this paper we have proposed a brand-new CALA algorithm, which includes a sampling process and an iteration process. After a small-step sampling phase, in which the variance of action selection probability is large so that the under-sampled action can be further explored, the learning process turns to a large-step iterative phase and the corresponding variance gradually decreases. The experiments with regard to function optimization evidently show that our proposed method outshines the original algorithm in any case with different initial parameters. Besides, our algorithm is also robust to the noise added to the function.

However, although a majority of experiments could converge to the global optimum in our experiment, there still exist points that get stuck in the local optimal area. For future work, we believe a combination of our algorithm with some classical technologies to getting out of undesirable local optima can be a promising field. And further study of CALA is still open.

ACKNOWLEDGEMENT

This research work is funded by the National Science Foundation of China (61271316), 973 Program of China (2013CB329605), Shanghai Key Laboratory of Integrated Administration Technologies for Information Security.

REFERENCES

Beigy, H. & Meybodi, M. 2006. A new continuous action-set learning automaton for function optimization. *Journal of the Franklin Institute*, 343, 27–47.

Kumar, N., Lee, J.H. & Rodrigues, J.J. 2014. Intelligent mobile video surveillance system as a bayesian coalition game in vehicular sensor networks: learning automata approach. *Intelligent Transportation Systems, IEEE Transactions on*,16, 1148 - 1161.

Thathachar, M.L. & Sastry, P.S. 1985. A new approach to the design of reinforcement schemes for learning automata. *Systems, Man and Cybernetics, IEEE Transactions on*, 168–175.

Misra, S., Krishna, P.V., Saritha, V., Agarwal, H., Shu, L. & Obaidat, M. S. 2015. Efficient medium access control for cyber C physical systems with heterogeneous networks. *Systems Journal, IEEE*, 9, 22–30.

Oommen, B.J. & Hashem, M.K. 2013. Modeling the Learning Process of the Teacher in a Tutorial-Like System Using Learning Automata. *Cybernetics, IEEE Transactions on*, 43, 2020–2031.

Papadimitriou, G.I., Sklira, M. & Pomportsis, A.S. 2004. A new class of ε-optimal learning automata. *Systems, Man, and Cybernetics, Part B: Cybernetics, IEEE Transactions on*, 34, 246–254.

Santharam, G., Sastry, P. & Thathachar, M. 1994. Continuous action set learning automata for stochastic optimization. *Journal of the Franklin Institute*, 331, 607–628.

Thathachar, M. & Oommen, B. 1979. Discretized reward-inaction learning automata. *J. Cybern. Inf. Sci*, 2, 24–29.

Thathachar, M. & Sastry, P.S. 2002. Varieties of learning automata: an overview. *Systems, Man, and Cybernetics, Part B: Cybernetics, IEEE Transactions on*, 32, 711–722.

Zhang, J., Wang, C. & Zhou, M. 2014. Last-position elimination-based learning automata. *Cybernetics, IEEE Transactions on*, 44, 2484–2492.

Zhang, X., Granmo, O.C. & Oommen, B.J. 2013. On incorporating the paradigms of discretization and Bayesian estimation to create a new family of pursuit learning automata. *Applied intelligence*, 39, 782–792.

Zhong, W., Xu, Y., Wang, J., Li, D. & Tianfield, H. 2014. Adaptive mechanism design and game theoretic analysis of auction-driven dynamic spectrum access in cognitive radio networks. *EURASIP Journal on Wireless Communications and Networking*, 2014, 1–14.

Signal and Information Processing, Networking and Computers – Chen & Huang (Eds)
© *2016 Taylor & Francis Group, London, ISBN 978-1-138-02881-4*

Full-Duplex Relay based on GMD block diagonalization in MIMO relay systems

Yaxin Wang, Xin Zhang & Dacheng Yang
Beijing University of Posts and Telecommunications, Beijing, P.R. China

ABSTRACT: This paper proposed a modified Full-Duplex Relay (FDR) based on Block Diagonalization (BD) over Multiple-Input-Multiple-Output (MIMO) channel. Two operating modes are considered for two FDR scenarios to achieve capacity and Bit Error Ratio (BER)improvement over the conventional FDR with BD. Geometric Mean Decomposition (GMD) is employed instead of Singular Value Decomposition (SVD) used in conventional BD method, which simplifies the computation of power allocation process significantly. The employment of GMD also obtains a further improvement over BER performance. Simulation results demonstrate the effectiveness of the proposed algorithm compared to conventional FDR.

Keywords: FDR; block diagonalization; two operating modes; GMD

1 INTRODUCTION

In recent years, full-duplex transmission has become one of the topics of most concern since it's a keypoint of next gerneration wireless communication technology. As a result, Full-Duplex Relay (FDR) comes back in the spotlight. FDR allows Relay Station (RS) to simultaneously transmit and receive at the same frequency bands, but it's important to reduce the self-interference induced by the RS's transmitter to its own receiver (Wang, B. et al. 2005).

A FDR based on block diagonalization (BD) is proposed in Lee, C. H. et al. (2010), which can serve multiple users in MIMO channel at the same time. In Lee, J. H. & Shin, O. S. (2010) it has been proved that the FDR based on BD outperforms the Zero-Forcing beamforming (ZF) FDR. However, the underutilization of relay exists since the signal transmitted by the RS will cause interference to the Mobile Station (MS) that is not served by the RS. The Singular Value Decomposition (SVD) decomposes the channel into several different component channels. The component channel with low power allocated suffers from high bit error ratio for the Signal Noise Ratio (SNR) of the received signal can be very low. In that case, the overall BER performance is not satisfied.

In this paper, a modified FDR based on BD over the MIMO channel is proposed. The proposed FDR employs two different operating modes for two scenarios to make the best use of the relay. Geometric Mean Decomposition (GMD) (Jiang, Y. et al. 2005) is employed instead of SVD to obtain several identical component channels, so that the power can be allocated equally to the component channels. It means that we can solve the power allocation problem with only one variable which represents the power of every component channel instead of several ones. And the overall BER performance will be improved since the power is allocated to the component channels equally and no component channel works in very low SNR. Using the Karush-Kuhn-Tucker (KKT) conditions (Boyd, S. & Vandenberghe, L. 2004), we derive the optimal power allocation scheme for the MIMO FDR system that maximises the system capacity under the given power constraints of BS and RS. Numerical results show that the modified FDR with BD provides better capacity performance and significant BER improvement over the conventional FDR based on BD.

Notations: Vectors and matrices are denoted by bold symbols. The operators $(\cdot)^T$ and $(\cdot)^H$ denote transpose and conjugate transpose respectively. The operator diag($\cdot$) can convert a vector into a square matrix with the vector elements along the diagonal.

2   SYSTEM MODEL

A MIMO relaying downlink system is considered. The BS and RS have $2N$ transmit antennas, while the RS and MS are equipped with N receive antennas. In this system, we use Decode-and-Forward (DF) relay proposed in Nabar, R. U. et al. (2004), Nosratinia, A. et al. (2004). The MSs are categorised into 3 groups according to the reachability of the signals from the BS and RS:

i. MSs in the first group are located only in the BS coverage, and we note them as MS-1.
ii. MSs noted as MS-2 are in the second group that are located out of the BS coverage but in the RS coverage.
iii. MS-3 notes the MSs in the third group that are located in both the BS and RS coverage.

In operating mode 1, MS-1 and MS-2 are served at the same time, and in operating mode 2 only MS-3 is served.

The Channel State Information (CSI) can be achieved at the BS in use of CSI feedback schemes (Marzetta, T. L. & Hochwald, B. M. 2006), then the MSs are distributed into the three groups according to the channel gain of BS to MS and RS to MS link. Whether a MS is in coverage of BS or RS is decided by the channel gain and a threshold that is set in advance at the BS. The number of MSs in each group will decide the choice of operating mode and the shifting between the two operating modes. The distribution of the MSs and the shifting algorithm are problems in system level, so we won't discuss them in this paper. In the following parts, we assume that there are enough MSs for every operating mode, and the FDR performance in each operating mode is studied in link level.

2.1   *Operating mode 1*

Figure 1 shows the MIMO FDR system model operating in Mode 1.

We use y_i to denote the $N \times 1$ received signal at the MS-i, $i = 1, 2$ respectively.

$$y_1 = H_1^{BS}s^{BS} + z_1$$
$$y_2 = H_2^{BS}s^{BS} + z_2 \qquad (1)$$

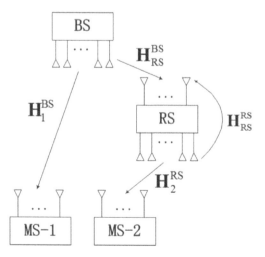

Figure 1.   MIMO relay system model for operating mode 1.

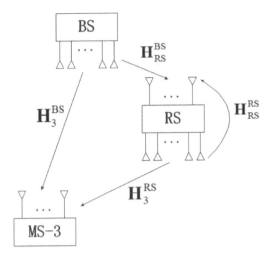

Figure 2.   MIMO relay system model for operating mode 2.

where the $N \times 2N$ channel matrix between the BS and MS-i is denoted as H_i^{BS}, while the $N \times 2N$ channel matrix between the RS and MS-i is denoted as H_i^{RS}. The $2N \times 1$ transmit signals from the BS and RS are denoted by s^{BS} and s^{RS} respectively. And z_i denotes the zero-mean additive white Gaussian noise (AWGN) whose covariance matrix is $\sigma^2 I$. We can also express the received signal at the RS as follows:

$$y_{RS} = H_{RS}^{BS} s^{BS} + H_{RS}^{RS} s^{RS} + z_{RS} \tag{2}$$

where H_{RS}^{BS} and H_{RS}^{RS} denote the $N \times 2N$ channel matrixes of BS-RS and RS-RS respectively. z_{RS} is the AWGN whose covariance matrix is $\sigma^2 I$. In the right side of (2), the second term denotes the self-interference of the RS.

2.2   *Operating mode 2*

For operating mode 2, the FDR system model is illustrated in Figure 2.
   The $N \times 1$ received signal vector y_3 at MS-3 is respectively given as

$$y_3 = H_3^{BS} s^{BS} + H_3^{RS} s^{RS} + z_3 \tag{3}$$

The received signal at the RS can be expressed the same as (2).

3   PRECODING AND POWER ALLOCATION METHOD OF MODIFIED FDR WITH BD

In this section, we introduce the proposed BD precoding method for FDR. We use P_T^{BS} and P_T^{RS} to denote the total transmit power of the BS and RS respectively. In use of CSI feedback schemes, the BS can get all the CSI of the BS-RS and BS-MS links. Using the method proposed in Marzetta, T. L. & Hochwald, B. M. (2006), the CSI of RS-MS link is also assumed to be available for BS.

3.1   *Precoding matrix of operating mode 1*

In operating mode 1, the transmit signal in (1) and (2) can be expressed as

$$s^{BS} = [W^{BS:1} W^{BS:RS}] P^{BS} \begin{bmatrix} x^{(1)} \\ x^{(2)} \end{bmatrix} \qquad s^{RS} = [W^{RS:2}] P^{RS} \tilde{x}^{(2)} \tag{4}$$

where $x^{(i)} = [x_1^{(i)} \dots x_N^{(i)}]^T$ denotes the $N \times 1$ data symbol transmitted to the MS-i. The data symbol $x^{(2)}$ at the BS is first transmitted to the RS. The RS decodes $x^{(2)}$ and forwards $\tilde{x}^{(2)} = [\tilde{x}_1^{(2)} \dots \tilde{x}_N^{(2)}]^T$ to the MS-2. It's a delayed version of $x^{(2)}$.

Based on BD principle, the precoding matrix $W^{BS:1}$ corresponding to the MS-1 should be able to nullify the interference so that

$$H_{RS}^{BS} W^{BS:1} = 0 \tag{5}$$

The SVD of the $N \times 2N$ channel matrix H_{RS}^{BS} is derived as

$$H_{RS}^{BS} = \tilde{U}_1 \tilde{\Lambda}_1 [\tilde{V}_1^{(1)} \tilde{V}_1^{(0)}]^H \tag{6}$$

where $\tilde{\Lambda}_1$ consists of the singular values of H_{RS}^{BS} and $\tilde{U}_1$ is the left singular matrix. $\tilde{V}_1^{(1)}$ and $\tilde{V}_1^{(0)}$ represents the right singular vectors corresponding to non-zero and zero singular values, respectively. It is shown that

$$H_{RS}^{BS} \tilde{V}_1^{(0)} = 0 \tag{7}$$

If we define

$$\hat{H}_1 = H_1^{BS} \tilde{V}_1^{(0)} \tag{8}$$

the SVD of $\hat{H}_1$ becomes

$$\hat{H}_1 = \hat{U}_1 \tilde{\Lambda}_1 \hat{V}_1^H \tag{9}$$

where $\tilde{\Lambda}_1 = \text{diag}[\lambda_1^{BS:1}, \dots, \lambda_N^{BS:1}]$. From (7) and (9), the precoding matrix $\hat{W}^{BS:1}$ can be derived as

$$\hat{W}^{BS:1} = \tilde{V}_1^{(0)} \hat{V}_1 \tag{10}$$

We can compute $\hat{W}^{BS:RS}$, $\tilde{\Lambda}_{RS} = \text{diag}[\lambda_1^{BS:RS}, \dots, \lambda_N^{BS:RS}]$ and $\hat{U}_{RS}$ corresponding to the RS, $\hat{W}^{RS:2}$, $\tilde{\Lambda}_{RS:2} = \text{diag}[\lambda_1^{RS:2}, \dots, \lambda_N^{RS:2}]$, and $\hat{U}_2$ corresponding to the MS-2 similarly. The precoding matrices can be expressed as

$$\begin{aligned} \hat{W}^{BS:1} &= [\hat{w}_1^{BS:1} \quad \dots \quad \hat{w}_N^{BS:1}] \\ \hat{W}^{BS:RS} &= [\hat{w}_1^{BS:RS} \quad \dots \quad \hat{w}_N^{BS:RS}] \\ \hat{W}^{RS:2} &= [\hat{w}_1^{RS:2} \quad \dots \quad \hat{w}_N^{RS:2}] \end{aligned} \tag{11}$$

where $\hat{w}_n^{BS:1}$, $\hat{w}_n^{BS:RS}$ and $\hat{w}_n^{RS:2}$ denotes the $2N \times 1$ precoding vectors for the nth data stream with $n = 1, \dots, N$.

As derived in (9) the SVD decomposes the BS-to-MS1 channel into N different component channels, where $\lambda_i^{BS:1}, i = 1, \dots, N$ may have a great disparity in numerical value. Then the data streams transmitted will be constructed with different modulation and coding scheme (MCS). Moreover, the component channel with small $\lambda_i^{BS:1}$ may suffer from high BER, so that the overall BER performance won't be satisfied. So we employed GMD instead of SVD to obtain N same component channels.

The GMD of $\hat{H}_1$ is

$$\hat{H}_1 = \hat{Q}_1 \hat{R}_1 \hat{P}_1^H \tag{12}$$

where

$$\hat{R}_1 = \begin{bmatrix} \overline{\lambda}_1 & r_{1(1,2)} & \cdots & r_{1(1,N)} \\ 0 & \overline{\lambda}_1 & \ddots & \vdots \\ \vdots & \ddots & \ddots & r_{1(N-1,N)} \\ 0 & \cdots & 0 & \overline{\lambda}_1 \end{bmatrix} \qquad (13)$$

From (7) and (12), the precoding matrix with GMD $W^{BS:1}$ can be derived as

$$W^{BS:1} = \tilde{V}_1^{(0)}\hat{P}_1 \qquad (14)$$

We can also compute $W^{BS:RS}$, $\hat{R}_{RS}$ and $\hat{Q}_{RS}$, $W^{RS:2}$, $\hat{R}_2$, and $\hat{Q}_2$ similarly.

3.2 Precoding matrix of operating mode 2

In operating mode 2, the transmit signal vector in (2) and (3) can be expressed as

$$\begin{aligned} \underline{s}^{BS} &= [W^{BS:3} W^{BS:RS}]P^{BS}[x^{(3)} \quad \tilde{x}^{(3)}]^T \\ \underline{s}^{RS} &= [W^{RS:3}]P^{RS}\tilde{x}^{(3)} \end{aligned} \qquad (15)$$

The data symbol $x^{(3)}$ at the BS is first transmitted to the RS. Then, the RS decodes $x^{(3)}$ and forwards $\tilde{x}^{(3)}$ to the MS-3, which is a delayed version of $x^{(3)}$. The MS-3 receives and decodes $\tilde{x}^{(3)}$, which comes from both BS and RS, to achieve spatial diversity gain.

Using BD, the SVD precoding matrix $\hat{W}^{BS:3}$ and GMD precoding matrix $W^{BS:3}$ can be given as

$$\begin{aligned} H_{RS}^{BS} &= \tilde{U}_1 \tilde{\Lambda}_1 [\tilde{V}_1^{(1)} \tilde{V}_1^{(0)}]^H \\ \hat{H}_1 &= H_3^{BS} \tilde{V}_1^{(0)} \\ \hat{H}_1 &= \hat{Q}_1 \hat{R}_1 \hat{P}_1^H \\ W^{BS:3} &= \tilde{V}_1^{(0)} \hat{P}_1 \end{aligned} \qquad (16)$$

where $\tilde{\Lambda}_1 = \text{diag}[\lambda_1^{BS:3}, \ldots, \lambda_N^{BS:3}]$ and $\hat{R}_1$ is the same as shown in (13).

The SVD precoding matrix $\hat{W}^{BS:RS}$ and GMD precoding matrix $W^{BS:RS}$ can be given as

$$\begin{aligned} H_3^{BS} &= \tilde{U}_{RS} \tilde{\Lambda}_{RS} [\tilde{V}_{RS}^{(1)} \tilde{V}_{RS}^{(0)}]^H \\ \hat{H}_{RS} &= H_{RS}^{BS} \tilde{V}_{RS}^{(0)} \\ \hat{H}_{RS} &= \hat{Q}_{RS} \hat{R}_{RS} \hat{P}_{RS}^H \\ W^{BS:RS} &= \tilde{V}_{RS}^{(0)} \hat{P}_{RS} \end{aligned} \qquad (17)$$

where $\tilde{\Lambda}_{RS} = \text{diag}[\lambda_1^{BS:RS}, \ldots, \lambda_N^{BS:RS}]$

The SVD precoding matrix $\hat{W}^{RS:3}$ and GMD precoding matrix $W^{RS:3}$ can be given as

$$\begin{aligned} H_{RS}^{RS} &= \tilde{U}_3 \tilde{\Lambda}_3 [\tilde{V}_3^{(1)} \tilde{V}_3^{(0)}]^H \\ \hat{H}_3 &= H_3^{RS} \tilde{V}_3^{(0)} \\ \hat{H}_3 &= \hat{Q}_3 \hat{R}_3 \hat{P}_3^H \\ W^{RS:3} &= \tilde{V}_3^{(0)} \hat{P}_3 \end{aligned} \qquad (18)$$

where $\tilde{\Lambda}_3 = \text{diag}[\lambda_1^{RS:3}, \ldots, \lambda_N^{RS:3}]$

3.3 *Power allocation of operating mode 1*

The power allocation matrix $\mathbf{P}^{BS}$ and $\mathbf{P}^{RS}$ are

$$\mathbf{P}^{BS} = \text{diag}[\ \mathbf{p}^{BS:1}\quad \mathbf{p}^{BS:RS}\ ]$$
$$\mathbf{P}^{RS} = \text{diag}[\ \mathbf{p}^{RS:2}\ ] \tag{19}$$

where

$$\mathbf{p}^{BS:1} = \left[\ \sqrt{p_1^{BS:1}},\ldots,\sqrt{p_N^{BS:1}}\ \right]$$
$$\mathbf{p}^{BS:RS} = \left[\ \sqrt{p_1^{BS:RS}},\ldots,\sqrt{p_N^{BS:RS}}\ \right] \tag{20}$$
$$\mathbf{p}^{RS:2} = \left[\ \sqrt{p_1^{RS:2}},\ldots,\sqrt{p_N^{RS:2}}\ \right]$$

From (4) and (19), the transmit signals at the BS and RS can be written as

$$\mathbf{s}^{BS} = \sum_{n=1}^{N} \left[\mathbf{w}_n^{BS:1}\sqrt{p_n^{BS:1}}\,x_n^{(1)} + \mathbf{w}_n^{BS:RS}\sqrt{p_n^{BS:RS}}\,x_n^{(2)} \right] \tag{21}$$

$$\mathbf{s}^{RS} = \sum_{n=1}^{N} \mathbf{w}_n^{RS:2}\sqrt{p_n^{RS:2}}\,\tilde{x}_n^{(2)} \tag{22}$$

The total power constraints of the BS and RS can be expressed as

$$\sum_{n=1}^{N}\left[\ \|\mathbf{w}_n^{BS:1}\|^2\, p_n^{BS:1} + \|\mathbf{w}_n^{BS:RS}\|^2\, p_n^{BS:RS}\ \right] = P_T^{BS} \tag{23}$$

and

$$\sum_{n=1}^{N}\|\mathbf{w}_n^{RS:2}\|^2\, p_n^{RS:2} \le P_T^{RS} \tag{24}$$

Since each column of the BD precoder matrix is extracted from right singular vectors as shown in (14), it has unit norm

$$\|\mathbf{w}_n^{BS:1}\|^2 = \|\mathbf{w}_n^{BS:RS}\|^2 = \|\mathbf{w}_n^{RS:2}\|^2 = 1$$
$$n = 1,\ldots,N \tag{25}$$

Substituting (21) into (1), using $\hat{\mathbf{Q}}_1$ in (12) as a receive filter at the MS-1, after applying the V-BLAST (Wolniansky, P. et al. 1998) receiving method, we can get the received signal at the MS-1 as

$$\bar{\mathbf{y}}_1 = \begin{bmatrix} \bar{\lambda}_1\sqrt{p_1^{BS:1}}\,x_1^{(1)} \\ \bar{\lambda}_1\sqrt{p_2^{BS:1}}\,x_2^{(1)} \end{bmatrix} + \bar{\mathbf{z}}_1 \tag{26}$$

Similarly, the received signals at the RS and MS-2 can be expressed as

$$\bar{\mathbf{y}}_{RS} = \begin{bmatrix} \bar{\lambda}_{RS}\sqrt{p_1^{BS:RS}}\,x_1^{(2)} \\ \bar{\lambda}_{RS}\sqrt{p_2^{BS:RS}}\,x_2^{(2)} \end{bmatrix} + \bar{\mathbf{z}}_{RS} \tag{27}$$

$$\bar{\mathbf{y}}_2 = \begin{bmatrix} \bar{\lambda}_2\sqrt{p_1^{RS:2}}\,\tilde{x}_1^{(2)} \\ \bar{\lambda}_2\sqrt{p_2^{RS:2}}\,\tilde{x}_2^{(2)} \end{bmatrix} + \bar{\mathbf{z}}_2 \tag{28}$$

We can derive the relationship between the transmit powers as

$$
\begin{aligned}
(\overline{\lambda}_{RS})^2 \, p_1^{BS:RS} &= (\overline{\lambda}_2)^2 \, p_1^{RS:2} \\
(\overline{\lambda}_{RS})^2 \, p_2^{BS:RS} &= (\overline{\lambda}_2)^2 \, p_2^{RS:2}
\end{aligned}
\tag{29}
$$

and

$$
\begin{aligned}
(\overline{\lambda}_{RS})^2 \, p_2^{BS:RS} &= (\overline{\lambda}_2)^2 \, p_1^{RS:2} \\
(\overline{\lambda}_{RS})^2 \, p_1^{BS:RS} &= (\overline{\lambda}_2)^2 \, p_2^{RS:2}
\end{aligned}
\tag{30}
$$

Therefore, the capacity of the **FDR** in operating mode 1 can be computed as

$$
\begin{aligned}
C_1 &= \log_2\!\left(1 + \frac{(\overline{\lambda}_1)^2 \, p_1^{BS:1}}{\sigma^2}\right) + \log_2\!\left(1 + \frac{(\overline{\lambda}_1)^2 \, p_2^{BS:1}}{\sigma^2}\right) \\
&\quad + \log_2\!\left(1 + \frac{(\overline{\lambda}_{RS})^2 \, p_1^{BS:RS}}{\sigma^2}\right) + \log_2\!\left(1 + \frac{(\overline{\lambda}_{RS})^2 \, p_2^{BS:RS}}{\sigma^2}\right)
\end{aligned}
\tag{31}
$$

The power constraints of the BS and RS in (23) and (24) can be written as

$$
p_1^{BS:1} + p_2^{BS:1} + p_1^{BS:RS} + p_2^{BS:RS} = P_T^{BS}
\tag{32}
$$

$$
\phi_1^{BS:RS} \, p_1^{BS:RS} + \phi_2^{BS:RS} \, p_2^{BS:RS} \leq P_T^{RS}
\tag{33}
$$

where

$$
\phi_1^{BS:RS} = \phi_2^{BS:RS} = \frac{(\overline{\lambda}_{RS})^2}{(\overline{\lambda}_2)^2}
\tag{34}
$$

3.4 *Power allocation of operating mode 2*

The power allocation matrix $\underline{P}^{BS}$ and $\underline{P}^{RS}$ are

$$
\begin{aligned}
\underline{P}^{BS} &= \mathrm{diag}[p^{BS:3} \quad p^{BS:RS}] \\
\underline{P}^{RS} &= \mathrm{diag}[p^{RS:3}]
\end{aligned}
\tag{35}
$$

Similar with operating mode 1, the received signals at the RS can be expressed as

$$
\overline{y}_{RS} = \begin{bmatrix} \overline{\lambda}_{RS}\sqrt{p_1^{BS:RS}}\,x_1^{(3)} \\ \overline{\lambda}_{RS}\sqrt{p_2^{BS:RS}}\,x_2^{(3)} \end{bmatrix} + \overline{z}_{RS}
\tag{36}
$$

Using $\hat{Q}_1$ and $\hat{Q}_3$ as receive filters, the received signals at MS-3 are expressed as

$$
\overline{y}_3 = \begin{bmatrix} (\overline{\lambda}_1\sqrt{p_1^{BS:3}} + \overline{\lambda}_3\sqrt{p_1^{RS:3}})\tilde{x}_1^{(3)} \\ (\overline{\lambda}_1\sqrt{p_2^{BS:3}} + \overline{\lambda}_3\sqrt{p_2^{RS:3}})\tilde{x}_2^{(3)} \end{bmatrix} + \overline{z}_3
\tag{37}
$$

where $\overline{z}_3$ includes the noise and interference after the receive filters. The capacity of the FDR in operating mode 2 can be derived as

$$C_2 = \log_2\left(1 + \frac{(\overline{\lambda}_1)^2 p_1^{BS:3}}{\sigma^2}\right) + \log_2\left(1 + \frac{(\overline{\lambda}_1)^2 p_2^{BS:3}}{\sigma^2}\right)$$
$$+ \log_2\left(1 + \frac{(\overline{\lambda}_{RS})^2 p_1^{BS:RS}}{\sigma^2}\right) + \log_2\left(1 + \frac{(\overline{\lambda}_{RS})^2 p_2^{BS:RS}}{\sigma^2}\right) \tag{38}$$

The transmit power constraints of the BS and RS can be written as

$$p_1^{BS:3} + p_2^{BS:3} + p_1^{BS:RS} + p_2^{BS:RS} = P_T^{BS} \tag{39}$$

$$\phi_1^{BS:RS} p_1^{BS:RS} + \phi_2^{BS:RS} p_2^{BS:RS} \leq P_T^{RS} \tag{40}$$

where

$$\phi_1^{BS:RS} = \phi_2^{BS:RS} = \frac{(\overline{\lambda}_{RS})^2}{(\overline{\lambda}_3)^2} \tag{41}$$

The power allocation method for SVD FDR and GMD FDR using KKT conditions is carried out as shown in Lee, J. H. & Shin, O. S. (2010) similarly.

4   SIMULATION RESULTS

In this section, we investigate the ergodic system capacity and bit error rate (BER) performance of our proposed SVD FDR, GMD FDR and the conventional FDR in Lee, J. H. & Shin, O. S. (2010). We assume $N = 2$, and $\sigma^2 = 1$. For each MS, channel coefficients of the BS-MS and RS-MS links follow independent and identically distributed (i.i.d.) complex Gaussian distribution which has zero mean and unit variance. We assume that the number of MSs in the three groups is the same, and we denote the number of MSs in every group as M. When we select an MS in every group out of M MSs, the MS that maximises the system capacity is chosen after the ergodic process. The SNR represents σ_s^2/σ^2 and units in dB, where σ_s^2 is the variance of source signal s^{BS}. The power constraint of P_T^{BS} and P_T^{RS} mentioned in this section units in unit variance, not in dB.

In Figure 3, system capacity performance versus M is investigated for the proposed FDR in different operating mode. The performance of conventional SVD FDR in Lee, J. H. & Shin, O. S. (2010) is also investigated as a reference.

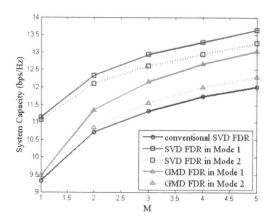

Figure 3.   system capacity performance, $P_T^{BS} = 20$, $P_T^{RS} = 15$, $SNR = 0\ dB$.

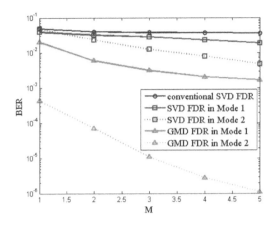

Figure 4.　BER performance versus M, $P_T^{BS} = 20, P_T^{RS} = 15$, SNR=0dB.

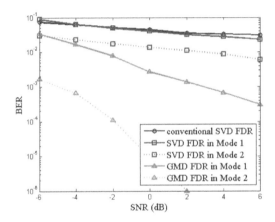

Figure 5.　BER performance versus SNR, $P_T^{BS} = 20$, $P_T^{RS} = 15$, $M = 3$.

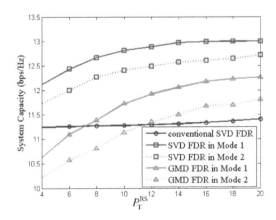

Figure 6.　System capacity performance with different power constraint of the relay, $P_T^{BS} = 20, M = 3$.

Figure 4 shows the BER performance versus M corresponding to the system in Figure 3.

It is observed in Figure 3 and Figure 4 that the proposed SVD FDR of both operation modes outperforms the conventional FDR in system capacity. The performance gap is about 1.5 bps/Hz. There is also a significant improvement in BER performance against the conventional FDR.

Figure 5 shows the overall BER performance versus SNR. Figure 6 shows the system capacity performance with different power constraint of the relay. We observe that the proposed FDR enjoys more processing gain with increasing SNR and P_T^{RS}.

5 CONCLUSIONS

In this paper, a modified FDR based on BD over MIMO channel is proposed. The proposed FDR employs two different operating modes for two scenarios to make the best use of the relay. GMD is employed instead of SVD to obtain better BER performance and simplify the power allocation algorithm. Numerical results show that the proposed FDRs provide better capacity performance and significant BER improvement over the conventional FDR and make better use of the relay.

ACKNOWLEDGEMENT

The research is sponsored by Project 61471066 supported by NSFC, and by National Science and Technology Major Project of the Ministry of Science and Technology under project name as "Wireless mobile spectrum research and verification for WRC15" with 2014ZX03003013-004.

REFERENCES

Boyd, S. & Vandenberghe, L. 2004. *Convex optimization*. New York: Cambridge University Press.

Jiang, Y., Hager, W. & Li, H. 2005. The geometric mean decomposition. *Linear algebra and its applications*, no. 1, vol. 396: 373–384.

Lee, C.H., Lee, J.H., Shin, O.S. & Kim, S.C. 2010. Sum rate analysis of multiantenna multiuser relay channel. *Communications, IET*, no. 17, vol. 4: 2032–2040.

Lee, J.H. & Shin, O.S. 2010. Full-duplex relay based on block diagonalisation in multiple-input multiple output relay systems. *Communications, IET*, no. 15, vol. 4: 1817–1826.

Marzetta, T.L. & Hochwald, B.M. 2006. Fast transfer of channel state information in wireless systems. *Signal processing, IEEE transactions on*, no. 4, vol. 54: 1268–1278.

Nabar, R.U., Bolcskei, H. & Kneubuhler, F.W. 2004. Fading relay channels: performance limits and space-time signal design. *Selected areas in communications, IEEE journal on*, no. 6, vol. 22: 1099–1109.

Nosratinia, A., Hunter, T.E. & Hedayat, A. 2004. Cooperative communication in wireless networks. *Communications Magazine, IEEE*, no. 10, vol. 42: 74–80.

Wang, B., Zhang, J. & Host-Madsen, A. 2005. On the capacity of MIMO relay channels. *Information theory, IEEE transactions on*, no. 1, vol. 51: 29–43.

Wolniansky, P., Foschini, G., Golden, G. & Valenzuela, R. 1998. V-blast: an architecture for realizing very high data rates over the rich-scattering wireless channel. *ISSSE 98. 1998 URSI International symposium on*: 295–300.

Signal and Information Processing, Networking and Computers – Chen & Huang (Eds)
© 2016 Taylor & Francis Group, London, ISBN 978-1-138-02881-4

An effective and scalable algorithm for hybrid recommendation based on Learning To Rank

Pingfan He, Hanning Yuan, Jiehao Chen & Chong Zhao
School of Software Engineering, Beijing Institute of Technology, Beijing, China

ABSTRACT: Recently, learning to rank in the domain of recommendation has drawn intensive attention. Though many approaches have been proposed, and proved their effectiveness in providing accurate recommendations, they lack emphasis on diversity. However, the predictive accuracy is not enough to judge the performance of a recommended system and diversity has been regarded as a quality dimension for recommendation. In this paper, we propose a formal model based on learning to rank for hybrid recommendation which integrates diversity. We also propose the representation of diversity features by using entropy based on attributes of users and items. Experimental results in the movie domain show the advantages of our proposal in both accuracy and diversity.

Keywords: recommender systems; learning to rank; matrix factorization; entropy; diversity

1 INTRODUCTION

With the coming of the information age, people are unprecedentedly faced with the problem of information overload (Amatriain 2014), where people have to handle a plethora of contents every day and choice has become an affliction in buying a book, watching a movie, etc. In Anderson (2013), Chris Anderson refers that we are entering the age of recommendation, and the search engine no longer meets the needs of discovery that may bring people wonderful experiences. Thus, recommender systems taking into account accuracy and diversity have been the subject of intensive research efforts (Zhang 2008).

Collaborative Filtering (CF) has been regarded as one of the most effective and successful recommendation technologies (Adomavicius et al. 2005), which recommends items only based on the users' past behaviors. Recently, the research attention in the area of CF has shifted away from rating prediction, and focuses more on the quality of the ranking of recommendation lists (McNee et al. 2006). The main intuition in philosophy could be crucial, that an item with a high score may not interest you due to your unique preferences, but a low-rated item may raise your curiosity and lead to consumption. Among recent novel approaches, Learning To Rank (LTR) treats recommendation as a ranking problem (Karatzoglou et al. 2013), which contributes a lot to improving the rank of top-N recommendations for collaborative filtering. Weston et al. (2013) present the k-order statistic loss function and deploy their LTR approach to two real-world systems, Google Music and YouTube video recommendations, where recommendation effectiveness and improvements have been obtained. In some e-commerce websites, e.g. Taobao in China, offline models based on LTR have been developed for mobile recommendation with respect to different business purposes, which have brought huge economic benefits. As a cutting-edge research topic, LTR is also adapted or extended to other model-based collaborative filtering methods (Shi Y et al. 2010), e.g. Matrix Factorization (MF). These approaches focus on ranking and could output recommendation lists with high accuracy, but without enough attention to diversification in order to avoid monotony, i.e. diversity is not embodied as an element in the solution procedure without respect to re-rank the recommendation lists to achieve diversification.

Diversity as a quality dimension of recommender systems should be considered as one of the key factors in recommender techniques (Vargas et al. 2013), which can increase the chances of providing users serendipity. Lathia et al. (2010) show that temporal diversity is an important facet of recommender systems and explore how user rating patterns affect diversity. In Di Noia et al. (2014), the authors investigate the concept of individual diversity, from which we can learn the users' multiple needs in recommendation lists.

In this paper, we propose an effective and scalable recommender algorithm based on LTR. Our approach also focuses on ranking or recommendation list and makes use of the advantages of list-wise matrix factorization to guarantee accuracy, at the same time taking into account the diversity. The main contributions of this paper are:

- In this paper, we propose a formal model based on LTR, which can be effectively used to fuse features and dynamically enhance the trade-off between accuracy and diversity for hybrid recommendation. Features extracted from the user profile et al. are classified into two types: accuracy features and diversity features. The model has been experimentally evaluated in the movie domain proving its advantages in both accuracy and diversity.
- We explore the combination with list-wise matrix factorization for our approach, in order to extract effective features for accuracy.
- We propose a novel method for diversity feature representation by using the concept of entropy based on the attributes of users and items.

The rest of the paper is structured as follows: after presenting our proposed approach in Session 2, we introduce accuracy feature extraction in Session 3. Then, we show how to use entropy to represent diversity features in Session 4. Session 5 gives an introduction for feature fusion. Experimental results are presented in Session 6, followed by the conclusion in Session 7.

2 PROPOSED MODEL AND APPROACH

Learning to rank for document retrieval was presented in Hang Li (2011); we make some adjustments and propose our LTR approach for recommendation. First of all, we introduce some notations used in this paper, shown in Table 1.

The framework of our proposed model is illustrated in Figure 1. First of all, we randomly split the dataset into training dataset (e.g. 80%) and test dataset (e.g. 20%), and then extract features from the training dataset. In the training dataset, if rating scores are available, the scores can be used as grade labels. Otherwise we should analyze the training data and set reasonable labels, e.g. using binary relevance obtained from user implicit feedback.

Accuracy features. These features aim to improve the accuracy of recommender systems, including rating features, latent features by MF, features extracted from user profiles or item contents, demographic features et al. Since content-based features are associated with specific

Table 1. List of notations and descriptions.

Notation	Explanation
U	User set
I	Item set
R	The user-item rating matrix
u_i	The ith user
i_j	The jth item
$x_{i,j}$	Feature vector created from (u_i, i_j)
x_i	Feature vector set of the ith user, i.e. $\{x_{i,1}, x_{i,2}, x_{i,3}, \ldots, x_{i,n_i}\}$
$y_{i,j}$	The relevance of u_i with respect to i_j, or the grade label for $x_{i,j}$
y_i	The grade label set correspond to x_i, i.e. $\{y_{i,1}, y_{i,2}, y_{i,3}, \ldots, y_{i,n_i}\}$

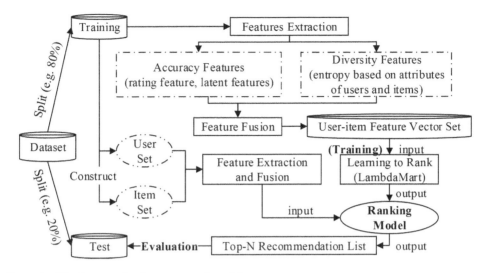

Figure 1.   The framework of our proposed model.

scenes, we only experiment on the more generic features without any concern for the contents of accuracy, i.e. rating features and latent features by MF.

Diversity features. We make use of the concept of entropy to represent the user's diversity needs and characteristics, as well as the item's diversity satisfaction and suitable degree, based on the attributes of both users and items

Feature fusion. We first represent the accuracy feature vector set as x_i^a and the diversity feature vector set as x_i^d of u_i. And we use a simple strategy to perform feature fusion:

$$x_i = (1 - \omega_i) \cdot x_i^a + \omega_i \cdot x_i^d \qquad (1)$$

where ω_i denotes the fusion weight of u_i, different from each other and calculated according to user diversity characteristics (see Session 5).

After feature fusion, the ranking model can be learned from the feature vector set of all users by adopting any learning to rank tools. In this paper, we adopt LambdaMART (Burges et al. 2010) to learn the ranking model, because of its success in solving the real world ranking problem. Finally, an evaluation step is taken to judge the ranking model.

3   ACCURACY FEATURE EXTRACTION

3.1   *Latent features by MF*

In the approaches of matrix factorization (Koren et al. 2009), the given user-item rating matrix R is factorized into two low-rank matrices, P and Q. A d-dimensional set of latent features is used to represent both users (P) and items (Q). And we use P_i to denote the d-dimensional column latent features of u_i and Q_i to denote the d-dimensional column latent features of i_j. The value of $P_i^T \cdot Q_j$ represents the relevance between u_i and i_j, which can be used to rank items for u_i. However, in this paper, we only exploit the significance of the latent features and use both P_i and Q_j to represent the features for the user-item pair (u_i, i_j). We represent the latent features vector set as x_i^{al} of u_i, and the latent features vector of u_i and i_j is denoted as $x_{i,j}^{a,l} = (P_i^T, Q_j^T)$, and we have $x_i^{a,l} = \{x_{i,1}^{a,l}, x_{i,2}^{a,l}, x_{i,3}^{a,l}, \ldots, x_{i,n_i}^{a,l}\}$.

As for the matrix factorization approaches, any type of algorithm fits our proposed model, theoretically. We make use of list-wise matrix factorization (ListRank-MF) (Shi Y et al. 2010), for its quality of ranking and effectiveness. ListRank-MF uses the cross entropy of the top one probabilities of the items in training lists and ranking lists, output by the ranking model to formulate the loss function, shown as follows:

$$L(P,Q) = \sum_{i=1}^{m} \left\{ -\sum_{j=1}^{n} I(r_{i,j}) \cdot \frac{\exp(r_{i,j})}{\sum_{k=1}^{n} I(r_{i,k}) \cdot \exp(r_{i,k})} \cdot \log \left(\frac{\exp\left(g\left(P_i^T \cdot Q_j\right)\right)}{\sum_{k=1}^{n} I(r_{i,k}) \cdot \exp(P_i^T \cdot Q_k)} \right) \right\}$$
$$+ \frac{\lambda}{2} \left(\sum_{i=1}^{m} \|P_i\|_d^2 + \sum_{j=1}^{n} \|Q_j\|_d^2 \right) \tag{2}$$

where $I(\cdot)$ is an indicator function that equals 1 if $r_{i,j} > 0$, and 0 otherwise. $g(\cdot)$ is a logistic function, i.e. $g(x) = 1/(1 + \exp(-x))$.

The loss function $L(P, Q)$ is optimized by gradient descent with alternatively fixed P and Q, and the gradients of $L(P, Q)$ with respect to P and Q can be computed as follows.

$$\frac{\partial L}{\partial P_i} = \sum_{j=1}^{n} I(r_{i,j}) \cdot \left(\frac{\exp\left(g\left(P_i^T \cdot Q_j\right)\right)}{\sum_{k=1}^{n} I(r_{i,k}) \cdot \exp(P_i^T \cdot Q_k)} - \frac{\exp(r_{i,j})}{\sum_{k=1}^{n} I(r_{i,k}) \cdot \exp(r_{i,k})} \right) \cdot g'\left(P_i^T \cdot Q_j\right) \cdot Q_j + \lambda P_i$$

$$\tag{3}$$

$$\frac{\partial L}{\partial Q_j} = \sum_{i=1}^{m} I(r_{i,j}) \cdot \left(\frac{\exp\left(g\left(P_i^T \cdot Q_j\right)\right)}{\sum_{k=1}^{n} I(r_{i,k}) \cdot \exp(P_i^T \cdot Q_k)} - \frac{\exp(r_{i,j})}{\sum_{k=1}^{n} I(r_{i,k}) \cdot \exp(r_{i,k})} \right) \cdot g'\left(P_i^T \cdot Q_j\right) \cdot P_i + \lambda Q_j$$

$$\tag{4}$$

Note that $g'(x)$ is the derivative of $g(x)$, i.e. $g'(x) = g(x) \cdot (1 - g(x))$.

3.2 Rating features

In this paper, we design rating features for each user and item. As for the user, we use the user's average normalized rating score and normalized rating times as the rating features. The ith user (u_i)'s average normalized rating score is defined as:

$$\bar{R}_{u_i} = \sum_{k=1}^{|R_{u_i}|} \left(0.5 + \frac{r_{u_i,k} - r_{u_i,min}}{r_{u_i,max} - r_{u_i,min}} \right) / |R_{u_i}| \tag{5}$$

where $|R_{u_i}|$ denotes the rating times of u_i and $r_{u_i,max}, r_{u_i,min}$ are the maximum and minimum rating score of u_i respectively. Particularly, if $r_{u_i,max}$ is equal to $r_{u_i,min}$, $\bar{R}_{u_i}$ is 0.5, and if there is no rating for u_i, $\bar{R}_{u_i}$ is 0. The ith user's normalized rating times is defined as:

$$\bar{N}_{R_{u_i}} = |R_{u_i}| / \sum_{k=1}^{m} |R_{u_k}| \tag{6}$$

In analogy, as for the item, we define the jth item (i_j)'s average normalized rating score and normalized rating times as the rating features. The definitions are shown as follows:

$$\bar{R}_{i_j} = \sum_{k=1}^{|R_{i_j}|} \left(0.5 + \frac{r_{i_j,k} - r_{i_j,min}}{r_{i_j,max} - r_{i_j,min}} \right) / |R_{i_j}| \tag{7}$$

$$\bar{N}_{R_{i_j}} = |R_{i_j}| / \sum_{k=1}^{n} |R_{i_k}| \tag{8}$$

For the user-item pair (u_i, u_j), we use $x_{i,j}^{a,r}$ to denote the respective rating features vector, i.e. $x_{i,j}^{a,r} = (\bar{R}_{u_i}, \bar{N}_{R_{u_i}}, \bar{R}_{i_j}, \bar{N}_{R_{i_j}})$, and the rating features vector set of u_i is denoted as $x_i^{a,r}$, specifically

$x_i^{a,r} = \{x_{i,1}^{a,r}, x_{i,2}^{a,r}, x_{i,3}^{a,r}, ..., x_{i,n_i}^{a,r}\}$. Generally, for each user-item pair (u_i, i_j), we have the accuracy feature vector $x_i^{a,r}$, and $x_{i,j}^a = (x_{i,j}^{a,l}, x_{i,j}^{a,r}) = (P_i^T, Q_j^T, \bar{R}_{u_i}, \bar{N}_{R_{u_i}}, \bar{R}_{i_j}, \bar{N}_{R_{i_j}})$.

4 DIVERSITY FEATURE REPRESENTATION BY ATTRIBUTE-BASED ENTROPY

In earlier work, Di Noia et al. (2014) gave a representation of the user's propensity in diversification and proposed an adaptive attribute-based re-ranking approach. In our proposal we follow up on the idea and take a further step to represent the diversity features for users and items.

Suppose that A^i (e.g. year and genre in the movie domain) is the set of item attributes and A^u (e.g. age, gender, occupation in the movie domain) is the set of user attributes. And we use a_j^i and a_j^u to denote the jth attribute of A^i and A^u. $|A^i|$ and $|A^u|$ are the sizes of A^i and A^u respectively. For a generic user u, we measure the diversity need of the user on the attribute a^i with K values of items ($a^i \in A^i$) by entropy in information theory, defined as follows:

$$E_{a^i}(u) = -\sum_{k=1}^{K} p_k \cdot \log_K p_k \tag{9}$$

where p_k is diversity probability of the kth value of a^i considering all the rated items of user u (user profile). If the value of $E_{a^i}(u)$ is high, we infer that the user u has a strong diversity need with respect to the attribute a^i of items. Meanwhile, we define the satisfaction degree of an item i for the user's diverse need on the attribute a^i as follows:

$$S_{a^i}(i) = \sum_{k=1}^{N_{R_i}} E_{a^i}(u_k) / (N_{R_i} \cdot \log(1 + N_{R_i})) \tag{10}$$

where N_{R_i} denotes the number of users' ratings on the item i. The term $1/\log(1 + N_{R_i})$ punishes the satisfaction degree of popular items by taking into account the hypothesis that an item, known by fewer users, is more likely to increase the diversity.

Analogously, for an item i, we define the diversity suitable degree of the item for a user with the attribute a^u with L values ($a^u \in A^u$), shown as follows:

$$H_{a^u}(i) = -\sum_{l=1}^{L} p_l \cdot \log_L p_l \tag{11}$$

where p_l denotes the diversity probability of the lth value of a^u considering all the users rated the item i (item profile). The high value of $H_{a^u}(i)$ is interpreted as the item i is widely suitable for the users with different values of attribute a^u. And the diversity characteristic of a user u on the attribute a^u is defined as:

$$C_{a^u}(u) = \left(\sum_{k=1}^{N_{R_u}} H_{a^u}(i_k) / N_{R_u}\right) \cdot \varphi(N_{R_u}) \tag{12}$$

where N_{R_u} is the number of items rated by the user u. $\varphi(\cdot)$ is a monotonically increasing function of N_{R_u}, used to tune the $C_{a^u}(u)$ and defined as $\varphi(x) = 1 - 1/\log(1 + x)$. And here we assume that a user u with a high value of N_{R_u}, can be regarded as an active user, who tends to view an item with a high diversity suitable degree $H_{a^u}(i)$.

We use $x_{i,j}^d$ to denote the diversity features represented by the attribute-based entropy for a user-item pair (u_i, i_j), thus $x_{i,j}^d = \{E_{a^i}(u_i), S_{a^i}(i_j), H_{a^u}(i_j), C_{a^u}(u_i) | a^i \in A^i, a^u \in A^u\}$.

5 FEATURE FUSION

Once the accuracy and diversity features have been extracted for each user-item pair, the final step is to perform feature fusion. This is done by calculating the fusion weight ω_i in Equation 1. Specifically, we use the diversity need $E_{a^i}(u)$, defined in Session 4, and the diversity feature weight θ to represent ω_i for the user u_i. The fusion method is defined as follows:

$$\omega_i = \theta \cdot \sum\nolimits_{a^i \in A^i} E_{a^i}(u_i) / (|A^i| \cdot E_{a^i_{max}}(u_i)) \tag{13}$$

where $E_{a^i_{max}}(u_i) = \max\{E_{a^i}(u_i) \mid a^i \in A^i\}$ is the normalized term and θ is used to bind the maximum proportion of diversity features, which can be dynamically set to balance the trade-off between accuracy and diversity. According to Equation 13, we can see that ω_i is different from user to user. As for the users with a high value of diversity need, the ω_i will be large.

6 EXPERIMENTS AND EVALUATION

6.1 Experimental setup

Our experiments are carried out on the MovieLens 1M[3] dataset,[1] which contains 1 million ratings from 6,000 users on 4,000 movies. Ratings are on a scale from 1 to 5 and each user has at least 20 ratings. Slightly in contrast to other approaches, we randomly selected 60% of the rated items as the training dataset and the rest as the test dataset, in which the users with less than 10 rated items in the test dataset were removed. Gender, age, and occupation are selected as the users' attributes and genres and years (divided into decades) are selected for the items. For measuring the accuracy we chose Normalized Discounted Cumulative Gain (Hang Li 2011) at position k (NDCG@k) as the evaluation metric. And we focused on NDCG@1~10 in this paper. As for diversity, we defined an evaluation metric by using the similarity between items for a recommendation list $l(u)$ for the user u:

$$Diversity(l(u)) = 1 - \sum\nolimits_{i,j \in l(u), i \neq j} sim(i,j) / \left(\frac{1}{2} \cdot |l(u)| \cdot (|l(u)| - 1) \right) \tag{14}$$

where $sim(i, j)$ is the similarity measure between the ith item and the jth item in $l(u)$, calculated by Jaccard index, and the overall diversity of recommendation lists is defined as:

$$Diversity = \sum\nolimits_{u \in U} Diversity(l(u)) / |U| \tag{15}$$

We report the average results, attained from all users, by repeating 5 runs of experimental procedures. The learning rate is set to 0.01 for the latent feature extraction by list-wise matrix factorization and we use the latent feature dimension of 5, the same as ListRank-MF. The LTR algorithm (i.e. LambdaMart) is implemented by an open source toolkit called RankLib.[2]

6.2 Effectiveness of our proposal

We first set $\theta = 0$ in Equation 13, to eliminate the diversity features when performing feature fusion, and thus we can only evaluate if our proposed model could provide accurate recommendations. We demonstrate the result of NDCG@1~10 as shown in Figure 2, from which we can observe that the results are empirically reasonable. Then, we investigate the impact of the diversity feature weight θ, from which we can see the effect of proposed diversity features and how the trade-off between accuracy and diversity is made. We only demonstrate NDCG@10 for accuracy during the change of θ and the result is shown in Figure 3. As we can see, with the increase of θ, the NDCG@10 shows a downward trend

---

[1] http://grouplens.org/datasets/movielens/
[2] http://sourceforge.net/p/lemur/wiki/RankLib/

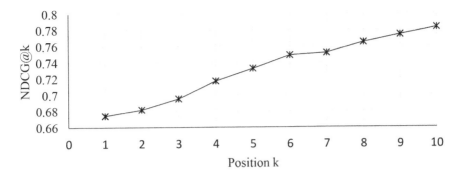

Figure 2. The effectiveness for accuracy with $\theta = 0$.

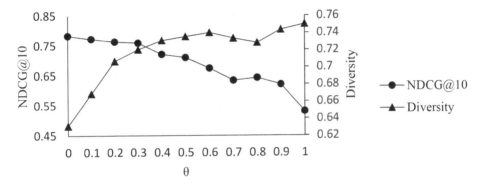

Figure 3. The impact of the diversity feature weight θ.

and simultaneously the diversity is going up when $\theta < 0.4$ and then shows a slow and frustrating rising tendency as a whole. When θ comes to 1, the overall diversity features are used to make the recommendation, and it seems that the recommendation lists are randomly selected.

6.3 Performance comparison

In this part, we choose the well-known item-based collaborative filtering (Item-CF) (Deshpande et al. 2004), Simon-SVD and the state-of-the-art ListRank-MF as the baselines to compare to our proposal. As for accuracy, we compare the values of NDCG@1~10, and similarly θ is set to 0 for our approach. The results are shown in Figure 4. Without considering the diversity, our approach achieves a performance improvement of ca. 3% over ListRank-MF, ca. 8% over Simon-SVD and ca. 13% over Item-CF, with respect to accuracy at NDCG@10. ListRank-MF slightly outperforms our proposal at NDCG@1~3. One possible reason is perphaps that LambdaMart directly optimizes the whole ranking list and the metric optimized is different to ListRank-MF, the other is the impact of rating features.

In order to measure the balance between accuracy (i.e. using NDCG@10) and diversity, we use a hybrid method, proposed in Zhou et al. (2010), to normalize the metrics and then make a combination. The method is defined as follows:

$$\psi(X_\alpha, Y_\alpha) = (1 - \mu) \frac{X_\alpha}{\max_\beta X_\beta} + \mu \frac{Y_\alpha}{\max_\beta Y_\beta} \tag{16}$$

where X_α denotes the NDCG@10 and Y_α denotes the diversity. We first calculate the maximum NDCG@10 and diversity among all compared approaches and then use them as the

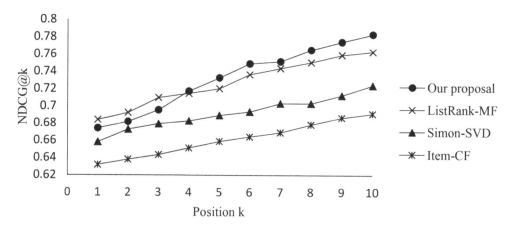

Figure 4.   NDCG@10 comparison.

Table 2.   Results of the balance between accuracy and diversity.

Approach	NDCG@10	Diversity	ψ (NDCG@10, Diversity)
Item-CF	0.6913	0.7012	0.9263
Simon-SVD	0.7241	0.6645	0.9411
ListRank-MF	**0.7629**	0.6812	0.9837
Our proposal	0.7592	**0.7203**	**0.9966**

normalizer for other values. The combination parameter μ is set as equal to the diversity feature weight θ in our proposal, to achieve a reasonable and fair trade-off between accuracy and diversity for the compared approaches. And in our experiments, we set $\mu = \theta = 0.3$. The results are shown in Table 2.

As can be observed, with the diversity feature weight $\theta = 0.3$, our proposal improves the diversity, the best among all, but with slightly decreasing NDCG@10. ListRank-MF outperforms others in accuracy, because it only focuses on the quality of ranking. Lacking emphasis on diversity, ListRank-MF gets a rather low value for diversity. Simon-SVD concentrates on accuracy and achieves a generic value for NDCG10, but with the lowest diversity. Particularly, Item-CF obtains a rather high value for diversity, because we choose a small collection of neighbors with 10 when calculating the similarity. Measured by Equation 16, our proposal gains the best balance between accuracy and diversity. Although, the measure should be further argued to be crucial, our proposal has shown its advantages in both accuracy and diversity.

6.4   Scalability analysis

In our proposal, the computational cost is mainly on three aspects of accuracy feature extraction by MF, diversity feature calculation and the training of the ranking model. As for accuracy feature extraction by MF, Shi Y et al. (2010) have indicated that the computational complexity is linear in the number of observed ratings. In terms of diversity feature calculation, the time complexity is $O(S \cdot (|A^i| + |A^u|) \cdot K)$, where S denotes the number of observed ratings and K is the maximum scale of attributes. Considering we often have $S \gg |A^i|, |A^u|$ and K is small, thus, the complexity can be regarded as linear with S. With respect to the training of the ranking model, the time complexity of centralized LambdaMART is $O(|S| \cdot |F|)$ (Bekkerman et al. 2011), where $|S|$ denotes the size of the training dataset for LambdaMART, i.e. the set of features vector, and $|F|$ is the number of features. When $|S| \gg |F|$, the complexity mainly depends on $|S|$. Moreover, the distributed LambdaMART can significantly reduce the complexity. All things considered, our proposal is computationally practical and can be scaled up for use on large-scale collections.

7 CONCLUSION

In this paper, we proposed a formal model based on LTR for hybrid recommendation, which focused on the balance between accuracy and diversity. First, we made use of the list-wise matrix factorization to combine with our model to extract accuracy features. In the meantime, some rating features were designed to improve the accuracy. Then we investigated the representation of diversity features, in order to incorporate diversity into our model. Attribute-based entropy representation was proposed for the diversity features. Experimental results in the movie domain showed that without considering diversity, our proposal outperformed the state-of-the-art ListRank-MF, Simon-SVD and the well-known Item-CF with respect to accuracy at NDCG@10 and as for the trade-off between accuracy and diversity, our proposal gained the best balance measured by a hybrid method. Moreover, according to scalability analysis, our proposal was computationally efficient and can be applied to large-scale cases. In future work, we will try to exploit deep learning to find better features and investigate more approaches or models to integrate with our proposal.

REFERENCES

Adomavicius G & Tuzhilin A. 2005. Toward the next generation of recommender systems: A survey of the state-of-the-art and possible extensions. Knowledge and Data Engineering, IEEE Transactions on. 17(6): 734–749.

Amatriain X. 2014. The recommender problem revisited. Proceedings of the 8th ACM Conference on Recommender systems. ACM, pp. 397–398.

Anderson, Chris. 2013. The long tail. Bertelsmann:Random House.

Bekkerman, Ron, Mikhail Bilenko & John Langford, eds. 2011. Scaling up machine learning: Parallel and distributed approaches. Cambridge: Cambridge University Press.

Burges C J C. 2010. From ranknet to lambdarank to lambdamart: An overview. Learning. 11: 23–581.

Deshpande M & Karypis G. 2004. Item-based top-n recommendation algorithms. ACM Transactions on Information Systems (TOIS). 22(1): 143–177.

Di Noia T, Ostuni V C & Rosati J, et al. 2014. An analysis of users' propensity toward diversity in recommendations. Proceedings of the 8th ACM Conference on Recommender systems. ACM, pp. 285–288.

Hang Li. 2011. A short introduction to learning to rank. IEICE Transactions on Information and Systems. 94(10): 1854–1862.

Karatzoglou A, Baltrunas L & Shi Y. 2013. Learning to rank for recommender systems. Proceedings of the 7th ACM Conference on Recommender Systems. ACM, pp. 493–494.

Koren Y, Bell R & Volinsky C. 2009. Matrix factorization techniques for recommender systems. Computer. (8): 30–37.

Lathia N, Hailes S & Capra L, et al. 2010. Temporal diversity in recommender systems. Proceedings of the 33rd international ACM SIGIR conference on Research and development in information retrieval. ACM, pp. 210–217.

Liu T Y. 2009. Learning to rank for information retrieval. Foundations and Trends in Information Retrieval. 3(3): 225–331.

McNee S M, Riedl J & Konstan J A. 2006. Being accurate is not enough: how accuracy metrics have hurt recommender systems. CHI' 06 extended abstracts on Human factors in computing systems. ACM, pp. 1097–1101.

Mnih A & Salakhutdinov R. 2007. Probabilistic matrix factorization. Advances in neural information processing systems. pp. 1257–1264.

Shi Y, Larson M & Hanjalic A. 2010. List-wise learning to rank with matrix factorization for collaborative filtering. Proceedings of the fourth ACM conference on Recommender systems. ACM, pp. 269–272.

Signal and Information Processing, Networking and Computers – Chen & Huang (Eds)
© 2016 Taylor & Francis Group, London, ISBN 978-1-138-02881-4

A Joint Source Channel Coding scheme based on environment vector clustering over WiFi channels

Tingting Huang, Songlin Sun, Yaoyao Guo, Na Chen & Zheng Zhou
Key Laboratory of Trustworthy Distributed Computing and Service (BUPT), Ministry of Education, Beijing University of Posts and Telecommunications, Beijing, China

ABSTRACT: A Joint Source Channel Coding (JSCC) scheme is a valid means, which guarantees a tradeoff between effectiveness and reliability of video transmission. This paper proposes a JSCC scheme by using the WiFi channel vector classification. Environment vectors are formed by channel environmental parameters, which are classified using k-means clustering and are further used to guide the strategy of resource optimization in real-time transmission. Experiment results show that the proposed algorithm could perceive the situation of the environment and adjust the bits allocated to the source and channel dynamically. Compared to other schemes that get optimal rate allocation through iterating all solutions, it saves a large amount of calculation.

Keywords: JSCC; HEVC; rate allocation; machine learning; WiFi

1 INTRODUCTION

With the development of communication technology, the channel environment is more and more complicated, especially the wireless channels. The wireless signals transmitted in the air are susceptible to noise and other factors. For example, when transmitted in WiFi channels, the compressed video will be greatly influenced by physical environmental factors such as noise, a fading effect caused by distance, obstacles, router placement, etc. Also, the number and behavior (different bandwidths occupy cases such as watching a movie, shopping, talking, and so on) of terminals that access the same WiFi hotspot will result in a limited bandwidth. Currently, for video compression, such as HEVC standard (Gary J. Sullivan, 2012), high compression ratio will cause sensitivity of the stream to the noise. At the same time, the bit stream transmitted through a channel is faced with a complex environment, which makes it very necessary to take source coding and channel coding into account together.

In the past, JSCC attracted a lot of attention because it is a comprehensive consideration of source compression efficiency and channel transmission reliability. At present, many research works focus on JSCC. For example, some articles consider overall end-to-end distortion, which is the sum of source distortion and channel distortion, as a measure of video transmission quality. The corresponding JSCC schemes are proposed in Yuan Zhang et al. (2007) and Ching-Hui Chen et al. (2011) based on the principle of minimizing end-to-end distortion. The JSCC scheme has many methods, such as efficient bit allocation scheme between source coding and channel coding proposed in some articles such as in Zhifeng Chen et al. (2012), Xinglei Zhu et al. (2010), Xuejuan Gao et al. (2008) and Qin Wang et al. (2013). Zhifeng Chen and Dapeng Wu (2012) designed a Cross Layer Rate Control (CLRC) algorithm based on R-D optimization. This algorithm can achieve the purpose of joint coding through a comprehensive selection of the quantization steps in source coding and coding rate in the channel coding. In Xinglei Zhu et al. (2010), the authors combined the concept of authentication and presented an optimal rate allocation scheme among source coding, channel coding, and authentication to get the best quality of end-to-end reconstructed video quality.

Apart from the above methods, there is another one that provides a different kind of channel protection for video bit streams with different importance in Yong Liu et al. (2008) and Zhou, C. et al. (2014). Based on the MIMO-OFDM system, Yong Liu et al. (2008) proposed a JSCC solution to provide different levels of error protection for different importance of the H.264 bit stream. Some key information in the H.264 stream file, such as picture parameter set and sequence parameter set, can be provided with more channel protection while general information is provided with common protection. More channel protection is provided not only for the importance of the general stream, but also for the hierarchical structure of video coding. More channel protection is provided to more important levels (e.g., basic layer) and important SU units in Chen Chi et al. (2010).

In this paper, a bit rate allocation scheme between source and channel is put forward according to different WiFi channel environments. First, we use advanced k-means clustering algorithm to cluster the channel environment, and then propose the optimal bit allocation scheme according to the different channel classes. When the user accesses the channel, a corresponding video transmission scheme is provided for users according to the real-time channel environment automatically. In addition, the video transmission scheme also will be adjusted with the change in the wireless channel environment.

The rest of this paper is organized as follows. Section 2 introduces related work. In Section 3, the proposed JSCC scheme was deduced in detail. The experiment and analysis of the proposed scheme is in Section 4. Finally, we conclude this paper in Section 5.

2 RELATED WORK

2.1 The factors influencing the WiFi channel environment

Let us assume that the video is transmitted in a WiFi environment. In general, the wireless signal transmission is more easily affected by environmental factors than wired signal transmission. Transmission distance, medium, obstacles, or other reasons (such as weather) can cause wastage of the wireless signal.

In addition to the physical factors, because the network requests of each user are independent, and all users accessing the same WiFi hotspots share limited bandwidth resources, the behavior of terminals will result in limited bandwidth. For example, the total number of terminals, the angle between the terminals' receiving antenna and the router, whether the usage of each terminal is a real-time request (video, web, game, or music), and so on will have a large impact on the usage of the users.

Compressed videos that were transmitted through the disturbed channel environment will suffer from errors, and may not even be able to correctly decode when these errors are serious. Now, there have already been several effective solutions to solve this kind of situation; the common strategies are Automatic Repeat Request (ARQ) and forward error correction (FEC). The proposed JSCC scheme in this paper uses FEC mechanism, which attaches a certain redundancy error correction code to the compressed video stream that shall be sent. The receiver can perform data error detection according to the error correcting codes, and this method improves the reliability of the transmission channel. To design a distribution rate to achieve the best anti-error performance, we need to take channel conditions into account.

2.2 K-means clustering

It is difficult to have an accurate understanding of a channel environment when user access to the hotspots, due to many factors, can influence the channel. Therefore, we need to choose a method to estimate the channel environment the video is transmitted in effectively.

The collection of physical or abstract objects can be divided into multiple classes according to similarity, and this process is called clustering. Clustering is an important unsupervised learning algorithm of machine learning and has been applied in many fields. Clustering is achieved by a variety of methods such as partitioning methods, hierarchical methods, density-based methods, grid-based methods, and model-based methods. K-means algorithm is one of the methods of partitioning.

As k-means algorithm is rapid, simple, scalable, and highly efficient for large data sets, we chose it for the classification of the channel environment in this paper.

2.3 The end-to-end distortion estimation

End-to-end distortion is an important measure to choose an effective strategy for videos transmitted in the wireless channel. First of all, we need define the necessary notation. Let $f_i(n)$ denote the value of pixel i in the frame n, $\hat{f}_i(n)$ the reconstructed pixel at the encoder, and $\tilde{f}_i(n)$ estimate of the reconstructed pixel at the decoder. Then the distortion $d(i,n)$ of the single pixel i in the frame n can be written as:

$$d(i,n) = E\{[f_i(n) - \tilde{f}_i(n)]^2\} = E\{[f_i(n) - \hat{f}_i(n)]^2\} + E\{[\hat{f}_i(n) - \tilde{f}_i(n)]^2\} \qquad (1)$$

Following a frame-level recursive approach, the overall distortion can be divided into the three items, namely source, error concealment, and error propagation distortion. Therefore, the overall distortion can now be expressed as:

$$d(i, n) = (1 - \rho)d_s(i, n) + (1 - \rho)d_{ep}(i, n) + \rho d_{ec}(i, n) \qquad (2)$$

where ρ is Packet Error Rate (PER). Error concealment is introduced by the block in the reference frame and is copied in the current frame. Error propagation caused by error in the reference frame and will spread to the frames which refer to it. Source distortions are based on source coding mode options, while error concealment and error propagation are related to the selection of the reference frame and channel FEC coding mode options. The overall distortion can be achieved by iteration (Yuan Zhang et al. 2007, Ching-Hui Chen et al. 2011).

3   THE PROPOSED JSCC SCHEME BASED ON ENVIRONMENT VECTOR

Our goal is to obtain a better experience of watching videos through bit allocation between source and channel with limited bandwidth and interference. The diagram of the whole system is as shown in Figure 1.

The design process of the scheme is mainly divided into three steps. First, according to different environmental parameters and user behaviors, the training samples of the environment vectors are classified into different categories by using k-means algorithm. Then a series of

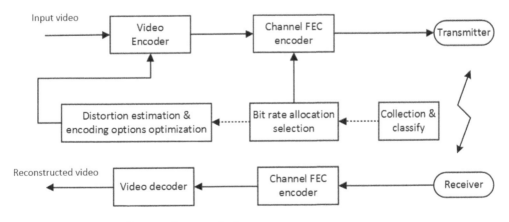

Figure 1.   The system diagram of proposed scheme.

corresponding rate allocation schemes (i.e., FEC coding mode) are set up to obtain an optimized tradeoff between compression efficiency and transmission robustness according to the different environmental categories. Finally, when users access the WiFi hotspot, bit rate allocation scheme is adjusted adaptively according to the real-time collection of environment data. Also, an available source encoding option will be implemented with this bit rate allocation. Specific details are in part A, B, and C of this section.

3.1 Environment vector clustering

In the wireless channel environment, there are many factors that can affect the wireless signal transmission, such as the emission source signal intensity, distance, terminals numbers, bandwidth, and other parameters. In addition to physical environmental factors, the users' behavior as a kind of environmental factors can also produce a great influence on the wireless signal transmission and receiving for each specific wireless environment. We need to classify the channel environment based on the environment vectors formed by user behaviors. K-means algorithm is used to cluster a large number of environmental vectors. That is to say, the input sample $x^{(i)}$, i = 1,2,...,n, is a series of environmental vectors as shown in Figure 2.

The basic process of K-means algorithm is as follows:

1. For a data set which contains n data objects $\{x^{(1)}, x^{(2)}, ..., x^{(n)}\}$, we choose $m_1, m_2, m_3, ..., m_K$ among them as the initial cluster centroids, randomly.
2. Calculate the distance of each object from the clustering center and classify each object to the category in which it should belong to according to the minimum distance. As shown in the following formula:

$$d(i) = \arg\min_j \| x^{(i)} - m_j \|^2 \tag{3}$$

where k is the number of clusters we set. $d(i)$ represents the class whose centroid has the shortest distance from the sample i in k classes, namely $d(i)$ is one number of 1 to k. m_j, j = 1,2,...,k, is the current centroid of the clusters.
3. Calculate again the centroid of the class for each class j:

$$m_j = \frac{\sum_{i=1}^{n} \gamma_j x^{(i)}}{\sum_{i=1}^{n} \gamma_j} \tag{4}$$

γ_j is an indicator function, when the data sample i belongs to the cluster j, $\gamma_j = 1$; otherwise, $\gamma_j = 0$.

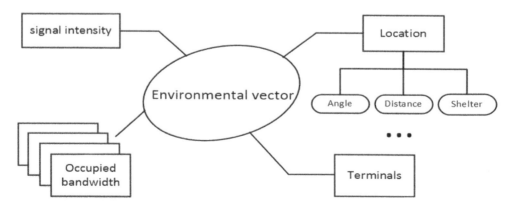

Figure 2.   The composition of environmental vector.

4. Repeat steps 2 and 3 until the algorithm converges, where, the size of k is related to the number of channel environment classes. If the value of k is greater, then there are more channel classes, and the corresponding rate allocation schemes are also more which can obtain approximate optimal bit allocation schemes based on the current channel environment compared with an algorithm iterating all schemes.

K-means algorithm mainly aims to minimize the distortion function as follows:

$$J(d,m) = \sum_{j=1}^{k} \sum_{i=1}^{n} \| x^{(i)} - m_{d(j)} \|^2 \tag{5}$$

J denotes sum of squared distances between each training sample and the cluster centroid $m_{d(i)}, j = 1,2,...,k$. First, while holding the cluster centroid $m_{d(i)}$ fixed, k-means algorithm minimizes J with respect to $d(i)$, and then while holding $d(i)$ fixed, minimizes J with respect to $m_{d(i)}$. Continue to repeat the above process, until J is decreased to the minimum. Reduction to the minimum of J means that clustering is complete.

We know that the k-means algorithm is extremely sensitive to the choice of the initial cluster center. If the choice of initial value is bad, the result can't get effective clustering effect. The proposed scheme chose a program with a minimal J as the final solution, by choosing different sets of initial clustering center and calculating their values of J.

3.2 *Corresponding rate allocation scheme*

In the wireless communication system, let us assume that the total network bandwidth is R, bit rate of the source encoder is R_s, r denotes the coding rate, so $R_s = r \times R$. Moreover, R_c is assigned to the bit rate of the channel encoder, $R_c = R - R_s = (1-r) \times R$.

In the current HEVC video coding standard due to the relatively high compression without corresponding code stream protection mechanism, it is very sensitive to channel error. Few bit errors can cause decoding failure extremely easily. The essence of JSCC is minimizing the total distortion by adjusting the coding rate r on the premise of limited channel bandwidth.

In this paper, we employ the general forward error correction (FEC) mechanism. Because of the sensitivity of HEVC to channel error, we select the worst channel condition in each channel environment class as the criterion of choosing the corresponding rate allocation scheme to guarantee that under each kind of environmental conditions in this class, the compressed bit stream can be transmitted correctly. After clustering for all training samples, we calculated the following formula:

$$s(i, j) = \arg\min_{x(i)} \{ x(i)\gamma_j - m_j \} \tag{6}$$

$s(i, j)$ is the sample of the worst channel condition in the class j.

We assumed that the k rate allocation scheme is corresponding to the code rate $r_1, r_2, ..., r_k$, $0 \approx r_1 < r_2 < ... < r_k < 1$, respectively.

When the wireless interference is serious, a lot of bandwidth is allocated to channel coding in order to increase its ability to resist error to ensure that users can watch video in real time. Because very few bits are assigned to source coding, the compressed video image is fuzzy. In addition, users can choose to pause watching and cache the video in a higher coding rate for better viewing experience.

By analogy, the better channel is, the more bits are used for source coding, and the lesser the bit rate used to resist error, the better the video reconstruction quality.

3.3 *Encoding mode selection*

The wireless channel is time-varying channel. Under the condition of limited channel bandwidth and certain channel interference, this scheme improves the overall performance of the system by bit rate adjustment between the source and the channel adaptively.

When a user accesses a certain wireless WiFi channel, the wireless router will automatically collect different environment parameters and form an environment vector. Then the current environment will be classified into its corresponding class, and the corresponding bit rate allocation scheme is set up. A reasonable source coding mode can be selected for video compression with this bit rate allocation in order to minimize end-to-end distortion.

After all, the environment data will be a "training data" to update the centroid of the class.

4   EXPERIMENT RESULTS AND ANALYSIS

4.1   *Average PSNR*

Because of the particular users' behaviors for each WiFi environment, we simulated a simple WiFi environment and clustered it into six classes, and k is equal to 6.

We implemented the experiments in HM15.0. The test sequences of "PartyScence" and "Race Horse" with the 832*480 resolution and "BQSqurare" and "BasketballPass" with 416*240 resolution are used in our experiments. We considered the transfer of 200 frames in additive white Gaussian noise (AWGN) channels with different classes. Each GOP contains four frames with a whole structure of {I,B,B,B,...}. We chose the PSNR luminance component of the reconstructed video as the measure of video quality. Since HEVC is strongly sensitive to channel error, and may cause source decoding failure when serious, we designed a different FEC coding rate r to ensure that the compressed video can be transmitted currently. The compressed video streams will not be transmitted under the worst channel conditions, namely "class 1". The other five classes, namely "class" 2–6, correspond to code rate r = 4/15, 3/7, 1/2, 4/ 7, 11/15, 1, respectively.

In order to describe the performance of our scheme, we compared it with the schemes of fixed rate coding as shown in Table 1. Let "fixed 1" denote the scheme with fixed coding rate 3/7 and "fixed 2", the scheme with fixed coding rate 1/2. Table 1 shows different PSNR under different channel classes. The "fail" denotes that the videos are not decoded or the reconstructed video quality is so poor that it can't be watched with that coding rate. In order to get accurate results, we repeated the transmission process 50 times and used the average results.

Table 1 shows the comparison results of the reconstructed quality of four sequences under our adaptive bit rate allocation scheme and two fixed channel coding rate schemes. It can be seen clearly from Table 1 that fixed channel coding rate schemes can't provide enough support for protection from poor channel environments, such as class 2, and result in decoding failure

Table 1.   PSNR-Y for the different bit rate allocation scheme employed under different channel classes.

Video sequence	Scheme	PSNR-Y(dB)				
		Class 2	Class 3	Class 4	Class 5	Class 6
PartyScence	Our scheme	27.00	28.70	29.28	29.80	30.65
	Fixed 1	Fail	28.70	28.70	28.70	28.70
	Fixed 2	Fail	Fail	29.28	29.28	29.28
RaceHorse	Our scheme	30.31	32.39	33.03	33.59	34.61
	Fixed 1	Fail	32.39	32.39	32.39	32.39
	Fixed 2	Fail	Fail	33.03	33.03	33.03
BQSqurare	Our scheme	29.39	31.08	31.71	32.26	33.24
	Fixed 1	Fail	31.08	31.08	31.08	31.08
	Fixed 2	Fail	Fail	31.71	31.71	31.71
BasketballPass	Our scheme	32.26	34.20	34.92	35.57	36.68
	Fixed 1	Fail	34.20	34.20	34.20	34.20
	Fixed 2	Fail	Fail	34.92	34.92	34.92

or very bad reconstructed video quality. However, when the channel environment is better, the fixed channel rate schemes can't obtain a higher performance of video quality, while our proposed one can achieve a higher reconstructed quality of 2–3 dB at the receiver side. The reason is that our proposed scheme provides more bits for channel protection in worse channel environments, while allocating more bits for source encoding in better channel environments.

Figure 3 shows the thirtieth frame in a reconstructed "BasketballPass" sequence with different schemes under channel environment class 3. It is clear from Figure 3(a), (b) that the performance with our scheme is almost the same as the fixed 1 scheme for using the same rate allocation 3/7, but in Figure 3(c), the fixed 2 scheme proposed a worse quality due to channel error.

In Figure 4, we can see that in class 5, the performance of the hundredth frame in the received "Race Horse" with our proposed scheme is more reliable than fixed 1 and fixed 2.

4.2 Complexity analysis

Due to the end-to-end distortion estimation method described in section 2.3, the objective of the system is to minimize decoder distortion:

$$\min_{\mu \in S, r \in C, ref \in M} E\left[d_{e2e}(\mu, r, ref)\right] \tag{7}$$

The vector of the candidate source and channel coding parameters for video compression and transmission are denoted as $u(k)$ and $r(k)$. Finally, let M be the set of reference frames.

General JSCC schemes perform rate allocation based on minimum end-to-end distortion and usually need to iterate all the bit rate allocation, which will cause quite a large amount of calculation. The worst case cost for general JSCC schemes is $O(|S| \times |C| \times |M|)$. In this paper, we preset the bit rate allocation for each environment class, that is to say that we have predefined the $r(k)$ before encoding only with simple classification calculation. The worst

| (a) Image with our scheme | (b) Image with fixed 1 scheme | (c) Image with fixed 2 scheme |

Figure 3.   The thirtieth frame in reconstructed "BasketballPass" sequence with different schemes in channel environment class 3.

| (a) Image with our scheme | (b) Image with fixed 1 scheme | (c) Image with fixed 2 scheme |

Figure 4.   The hundredth frame in a reconstructed "RaceHorse" sequence with different schemes in channel environment class 5.

case cost for our scheme will be $O(|S| \times |M|)$. However, the actual cost is even lower because not all the source encoding options can be used for a particular FEC code available in C.

As per the above analysis, each rate allocation scheme ensures that the compressed code streams can be transmitted without bit error under the corresponding class of the channel conditions as well as get an optimized video reconstruction quality. In addition, our scheme will save a lot of time because the complexity of classification calculation is much less than that of end-to-end distortion calculation iteratively. At the same time, we can know that the increase in the code rate will make the video quality better, so the increase in classifications' numbers can increase the number of the bit rate allocation schemes and achieve better results.

5 CONCLUSIONS

An efficient and concise JSCC scheme for allocating bit rate between source and channel according to different wireless WiFi channel environment classes has been presented in this paper. It is comprehensive considering the various physical environment factors as well as the channel usage. In particular, choosing the worst condition in each channel environment class as the criterion of choosing code rate r ensures that compressed bit stream can be transmitted correctly for each class. The scheme that allows bit rate allocation to be switched automatically in the wireless channel will get an optimized video reconstructed quality and transmission reliability. Further, the complexity issues have also been analyzed in this paper. Compared to other schemes which get the optimal rate allocation through iterating all solutions, this scheme can save us a large amount of calculation.

REFERENCES

Chen Chi, Yu Zhang, Yaosheng Fu, & Zhixing Yang. 2010. A new joint source and channel coding scheme for packet-based scalable multimedia streams. *GLOBECOM Workshops (GC Wkshps), 2010 IEEE, Miami, FL*: 954–959.

Ching-Hui Chen, Wei-Ho Chung, & Wang, Y.-C.F. 2011. Joint source-channel coding optimization with packet loss resilience for video transmission. *IEEE International Conference on Image Processing (ICIP), Brussels*: 2197–2200.

Gary J. Sullivan, Jens-Rainer Ohm, Woo-Jin Han, & Thomas Wiegand. 2012. Overview of the High Efficiency Video Coding (HEVC) standard. *IEEE Transactions on Circuits and Systems for Video Technology* 22:1649–1668.

Qin Wang, Wei Wang, Shi Jin, & Hongbo. 2013. Energy-aware joint source-channel coding control for quality-optimized wireless multimedia communications. *International Conference on Wireless Communications and Signal Processing (WCSP), Hangzhou, China*: 1–6.

Xinglei Zhu & Chang Wen Chen. 2010. A joint source-channel adaptive scheme for wireless H.264 video authentication. *IEEE International Conference on Multimedia and Expo (ICME), Suntec City*: 13–18.

Xuejuan Gao, Li Zhuo, Suyu Wang, & Lansun Shen. 2008. A H.264 based joint source channel coding scheme over wireless channels. *IIHMSP '08 International Conference on, Harbin, China*: 683–686.

Yong Liu, Qing-song Tong, Ai-Dong Men, & Zi-yi Quan. 2008. A joint source-channel coding scheme focused on unequal error protection for H.264 transmission over MIMO-OFDM system. *CCCM '08. ISECS International Colloquium on, Guangzhou, China*: 491–495.

Yuan Zhang, Wen Gao, Yan Lu, & Qingming Huang. 2007. Joint source-channel rate-distortion optimization for H.264 video coding over error-prone networks. *IEEE Transactions on Multimedia* 9:445–454.

Zhifeng Chen & Dapeng Wu. 2012. Rate-distortion optimized cross-layer rate control in wireless video communication. *IEEE Transactions on Circuits and Systems for Video Technology* 22:352–365.

Zhou, C., Lin, C.-W., Zhang, X., & Guo, Z. 2014. A novel JSCC scheme for UEP-based scalable video transmission over MIMO systems. *IEEE Transactions on Circuits and Systems for Video Technology*: 1002–1015.

Signal and Information Processing, Networking and Computers – Chen & Huang (Eds)
© 2016 Taylor & Francis Group, London, ISBN 978-1-138-02881-4

A robust topology control algorithm for channel allocation in the Cognitive Radio Ad Hoc networks

Yinghua Chen, Xiaojun Jing & Hai Huang
School of Information and Communication Engineering, Key Laboratory of Trustworthy Distributed Computing and Service (BUPT), Ministry of Education, Beijing University of Posts and Telecommunications, Beijing, China

ABSTRACT: Cognitive Radio Ad Hoc network, a multi-hop wireless communication etwork, is composed of Cognitive Radio (CR) techniques, which combines the advantage of CR with Ad Hoc network and makes a contribution of breaking the limit of the spectrum shortage for conventional Ad Hoc network. Thus, the channel allocation becomes a hot issue in this area. With topology control of the whole network, this paper proposes an algorithm which focuses on assigning channels to bring a high connectivity of the whole network, while maintaining a robust network at the same time. The simulation results reveal a better performance on the network connectivity with the proposed robust topology control algorithm. Meanwhile, the fairness between Primary User (PU) and Second User (SU) can be achieved using this algorithm.

Keywords: Cognitive Radio Ad Hoc Network; topology control; robust network

1 INTRODUCTION

Mobile Ad Hoc Network (MAHN) consists of a set of wireless nodes and is constructed without any infrastructure. Recently, with the development of wireless communication techniques, the limit of the spectrum resource is becoming a main barrier of it. In the cognitive Ad Hoc network, with the cognitive radio techniques, the Ad Hoc network is capable of sensing the spectrum from the surrounding and finding the available spectrum to use. Moreover, the cognitive Ad Hoc network can process the spectrum collaborations spontaneously based on the interference of external condition, the density of network nodes and user QoS ect (Akyildiz I. F. et al. 2009a). In the meantime, it solves the problem of the frequency spectrum resource scarcity and improves the spectrum efficiency and network capacity.

Compared with traditional self-organizing network, cognitive Ad Hoc network face more changeable spectrum environment and the fact is that the network will be more defensible to the spectrum allocated to the authorized users. In a cognitive Ad Hoc Network, the available spectrum distributes quite a wide range of frequency, so the choice of the channel is much more and the vacant channel always changes along with time. There is no wonder that the distribution of an authorized user's channel will somehow make an influence on the available channel for all the cognitive nodes. Due to the authorized users' activity, all the links in the network will be no longer fixed. When the channel is occupied by the PU, the rest of users should be able to change the channel and in a quick way.

Normally in a traditional Ad Hoc network, all the channel allocation problem will be solved with some algorithm or scheme based on topology control, and the control of topology always will be realized by some adjustment of the power of the network nodes. By changing the transmitted power of the nodes, the connectivity between all the network nodes can be predicted, so that the QoS on the network connectivity is satisfied. However, in the cognitive network, the relative location between network nodes and the transmitted power of all the nodes is not the only influence factors on the network topology. The available channels for all nodes can still affect the topology. In fact, two nodes can't communicate with each other if there is no common channel between these two nodes even its range of transmission and reception is overlapped. Thus, the channel allocation in a cognitive Ad Hoc network is more complicated than a conventional one. Only with a careful design of topology control algorithm can it meet the demand of reducing the network load while keeping the robust and capacity of the whole network.

In order to fix all the problems above, this paper focuses on the robust topology control algorithm which is used on the channel allocation in the cognitive Ad Hoc network. With this algorithm, the network will be more robust. The rest of this paper is consisted of four parts. Section 2 mainly gives the background of C-Ad Hoc network. And in Section 3, the system model will be analyzed. And section 4 is the mainly the analyses of the simulation result. Last, in the section 5, the conclusion of the entire paper will be exhibited.

2 BACKGROUND

2.1 Cognitive Ad Hoc network

The basic idea of cognitive radio is proposed for solving the problem of spectrum shortage and improving the spectrum efficiency. Most importantly, the cognitive radio authorizes the second user to work on the authorized spectrum opportunistically when it is provided with certain communication equipment. The key that the cognitive radio offers to the wireless communication equipment is the ability of sensing the spectrum hole and making use of it properly. In a cognitive radio network, there are three core concepts, which are spectrum sensing, spectrum sharing, and spectrum managing. Spectrum sharing is a process of solving the collision brought by the simultaneous accessing of multiple users on the same band of spectrum. This process is made by resources allocation, which will lead to the avoidance of the interference between the primary user and the cognitive user. And in this process, multiple cognitive nodes are accessing the spectrum opportunistically and sharing it.

Generally speaking, two points that distinguished the cognitive Ad Hoc network from the conventional self-organized network are the varying spectrum environments and the protection of the band authorized to the primary user. In a cognitive Ad Hoc network, the available band of the spectrum is always distributed on a wide range. Thus, there are multiple choices for setting available channels, and the specific channel situation is changing along with time. In this case, the distribution and the work condition of the primary user will somehow affect the available channels for the cognitive user. On the contrary, working bands for all nodes have already been set on a certain channel in the conventional self-organizing network, and all nodes are working on a fix channel.

Besides, in a cognitive Ad Hoc network, the end-to-end multi-hop communication might be processed through a different channel because of the fact that there is quite a large quantity of channels among the whole network, and the opportunistically accessing of the users. As a matter of fact, even the channels in the network are changing as the primary user will occupy some channels, which become totally useless for other users. From the above, we can see that combing the concept of cognitive radio network with Ad Hoc network does not only solve the problem of spectrum shortage, but also bright up some new problems to this system.

2.2 Topology control

Topology Control techniques have been successful in addressing such considerations in a distributed and efficient manner for traditional networks (Akyildiz I.F. et al. 2009b). Topology control is originally developed for Wireless Sensor Networks (WSNs) (P. Santi. 2005), MANETs (F. Dai & J. Wu. 2006), and wireless mesh networks to reduce energy consumption and interference.

In a self-organizing network, there is no control center for the contracture of nodes, so all nodes topology constructed depends on gathering the information of topology through the cooperation between one to other district of nodes. And the conventional Ad Hoc network exchanges all this topology construction information from one fix channel, which turns out to be totally different in the cognitive Ad Hoc network. The available channels for different districts of nodes aren't the same. So it's impossible for all the nodes to work on one single channel. And in this case, there might be some nodes in the network are blocked off from the topology information of the whole network, which is considered as one flaw in this system.

Topology control is usually done through power control, which is a scheme using power adjustment on one node to get the whole network topology under control and that satisfies the requirement for the network connectivity. However, in the cognitive Ad Hoc network, the factor influencing the topology is not only the power of each node, but also the availability to connect two nodes. Even if two nodes in the communication are allowed in the same area, if the available channels are occupied by the primary users, then the communication between these two nodes is not going to happen. So the topology in the cognitive Ad Hoc network is beyond complicated.

3 SYSTEM MODEL AND ALGORITHM

Among all the analyses of CR, the model of channel allocation can be divided into two types, which are the central one and the distributed one. A centralized system model manages the SU on taking the vacant spectrum by a center, which collects information on all SUs for building a spectrum "library" and then maximizing the channel allocation in the whole network. On the contrary, in a distributed system model, SU only senses the available channel from the neighbor nodes. The network is set up by the communication between neighbor nodes. Although, in this way, flexibility will be improved than in a centralized model, affected by a shadow effect, multipath effect, and the other hidden bug on the terminal, the performance of it is not expected. Moreover, the distributed model will cause many network segments, which should be avoided by a good network because the network will not be robust once there is a segment in it. In order to avoid this situation, a centralized system model is chosen here to keep the network robust to some extent.

Normally, the state of an authorized channel is a Markov model with two states, which means that the channel is in a vacant state or in a possessed state. The channel in the vacant means it's not occupied by PU and could be used by SU in data transmitting. The possessed state means a channel is used by PU, and it's not possible for a SU to use it again. There will be a collision if the SU can't sense and avoid the situation when a vacant channel changes the state into a possessed one.

3.1 *System model*

Assuming there are n SUs, and each of the nodes in the network are equipped with Q transceivers ($Q \geq 2$), each of the transceiver can work on a common channel, and the available accessing channel set is C. The network can be simulated as an undirected graph $G(V, E)$, in which v represents a user and $v \in V$. The $e = (u, v)$ stands for the link between users u and v, and these two nodes can transmit with each other, $e = (u, v) \in E$. G is a connected graph, each of the two nodes in it is connected directly or through a multi-hop link. Our target is to find a channel allocation scheme, which can allocate a channel for each node in the network with a limitation of the number of transceivers in each node. Potential Interference edge is defined as the link between the two nodes which in the transmitted range has been overlapped. As for the Interference edge that happens when the two set nodes of potential interference edge are allocated with same channel.

The transmitted range of a node is assumed to be r with an interference to be R. And R is usually 2–3 times bigger than r (Zhang W. Kandah F & Tang J. 2010). All nodes of the whole network could communicate with each other by this specific channel. By centralized control, each node has the same transmitted power. (Only when the nodes all have the same power, the link is bi-directed. Thus, the graph model can be turned into unidirectional graph, and the interference between the links in the network can be of symmetry). In order to avoid self-interfering, all the nodes work in a half-duplex mode, which means one transceiver in the node can either transmit or receive a signal at the same time. But, it allows one node to work on one channel while another node works at another channel in the meantime. For convenience to simulate the whole cognitive Ad Hoc network, the interference caused by PU is assumed to have the influence on the whole network, which makes the available accessing channel set C as same for all the nodes in the whole network.

As for the time for both channels in a vacant state or possessed state, it is assumed that the two times that it obeys are the exponential distributions within the parameters equal to λ_c and μ_c. Then the average time in which the vacant state can be sustained to the average time when the possessed time can be lasted in a channel ratio can be represented by R_c.

$$R_c = \frac{1/\lambda_c}{1/\lambda_c + 1/\mu_c} = \frac{\mu_c}{\lambda_c + \mu_c} \tag{1}$$

The interference parameter is defined as the ratio between the times for which the current channel and the whole channels are allocated. Assuming k_c is the number for how many times the channel c is allocated. Then, we can deduce the interference parameter I_c equals to

$$I_c = \frac{k_c}{\sum_{i=1}^{m} k_i} \tag{2}$$

m stands for the total number of times that the channels have been assigned, in this network.

Apparently, the channel is more suitable to be accessed when R_c turns bigger. And the accessing probability for channel c is

$$p_c = \beta R_c + (1-\beta)(1-I_c) \tag{3}$$

β calls the channel coordinated factor. For SU, β should be smaller as it can give a relatively high p_c. And for PU, β should be chosen from the channel set C with the maximum values.

The whole network is always evaluated by the robustness, and it mainly analyzed the connectivity of the whole nodes in the network, which is defined as:

$$p_d = \frac{\sum_{i,j=1}^{n} D_{i,j}}{2C_n^2} = \sum_{i,j=1}^{n} D_{i,j} \Big/ n(n-1) \tag{4}$$

where $i \neq j$, and when $D_{i,j} = 0$, the node i and j can reach for each other. $D_{i,j} = 1$ gives the opposite situation between node and i and j.

3.2 Algorithm

When it comes to the algorithm, first we should adjust the power for each node into a profit and same value which will assure the connectivity of the whole network. Besides, when some areas are intensively distributed with nodes, the potential interference edge will be more along with the increasing number of edge. According to the rules, we choose this part of the graph G, which is called G' as the target area by electing some redundant link in G. In this G', the nodes have less connectivity in the graph G. Then the robust topology control can be carried on with the following steps.

1. Computing the set of interference edge $\Omega(e)$ and the potential interference edge $N(e)$ for each edge e. And then ranking the edge in a decreasing order of $N(e)$.

2. Assigning channels to link $e = (u, v)$, which is finding the proper working channel for the transceiver in each node. Then initializing each node with turning the available channel set $\Gamma(u)$ into a blank set.
3. After processing step one, $e = (u, v)$, if $\Gamma(u) \cap \Gamma(v) = \Phi$, then choose the channel c with smallest p_c among $\Omega(e)$ for $\Gamma(u) \cap \Gamma(v)$; if the available channel $|\Gamma(u)|$ is less than the transceivers in the node, then choose the channel c with the smallest p_c among $\Omega(e)$ for e.
4. Testing the robustness of the network after assigning the channels to each node according to the three steps above. It is found whether node u and node v connected to each other would delete one working channel on the link $e = (u, v)$.
5. If the robust test passes, for those nodes that are out of the area of G', choosing the used channel from the neighbor nodes with the smallest possessed times.

4   SIMULATION RESULTS

The whole simulation for the system model is built in an area which is 800 m × 800 m with random 25 nodes. And the transmitting radius is 250 m, and the interference radius is 500 m. The λ_c and μ_c are chosen from [3, 20] randomly. The refreshing time for the channel statement is 100 ms, and sensing time is 10 ms.

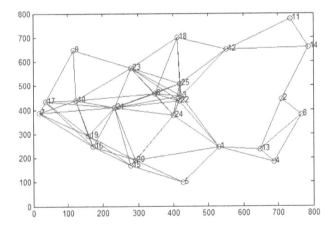

Figure 1.   Topology of the whole network and the potential interference edge on each node.

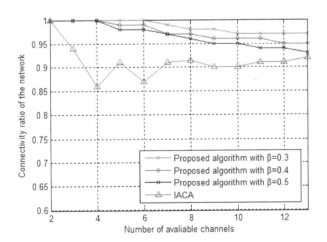

Figure 2.   Performance on network connectivity with proposed algorithm.

81

Each user owns multiple radio access with flow rate equals to 1 Mbit $\cdot$ s^{-1}. After 1,000 times simulation, the average topology is showed in Figure 1. Based on the potential interference, $N(e)$ can be evaluated and it's shown in the Figure 1.

In order to give a comparison between the performance of the proposed algorithm and other ordinary one, we here import an IACA (Interference Avoid Channel Allocation algorithm) which focuses on avoiding the interference when it comes to channel allocation. According to Figure 2, in the same number of channels, with the increasing β, the collision rate is reduced. With the same β, along with the increasing number of channels, the collision rate is reduced and the network is more robust. In a real network, the particular value of β can be settled by balancing the collision rate and the sensitivity of interference. The proposed algorithm can give a better performance on network stability than the IACA.

5 CONCLUSION

In this paper, a framework has to be proposed to analyze the stability of CR Ad Hoc network, which has a more complicated processing when dealing with the channel allocation problems. Moreover, an algorithm has been proposed to solve this type of problems by using topology control while assuring a robust network and bringing more fairness to PU and SU. In this paper, this algorithm has been modeled and simulated. The performance shows the whole network can be more robust with this algorithm. Furthermore, with the simulation result, the fact that a channel allocation algorithm brings a more robust network can be proved in a CR Ad Hoc Network.

ACKNOWLEDGMENTS

This work is supported by project NSFC 61471066 and the open funding project of State Key Lab of Virtual Reality Technology and Systems at Beihang University under Grant No. BUAA-VR-15 KF–19.

REFERENCES

Akyildiz, I.F. & Lee, W.Y. 2009b. Topology control in cooperative Ad Hoc networks. *IEEE* 23(4): 6–12.

Akyildiz, I.F., Lee W.Y. & Chowdhury, K.R. 2009a. Spectrum management in cognitive radio Ad Hoc networks [J], *Network, IEEE,* 23(4): 6–12.

Dai, F. & Wu, J. 2006. Mobility-sensitive topology control in mobile Ad Hoc networks, *IEEE Trans. Parallel Distrib. Syst.,* vol. 17, no. 6, pp. 522–535.

Santi, P. 2005. Topology control in wireless Ad Hoc and sensor networks, *ACM Comput. Surv.,* vol. 37, no. 2, pp. 164–194.

Zhang, W., Kandah, F. & Tang, J. 2010. Interference-aware robust topology design in multi-channel wireless mesh networks [C], *Consumer Communications and Networking Conference (CCNC), IEEE,* 1–5.

Signal and Information Processing, Networking and Computers – Chen & Huang (Eds)
© 2016 Taylor & Francis Group, London, ISBN 978-1-138-02881-4

Adaptive spread spectrum audio watermarking based on data redundancy analysis

Rangkun Li & Shuzheng Xu
Department of Electronic Engineering, Tsinghua University, Beijing, China

Bo Rong
Communications Research Centre, Nepean, Canada

Huazhong Yang
Department of Electronic Engineering, Tsinghua University, Beijing, China

ABSTRACT: In this paper an adaptive spread spectrum method is proposed for effective embedding and detection of audio watermarking. The method introduces audio data redundancy analysis at the watermark receiver to achieve enhanced extraction performance. In particular, audio signals tend to have large dynamic range, and data redundancy exists between neighboring audio samples. By removing data redundancy of the host audio signal, the host interference is suppressed. Furthermore, we introduce a double-sequence modulation technique at the watermark embedder. The modulation coefficients are adaptively selected to reduce the negative effect of perceptual shaping. The advantages of our method are verified through experiments.

Keywords: spread spectrum; redundancy analysis; adaptive; watermarking

1 INTRODUCTION

Digital watermarking is an advanced technique which has already found many useful applications like copyright protection, covert communication, broadcast monitoring, data authentication, etc (Mazurczyk et al. 2013). Three critical requirements for watermarking algorithms are transparency, robustness and capacity (Cvejic et al. 2009, Ghebleh et al. 2014). Which means the watermark signal should carry enough hidden data without damaging the perceptual quality of the host media. Also, the watermark extractor should be able to successfully decode the watermark information despite of various kinds of channel attacks.

Digital watermarking for audio signal is an important research subject which attracts a lot of attentions recently. Unlike video or image watermarking, audio watermarking is much more challenging for realization. Digital audio signals tend to have very low data rate with limited bandwidth, which leaves even less space for embedding additional information. Moveover, human ears are sensitive to even slight changes of audio signal. So extra care should be taken when altering the audio samples to embed watermark information. Nevertheless, researchers have already developed several effective audio watermarking schemes through different approaches. These efforts include phase modulation (Arnold et al. 2014, Djebbar et al. 2013, Garcia et al. 2013), echo hiding (Xiang et al. 2012), patchwork modulation (Natgunanathan et al. 2012) and quantization based methods (Fallahpour & Megias 2012). Certain locations in the audio signals are selected for watermark embedding: time-domain samples, spectral samples, DCT and DWT components, or other features like Empirical Mode Decomposition (EMD) (Khaldi & Boudraa 2013, Wang et al. 2014), Singular Value Decomposition (SVD) (Bhat et al. 2011) and log coordinate mapping (LCM) features (Kang et al. 2011), etc.

Among these watermarking methods, Spread Spectrum (SS) modulation is an advanced technique which achieves excellent properties of high robustness, easily achieved synchronization and low embedding distortion. However, SS based watermarking methods suffer from the host signal interference problem as the audio carriers have a negative influence to the correlative watermark extractor. To improve the SS watermarking performance, some researchers focus on modifying the watermark embedder using the informed embedding methods (Malvar & Florêncio 2003), and some others try to improve the performance at the extractor side (Gerek & Mihcak 2008).

This work endeavors to improve the SS audio watermarking performance by utilizing the statistical characteristics of audio signals. As stated in the following sections, the performance of SS watermarking has an inverse relationship with the host audio power, which is much higher than the watermark signal power. Therefore, we introduce data redundancy analysis at the watermark extractor to reduce the extra power of host audio and suppress the host interference. Then we analyze the matching factor losses caused by the perceptual shaping process at the watermark embedder. An adaptive double-sequence modulation method is further proposed to reduce the negative effect of the perceptual shaping.

2 SS BASED AUDIO WATERMARKING

In an SS watermarking system, the watermarked audio signal is created by linearly adding the host audio signal with an SS-modulated watermark signal. The SS modulator adopts a key-generated Pseudo-Noise (PN) sequence to represent one bit of watermark information. Using the symbols b, $\mathbf{u}$ and $\mathbf{x}$ to represent the bipolar watermark information bit, the PN sequence and the host audio signal respectively, the SS-watermarked audio signal can be expressed as $\mathbf{s} = \mathbf{x} + b\mathbf{u}$. If the PN sequence has a length of N, then one information bit is embedded into one audio frame s with N samples.

After a lossy communication channel with noise $\mathbf{n}$, the watermark decoder receives the watermarked and noise contaminated audio signal $\mathbf{y} = \mathbf{x} + b\mathbf{u} + \mathbf{n}$. The watermark extractor calculates the sufficient statistic r correlatively as (Malvar & Florêncio 2003):

$$r = \frac{\langle \mathbf{y}, \mathbf{u} \rangle}{\langle \mathbf{u}, \mathbf{u} \rangle} = \frac{\langle \mathbf{x} + b\mathbf{u} + \mathbf{n}, \mathbf{u} \rangle}{\| \mathbf{u} \|} = b + x + n \tag{1}$$

Where $x \overset{\Delta}{=} \langle \mathbf{x}, \mathbf{u} \rangle / \| \mathbf{u} \|$ and $n \overset{\Delta}{=} \langle \mathbf{n}, \mathbf{u} \rangle / \| \mathbf{u} \|$. Then the embedded watermark information b can be estimated with $\hat{b} = \text{sign}(r)$. As in (1), the accuracy of this estimation is influenced by x and n, which represent the host interference and the channel noise interference respectively.

Assume $\mathbf{x}$, $\mathbf{u}$ and $\mathbf{n}$ are all independent Gaussian signals with the variance of σ_x^2, σ_u^2 and σ_n^2 respectively, then the estimation accuracy can be derived using the Gaussian analysis method. The estimation accuracy is measured with the decoding error rate:

$$p = \frac{1}{2} \text{erfc} \left(\sqrt{\frac{N\sigma_u^2}{2(\sigma_n^2 + \sigma_x^2)}} \right) \tag{2}$$

The extraction performance of SS watermarking is positively related to the power ratio of the watermark signal and the host audio signal. Usually the watermark signal is suppressed at a much lower power level than the host audio signal to ensure the host fidelity. As a result, the decoding performance of SS watermarking is rather limited. Larger frame length N can be used to ensure more successful extraction. However, this will reduce the watermark embedding capacity.

Furthermore, keeping the watermark signal at low power level is not enough to guarantee its imperceptibility. As shown by studies of the Human Auditory System (HAS), an audio signal is imperceptible to the human ear only when it is shielded by a larger audio component in the frequency domain or in the time domain. This acoustical phenomena is called the simultaneous masking and the temporal masking (Brandenburg et al. 2013). The masking effect is widely adopted by lossy audio compression methods to calculate the allowable quantization

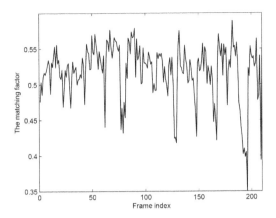

Figure 1.   The matching factor losses caused by perceptual shaping.

noises when storing the raw audio data into smaller compressed files. As in the case of SS watermarking, the watermark signal has to be perceptually shaped to make it imperceptible. The SS embedding process under perceptual shaping can be expressed as $\mathbf{s} = \mathbf{x} + b\mathbf{u}^s$, where $\mathbf{u}^s$ denotes the shaped watermark signal. Then the decoding error rate is calculated as:

$$p^s = \frac{1}{2}\text{erfc}\left(\sqrt{\frac{\lambda^2 N \sigma_{u^s}^2}{2(\sigma_n^2 + \sigma_x^2)}}\right) \tag{3}$$

Here, the parameter λ is defined as the matching factor, which is the normalized correlation between the shaped watermark signal and the original PN sequence:

$$\lambda \stackrel{\Delta}{=} \frac{\langle \mathbf{u}, \mathbf{u}^s \rangle}{\sqrt{\|\mathbf{u}\| \cdot \|\mathbf{u}^s\|}} \tag{4}$$

The physical meaning of λ is the measurement of correlation loss between the watermark signal and the PN sequence caused by the perceptual shaping process. The matching factors always satisfy $\lambda < 1$, and the equal sign is achieved when the shaping process is not applied. Figure 1 demonstrates the matching factor after the perceptual shaping. Real audio signal is used here and the matching factors are calculated frame by frame with frame length $N = 2048$. The correlations between the watermark signal and the PN sequence drop from 1 to 0.3 ~ 0.6 after the shaping process. This will cause performance downgrades at the correlative extractor and result in increase of decoding error rate.

From the above analysis, the performances of SS audio watermarking are negatively affected by two factors: the host audio interference and the perceptual shaping process. In the following sections, we attempt to reduce the performance losses by suppressing this two factors. Firstly, we adopt the data redundancy analysis at the watermark extractor to remove major power of the host audio. Then an adaptive double-sequence modulation method is introduced at the watermark embedder to minimize correlation losses caused by the shaping process.

3   DATA REDUNDANCY ANALYSIS AND ADAPTIVE MODULATION

3.1   *Data redundancy analysis for audio signal*

As discussed in the previous sections, the audio signals have much higher power than the watermark signals due to the fidelity requirements. This causes serious interference problems at the correlative watermark extractor. To reduce the negative impact of the host audio signal, one simple idea is to filter out the host power before the correlation with the decoding PN sequence. However, as explained in the previous section, the spectrum of the watermark signal overlaps with the host audio as the watermark signal uses spectral components of the host audio as maskers. As a result, simple filtering operations which attempt to remove the host audio power will cause damages to the watermark signal as well.

Nevertheless, the power of host signals can be suppressed through another way. Unlike random noise signals, audio signals contain large part of redundancy power. One sample from an audio signal can be predetermined by its neighboring samples only with a slight difference. Most speech coding standards are based on this fact and achieve high compression rate. To demonstrate this fact more clearly, we express one audio sample by linearly combine its M earlier samples. Let these $M+1$ adjacent samples be $\{x_{t-M}, x_{t-M+1} \cdots x_{t-1}, x_t\}$. Then the linear form of the audio data redundancy can be expressed as:

$$x_t = \sum_{k=1}^{M} w_k x_{t-k} + v_t \tag{5}$$

We can see that the audio samples follow the AR model. The coefficients $\mathbf{w} = \{w_k\}$ model the relationships between two samples which are k samples away. The residual signal $\mathbf{v} = \{v_t\}$ means the part of the audio sample which can not be predicted using the last M samples. From (5), if this redundancy analysis is applied to all samples of audio signal $\mathbf{x}$, then we can achieve the residual signal $\mathbf{v}$ which is much lower in power than $\mathbf{x}$.

The data redundancy analysis can be achieved by Linear Prediction Filtering (LPF). An M-order LPF calculates M prediction coefficients $\{w_k\}$ by minimizing the power of the residual signal: $\mathbf{v} = \mathrm{L}\{\mathbf{x}\}$. So the LPF calculation is a Minimum Squared-Error (MSE) searching problem. The Levinson-Durbin algorithm is an efficient tool for MSE searching. Rather than solving an M-order matrix function, the Levinson-Durbin algorithm calculates the prediction coefficients recursively with much lower computational complexity.

In our watermarking method, after receiving the watermarked audio signal at the watermark extractor, data redundancy removing is applied to the audio signal before the correlative extraction. As the audio signal is usually short-time stationary, the data redundancy analysis is performed over several frames.

3.2 *Adaptive embedding strategy*

As described in the earlier sections, SS watermarking uses a bipolar embedding method which modulates the watermark information bit b on one PN sequence $\mathbf{u}$. In our method, we adopt two orthogonal PN sequences $\mathbf{u}_1$ and $\mathbf{u}_2$ to carry the watermark information. In the same way as traditional SS watermarking, these two PN sequence should be perceptually shaped to guarantee their imperceptibility. The shaped PN sequences are denoted as $\mathbf{u}_1^s$ and $\mathbf{u}_2^s$ respectively.

Similar methods as in (Can et al. 2014) also adopt multiple PN sequences in their SS modulation and they are called the Frequency-Hopping Spread Spectrum (FHSS) methods, as opposed to the traditional Direct Sequence Spread Spectrum (DSSS). Nevertheless, unlike FHSS which selectively use only one watermarking sequence for embedding at a time, our method modulates the watermark information on two sequences adaptively and simultaneously:

$$\mathbf{s} = \mathbf{x} + \alpha_1 \mathbf{u}_1^s + \alpha_2 \mathbf{u}_2^s \tag{6}$$

Where α_1 and α_2 are the weighting coefficients for the two watermark sequences. The signs of α_1 and α_2 depend on the embedded information bit m:

$$\alpha_1 > 0, \alpha_2 < 0 \quad if \quad m = 0 \tag{7}$$

$$\alpha_1 < 0, \alpha_2 > 0 \quad if \quad m = 1 \tag{8}$$

At the watermark receiver, two sufficient statistics are calculated separatively. Assume the watermark sequences are normalized in power, $\| \mathbf{u}_1 \| = \| \mathbf{u}_2 \| = \| \mathbf{u}_1^s \| = \| \mathbf{u}_2^s \| = \| \mathbf{u} \|$, then

$$r_1 = \frac{\langle \mathbf{x} + \alpha_1 \mathbf{u}_1^s + \alpha_2 \mathbf{u}_2^s, \mathbf{u}_1 \rangle}{\langle \mathbf{u}_1, \mathbf{u}_1 \rangle} = \frac{\langle \mathbf{x}, \mathbf{u}_1 \rangle}{\| \mathbf{u} \|} + \alpha_1 \lambda_1 \tag{9}$$

$$r_2 = \frac{\langle \mathbf{x} + \alpha_1 \mathbf{u}_1^s + \alpha_2 \mathbf{u}_2^s, \mathbf{u}_2 \rangle}{\langle \mathbf{u}_2, \mathbf{u}_2 \rangle} = \frac{\langle \mathbf{x}, \mathbf{u}_2 \rangle}{\| \mathbf{u} \|} + \alpha_2 \lambda_2 \tag{10}$$

Where α_1 and α_1 mean the matching factors for the shaped watermark sequences $\mathbf{u}_1^s$ and $\mathbf{u}_2^s$. The extraction of watermark information can be performed through simple comparison:

$$\hat{m} = \begin{cases} 0 & if \quad r_1 > r_2 \\ 1 & if \quad r_1 < r_2 \end{cases}$$

As in (6), the power of embedded watermark signal should be limited, so $\| \alpha_1 \mathbf{u}_1^s + \alpha_2 \mathbf{u}_2^s \| = \| \mathbf{u} \|$, which can be further derived as:

$$\alpha_1^2 + \alpha_2^2 = 1 \tag{11}$$

From the above equations, it can be derived that minimum decoding error rate is achieved when α_1 and α_2 satisfy:

$$| \alpha_1 | = \frac{\lambda_2}{\sqrt{\lambda_1^2 + \lambda_2^2}}, \quad | \alpha_2 | = \frac{\lambda_1}{\sqrt{\lambda_1^2 + \lambda_2^2}} \tag{12}$$

This means the value of α_i is inversely proportional to λ_i. This is easy to understand because the watermark sequence should be given more weighing power if it is less affected by the perceptual shaping process.

Note here, if we let $\alpha_1 = 1, \alpha_2 = 0$ when the watermark embeds $m = 0$, and $\alpha_1 = -1, \alpha_2 = 0$ when embedding $m = 1$, then the embedding method is exactly the traditional bipolar spread spectrum watermarking method. And if we let $\alpha_1 = 1, \alpha_2 = 0$ when embedding $m = 0$, and $\alpha_1 = 0, \alpha_2 = 1$ when embedding $m = 1$, then this embedding method can be called the unipolar spread spectrum watermarking. Compared to the traditional SS methods, the proposed embedding method can assess the perceptual shaping effects on the watermark sequences, and adaptively select the weighting coefficients to reduce the negative influence of the shaping process.

4   PROPOSED WATERMARKING SCHEME

4.1   *Embedding algorithm*

The watermark embedding algorithm is demonstrated in Figure 2. Perceptual shaping is adopted to ensure the perceptual quality of the embedded watermark. The original message can be an image, a paragraph of text or other materials. During the preprocessing, the watermark data should be scrambled first using Arnold transform or other methods. This does not bring extra benefit to lower the extraction error rate. However, it helps to enhance the watermark security and prevent burst decoding errors. The embedding algorithm contains the following steps:

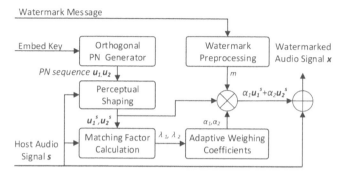

Figure 2. The adaptive watermark embedding structure.

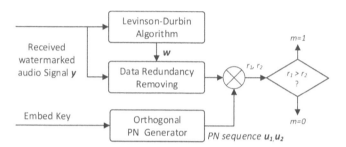

Figure 3. The watermark extraction structure.

Step 1: Input the host audio signal x and frame it by length N. N is chosen accordingly with different applications. Smaller N will result in higher embedding capacity, but with lower robustness.

Step 2: Generate two orthogonal PN sequences u_1 and u_2 using the embed key. The two PN sequences can be generated separately and then orthogonalized using the Gram-Schmidt algorithm.

Step 3: Apply perceptual analysis on the host audio using the psycho-acoustic model adopted by the MPEG standard. The audio signal x is transformed into the spectral domain using the FFT method. Then tonal and non-tonal maskers are extracted from the spectral samples which further generate the masking threshold. The masking threshold is the upper boundary below which the added watermark signals are imperceptible.

Step 4: Shape the PN sequences with the analyzed masking threshold. The shaping process is executed in the subband domain. The matching factors λ_1 and λ_1 are calculated using the host signal x and the shaped u_1^s, u_2^s

Step 5: Calculate the modulation coefficients α_1 and α_2 using the method in (12). Then modulate the watermark information $\hat{m}$ using (6) to complete the embedding.

4.2 Extraction algorithm

The structure of the extraction algorithm is illustrated in Figure 3. The algorithm adopts the data redundancy analysis before the correlative extraction. So that the host interference is reduced with most of its power removed. The extraction steps for our watermarking method is described as follows:

Step 1: Generate two PN-sequences using the same key and the same method as in the embedding process.

Step 2: Apply data redundancy analysis on the received watermarked audio signal using the Levinson-Durbin algorithm.

Table 1. Robustness tests under different channel conditions.

Channel	Parameters	ODG after attack	Decoding BER:%
Attack-free	None	−0.71	0.41
AWGN	20 dB	−3.03	4.42
Low-pass filtering	12 kHz	−1.33	2.02
Down-sampling	22.05 kHz	−1.21	1.58
Requantization	8 bit	−2.56	1.23
MP3 compression	64 kbps	−1.74	1.30
AAC compression	64 kbps	−1.53	0.97

Step 3: Remove the redundancy power of the audio signal power using the linear predictive filtering.

Step 4: Calculate the two sufficient statistics r_1 and r_2 using the correlation method, then the watermark information m is extracted through simple comparison.

5 EXPERIMENT RESULTS

Experiments are performed on real audio signals with different acoustical types. The audio signals are sampled at 44.1 kHz with 16 bit precision. The frame length N is set as 2048, which results in an embedding capacity of 21.5 bps per channel. The performances are evaluated in two aspects: the embedding performances which measure the embedding distortion and the perceptual quality, and the extraction performances which contain the robustness tests.

The watermark-to-signal ratio $WSR \triangleq 10\log(\|\mathbf{u}\|/\|\mathbf{x}\|)$ is used for embedding distortion control. In our experiments, the WSR is set as a fixed value of −28 dB, which is very low to keep the audio power structure unchanged. This guarantees the undetectability and safety of the embedded watermark.

ODG measurement is commonly used for evaluations of audio compression algorithms. It calculates the perceptual difference of a modified audio signal from the original reference signal using arithmetic method. So it is also suitable for the quality testof audio watermarking algorithms. In the watermark embedding experiments, an average ODG score of −0.71 is achieved, which is in the range of "Imperceptible" (Cvejic et al. 2009).

For robustness tests, the decoding error rates are measured under different types of channel attacks. Different signal processing methods are introduced here to simulate the channel conditions: (1) Attack-free channel; (2) Additive white noise channel with 20 dB SNR; (3) Low-pass filtering with 12 kHz cutoff frequency; (4) Down-sample the audio signal with 22.05 kHz; (5) Requantize the audio signal with 8 bit precision; (6) MP3 compression with 64 kbps data rate; (7) AAC compression with 64 kbps data rate.

Table 1 lists the robustness test results. It can be seen that the host audio qualities are damaged by the channel attacks, however, most watermark information can still be successfully extracted. This means the embedded watermark information can not be easily removed through these signal processing methods.

6 CONCLUSIONS

An adaptive spread spectrum watermarking method is presented in this paper. The proposed method focuses on the performance losses caused by host interference and perceptual shaping. A double-sequence modulation mechanic is introduced in the watermark embedding process. The modulation weighing coefficients are selected adaptively according to the perceptual shaping results. For watermark extraction, we import data redundancy analysis on the received watermarked audio signals to remove most of the host power, so that the correlative watermark extractor are less affected by the host interference. The performances of our method are proved by experiments on real audio signals.

REFERENCES

Arnold, M. Chen, X.M. Baum, P. Gries, U. & G. Doerr. 2014. A phase-based audio watermarking system robust to acoustic path propagation. *Information Forensics and Security, IEEE Transactions on* 9(3): 411–425.

Bhat, V. Sengupta, I. & Das, A. 2011. A new audio watermarking scheme based on singular value decomposition and quantization. *Circuits, Systems, and Signal Processing* 30(5): 915–927.

Brandenburg, K. Faller, C. Herre, J. Johnston, J.D. & Kleijn, W.B. 2013. Perceptual coding of high-quality digital audio. *Proceedings of the IEEE* 101(9): 1905–1919.

Can, Y.S. Alagoz, F. & Burus, M.E. 2014. A novel spread spectrum digital audio watermarking technique. *Journal of Advances in Computer Networks* 2(1): 6–9.

Cvejic, N. Drajic, D. & Seppänen T. 2009. Audio watermarking: More than meets the ear. In *Recent Advances in Multimedia Signal Processing and Communications*: pp. 523–550. Springer.

Djebbar, F. Ayad, B. Abed-Meraim, K. & Hamam, H. 2013. Unified phase and magnitude speech spectra data hiding algorithm. *Security and Communication Networks* 6(8): 961–971.

Fallahpour, M. & Megias, D. 2012. High capacity logarithmic audio watermarking based on the human auditory system. *Multimedia (ISM), 2012 IEEE International Symposium on*: 28–31. IEEE.

Garcia-Hernandez, J.J. Parra-Michel, R. Feregrino-Uribe, C. & Cumplido, R. 2013. High payload data-hiding in audio signals based on a modified OFDM approach. *Expert Systems with Applications* 40(8): 3055–3064.

Gerek, N. & Mihcak, M.K. 2008. Generalized improved spread spectrum watermarking robust against translation attacks. *Acoustics, Speech and Signal Processing (ICASSP), 2008 IEEE International Conference on*: 1673–1676. IEEE.

Ghebleh, M. Kanso, A. & Own, H.S. 2014. A blind chaos-based watermarking technique. *Security and Communication Networks* 7(4): 800–811.

Kang, X. Yang, R. & Huang J. 2011. Geometric invariant audio watermarking based on an LCM feature. *Multimedia, IEEE Transactions on* 13(2): 181–190.

Khaldi, K. & Boudraa, A. 2013. Audio watermarking via EMD. *Audio, Speech, and Language Processing, IEEE Transactions on* 21(3): 675–680.

Malvar, H.S. & Florêncio, D.A. 2003. Improved spread spectrum: a new modulation technique for robust watermarking. *Signal Processing, IEEE Transactions on* 51(4): 898–905.

Mazurczyk, W. Szczypiorski, K. Tian, H. & Liu, Y. 2013. Trends in modern information hiding: techniques, applications and detection. *Security and Communication Networks* 6(11): 1414–1415.

Natgunanathan, I. Xiang, Y. Rong, Y. Zhou, W. & Guo, S. 2012. Robust patchwork-based embedding and decoding scheme for digital audio watermarking. *Audio, Speech, and Language Processing, IEEE Transactions on* 20(8):2232–2239.

Wang, X.G. Niu, P.P. Yang, H.Y. Zhang, Y. & Ma, T.X. 2014. A robust audio watermarking scheme using higher-order statistics in empirical mode decomposition domain. *Fundamenta Informaticae* 130(4): 467–490.

Xiang, Y. Natgunanathan, I. Peng, D. Zhou, W. & Yu, S. 2012. A dual-channel time-spread echo method for audio watermarking. *Information Forensics and Security, IEEE Transactions on* 7(2): 383–392.

Signal and Information Processing, Networking and Computers – Chen & Huang (Eds)
© *2016 Taylor & Francis Group, London, ISBN 978-1-138-02881-4*

A graph coloring based resource allocation in Heterogeneous Networks

Zhengmao Ye, Xiaojun Jing & Hai Huang
School of Information and Communication Engineering, Key Laboratory of Trustworthy Distributed Computing and Service (BUPT), Ministry of Education, Beijing University of Posts and Telecommunications, Beijing, China

ABSTRACT: This paper proposes a novel resource allocation scheme applying improved graph coloring algorithm in LTE-A Heterogeneous Networks (HetNets). Based on our interference graph, a dynamic orthogonal spectrum sharing between macrocells and femtocells is designed to reduce the cross-tier interference. In proposed scheme, a cost function to quantize resource reuse condition and a resource estimation method are utilized to construct disjoint clusters for all Base Stations (BSs). The cluster-based resource allocation algorithm not only achieve the mitigation of interference to a large extent, but also strike a balance between spectrum reusability and resource requirements. It is demonstrated by simulation that significant performance can be achieved under our proposed scheme.

Keywords: HetNet; resource allocation; graph coloring

1   INTRODUCTION

Due to the constant pursuit for Quality of Service (QoS), each new generation of wireless communication system promises higher capacity and better coverage. Furthermore, the necessity for more energy efficient, or green technologies is growing. Meanwhile owing to the poor indoor coverage, its especially difficult to maintain the desirous and stable services (G. Mansfield, 2008). The conventional macro-cellular networks may fail to improve the QoS of mobile communication for indoor subscribers in a cost effective manner.

The Third Generation Partnership Project Long Term Evolution (3GPP LTE) has introduced low power nodes placed indoors and the coexistence between femtocell and macrocell has acquired more popularity in the scientific field of HetNets (3GPP TR 36.814 V9.0.0. 2010, S. Carlaw. 2008). In such case, the User Equipment (UE) would hand off from the cellular base station to the Femtocell Base Station (FBS) installed at home or an office when operating indoors or when it is close enough to the FBS. This idea lessens the gap between transmitters and receivers which creates the dual benefits of higher-quality links and more spatial reuse. As a result, it leads to enhancement of capacity and improvement of coverage.

As the saying goes, every coin has two sides. FBSs can ease the burden on Macrocell Base Stations (MBSs) but, in turn, bring more interference by sharing the same licensed spectrum (D. Knisely et al, 2009). In order to promote the large-scale deployment of femtocells, an appropriate interference management technique is essential. Radio resource allocation scheme that can effectively mitigate interference and achieve a sufficient radio resource utilization deserves to be identified as one of the major challenges considering the realization of Het-Nets in LTE-A and even future network (T. Zahir et al. c2012, D. Lpez-Prez et al. 2009, W. Yi et al. 2009, J. Xiang et al. 2010, C. Wei Tan. 2011). T. Zahir et al. (2012) and D. Lpez-Prez et al. (2009) provide good surveys of femtocell interference managements. In W. Yi et al. (2009) and J. Xiang et al. (2010), the authors focus on spectrum sharing algorithms in femtocell networks to promote network capacity. An optimal power control is proposed to minimize the MUEs outage probability in heterogeneous networks in C. Wei Tan (2011).

The problem of interference mitigation is first addressed by using graphic approach in R.Y. Chang (2009). Recently, graph theory is widely used on the reduction of interference in LTE network and the vertex, which generally is BS in the traditional interference graph modeling schemes, expands to UE and FBS now (S. Sadr & R. Adve. 2012, Q. Zhang et al. 2013).

In S. Sadr & R. Adve (2012), a weighted undirected graph can be found, in which the vertex of the graph is MUE or FBS. In this scheme, not only one sub-band is assigned to MUE and FBS, but also other available sub-bands are assigned to FBS under the interference constraint to improve the spectrum efficiency regardless of actual Resource Blocks (RBs) requirements.

In Q. Zhang et al. (2013), a self-organized resource allocation scheme based on a coloring method in graph theoretic and a clustering method in OFDMA femtocells is proposed to mitigate co-tier interference between femtocells. The femtocells which have least interference are grouped in one cluster so that they can reuse the same frequency band without interfering each other and the interference between clusters will be mitigated by using orthogonal spectrum allocations. Hence, the co-tier interference can be mitigated significantly, while the spectrum efficiency and the total system capacity of femtocell network can be improved as well. Whats more, RBs are allocated to each cluster adaptively according to the average requirements of cluster. But it failsto address the variety of RBs requirements among femtocells in the same clusters.

In this paper, we consider a macro-femto networks where a large quantity of femtocells are randomly deployed and focus on how to implement theory of graph into allocation and utilization of resource so that the system can achieve an efficiency and fair performance. A novel dynamic resource allocation scheme based on graph coloring is proposed. In our scheme, the interference graph needs to be modeled according to the interference tolerance of UEs in each macrocells or fametocells. And a modified coloring algorithm with a cost function to quantize resource reuse condition is utilized to construct disjoint clusters for all BSs. The BSs which have least interference are grouped in one cluster so that they can reuse the same frequency bands without interfering each other. Subsequently, each BS calculates its requirements and the BSs with more RBs requirements are allowed to join more than one cluster. Hence, the co-tier interference can be mitigated significantly, while the spectrum efficiency and the system capacity can be improved as well.

The rest of this paper is organized as follows. In section II, we describe the system model and graph formation. Then we discuss the conventional graph coloring in section III. The proposed method and its simulation are presented in Section IV and V, respectively.

2 CONSTRUCTION OF THE INTERFERENCE GRAPH

We consider, as shown in Figure 1, the downlink transmission of a two-tier multi-cellular system. Femtocells are located randomly and only cover in small cycle with a fix radius, while the rest area is served by MBS.

The total bandwidth is divided into N_{RB} RBs which is the minimum unit of the resource that can be allocated to one UE and the same RB can be reused among different UE (MUE and FUE) at a time. We assume that MBSs hold a fix number of RBs and share the rest with FBSs. Also assume the data rate requirements for UE u is $R_{u,min}$.

The Signal to Interference plus Noise Ratio (SINR) γ_u of UE u served by BS i is

$$\gamma_u = \frac{P_i H_{u,i}}{\sum_{j \in \bar{I}_i} P_j H_{u,j} + \sigma^2} \tag{1}$$

where $\bar{I}_i$ is the set of tolerable interfering neighbors of BS i, P_i is the transmit power of BS i, $H_{u,i}$ is the channel gain between serving UE u and BS i, which describes the combined effect of path loss and other loss such as shadowing, penetration loss and σ^2 accounts for thermal

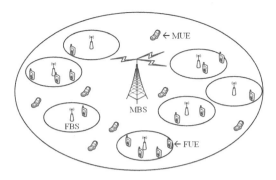

Figure 1.   A two-tier marco-femto system model.

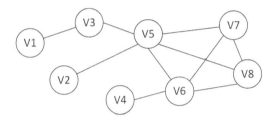

Figure 2.   Interference graph.

noise. Then UE can calculation the achievable data rate R_u with the allocated bandwidth B_u by using:

$$R_u = B_u \log_2(1 + \gamma_u)$$ (2)

To achieve the average data rates requirements of UE u, we define the set of tolerable interfering neighbors as

$$\bar{I}_i = I_i - I_i^{un}$$ (3)

where I_i is the set of all interfering BSs of BS i and I_i^{un} is the set of intolerable interfering BSs which must not use the same resource which have been allocated to BS i. If $R_u < R_{u,min}$, the strongest interfering BS among I_i is removed and put it into I_i^{un}. Then re-peat the process until satisfy the requirements.

Each BS collects the intolerable interfering BSs and lists the interfering source, then shares with the corresponding neighbor BSs to build an interference graph which indicates the interference relationship between the BSs. The graph can be represented as $G(V, E)$, where $V = \{v_1, v_2, \ldots, v_i\}$ is the vertex set which represents all N BSs, and $E = \{e_{i,j} \mid i, j \in V\}$ is the set of edges which represent interference collision.

Due to various channel conditions between BSs and their users, there are different tolerance of interference among BSs. The BSs whose UEs are on better channel condition can suffer more interference and maintain the service with expected date rate, which means they have a better capacity to tolerate interference and a smaller set of intolerable interfering BSs. So we can leverage their excess capacity to reduce the complexity of graph and improve reusability of resource without disappointing the expectation.

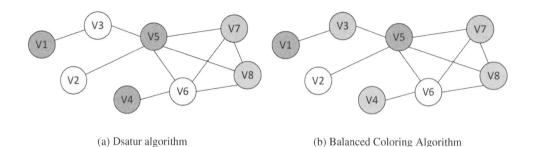

(a) Dsatur algorithm   (b) Balanced Coloring Algorithm

Figure 3.   Coloring results comparison.

3   COLORING ALGORITHM

In order to mitigate interference in downlink, the intolerable interference should not use the common frequency spectrum simultaneously, so the vertices connected by edges should be colored distinctively. Graph coloring algorithms color the vertices of a graph with minimum number of colors, such that no two connected vertices (the intolerable interfering BSs) have the same color. By assuming each color as a part of radio spectrum, graph coloring facilitates resource assignment while mitigates the frequency conflicts.

To achieve the minimum colors, several coloring methods have been proposed. Dsatur algorithm (D. Br elaz, 1979.) stands out from various graph coloring algorithms due to its computational efficiency and low complexity. The details of algorithm are listed in Algorithm 1, where the saturation degree of vertex i, ξ_i, indicates the total number of different colors to which the vertex is connected.

Algorithm 1

---

1. Initialize graph $G(V, E)$ and color set C
2. Form the uncolored vertices set V_{un}
3. **Repeat**
4. Select v_i having maximum ξ_i from V_{un}
      or having maximum number of connected vertices
      if there are vertices having same ξ_i values
5. Assign c with the initial color of C
6. **If** none of the connected of V_n are colored by c
      then color v_i with c
      else assign c with next color of C , and try again
7. Remove v_i from V_{un}
8. **Until** V_{un} is empty

The obtained set of colors, C, ermines the minimum number of required colors, $C = |C|$, to cancel strong conflicts in given interference graph.

It is noteworthy that, the objective of this algorithm is to find the minimum used number of colors which mitigate interference within limited resource. In this algorithm, a same sequence of colors are tried in each vertex coloring operation, which results in the different reuse rate among colors. For example, the initial color could be reused much higher than others while the last reused seldom. The unequal reuse of available colors translate into re-source allocation is unbalanced reuse of frequency. Whats more, the interference among unconnected vertices which are same-colored in interference graph still exists, and the sum of a large number tolerant interference could not be ignored, so this unbalance could degrade the performance of interference management and extend the fairness among UEs.

Whats more, in some way, all BSs assigned one same color would be allocated with a fixed number of resource regardless of the requirements variety among them, and in rest methods, each BS is assigned all available colors which lead to more interference and higher complexity.

These shortcomings of conventional graph coloring are addressed in our proposed scheme introduced in next section.

4 RESOURCE ALLOCATION ALGORITHM BASED ON GRAPHING AND COLORING

The drawback of conventional graph coloring is the inefficient resource utilization which consists of unbalanced frequency reuse and inefficient allocation. In order to improve resource utilization and flexibility of allocation management, a novel resource allocation algorithm based on graph coloring method is proposed and divided into three parts. In coloring scheme part, we propose a cost function to quantize resource utilization and implement our balanced coloring algorithm with its help. Then, each BS calculate their requirements and estimate the expected number of colors. Subsequently, in allocation scheme part, the RBs can be allocated adaptively to each BS based on requirements while mitigate the co-tier interference.

4.1 Balanced coloring algorithm

In this stage, first execute Dsatur algorithm to guarantee the minimum number of colors, then taking consideration of the tradeoff between efficiency and flexibility we modify the result by applying our coloring scheme.

Inspired by (S. Uygungelen, 2012), we define the cost function $H(c,i)$ of assigning color c to vertex i as:

$$H(c,i) = |V_c| \tag{4}$$

where V_c, with cardinality $|V_c|$, is the set of vertices which are colored by color c assigned to vertex i.

The pseudocode of our proposed coloring algorithm is given as following.

Algorithm 2
---
1. Execute Algorithm 1
2. Obtain the color pool, C and average cost $\bar{H}$
3. Form the high cost vertices set V_h ,
    where v_i satisfies $H(c,i) > \alpha \bar{H}$
4. **Repeat**
5. Select v_i having maximum $H(c,i)$
    or having maximum number of connected vertices
    if there are vertices having same cost value
6. Get available color set C^a
7. Recolor v_i with $c = \arg\min \{H(c,i), c \in C^a\}$
8. Remove v_i from V_h
9. **Until** V_h is empty

Considering the scenario in Figure 1, we draw the interference graph as shown in Figure 2. Figure 3(a) is the result according to conventional coloring algorithm, and the obtained set of colors, C = red, yellow, blue, green, determines the minimum number of required colors. Figure 3(b) illustrates how the vertices are colored in our coloring scheme. Compared with Figure 3(a), it is obvious that the colors of connected vertices in our colored graph are different. We achieve a more balanced utilization without increasing resource and the vertices assigned same color are appropriately grouped in one cluster. In this way, the spatial reuse of spectrum is improved.

4.2 Requirement estimate

In this stage, we pay more attention to the variety of requirements among BSs. Assuming the number of clusters is N_c, we equally partition the available RBs into N_c disjoint groups and allocate to related cluster, respectively. Its obvious the RBs in one cluster might not enough to meet some BSs which have greater requirements. To achieve a flexible resource manage method, we allow to color one vertex with more than one color to meet the requirement. And the number of resource groups that BS i de-sires can be calculated by (5).

$$N_i^d = \left\lceil \frac{\sum_{u \in U_i}^{U_i} N_{i,u}}{N_g} \right\rceil - N_i^a \tag{5}$$

where N_i, N_g and $N_{i,u}$ represent the number of colors have assigned to BS i, the number of RBs in one cluster which is one out of the total RBs and the actual RBs requirements of UE u, respectively, as U_i is the set of UEs served by BS i.

4.3 Multiple coloring algorithm

In this stage we search and assign more resource to BSs which have greater RBs requirements. Similarly, implement the coloring process with the help of cost function and the pseudocode is given in Algorithm 3.

Algorithm 3

---

1. Obtain the color pool, **C**
2. Search vertices satisfy $N_i^d > 0$ and group as V_d
3. **Repeat**
4. Select v_i having maximum N_i^d
   or having maximum number of connected vertices
   if there are vertices having same desire value
5. Get available color set C'^a
6. Assign with another

$$c = \arg\min \{H(c,i), c \in C^a\}$$

7. Remove v_i from V_d
8. **Until** V_d is empty

Finally, if all colors are occurred by the connected vertices and the required RBs of BS i is greater than the actual allocated RBs N_i, the number of RBs which can be assigned to vertex i is N_i. Then assign RBs to each UE according to the rate of UE requirements in its served BSs requirement. Otherwise, the number of RBs will be assigned to UEs depending on its requirements.

5 PERFORMANCE EVALUATION

The system simulation parameters are configured according to 3GPP LTE specifications, and the main simulation parameters are listed in Table 1. In our simulation, 7 macrocells are considered, in each of which the same number of femtocells are placed. The FUEs are randomly distributed in the coverage area of femtocells and the number of FUEs is 1.5 times as many as the number of femtocells. The probability of FUEs outdoor is 50% and set their penetration loss as 5db. The SINR threshold for construction the interference graph detailed in Section II is set to 3 dB.

Table 1. Simulation parameter.

Parameter	Value
System bandwidth	10 MHZ
Carrier frequency	2 GHZ
Macrocell radius	500 m
Femetocell radius	25 m
MBS transmitted power	46 dBm
FBS transmitted power	20 dBm
Thermal noise density	174 dBm/Hz

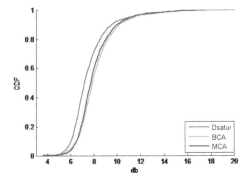

Figure 4.   CDF of Downlink SINR.   Figure 5.   CDF of Downlink throughput.

We compare our proposed scheme with conventional Dsatur algorithm. For specific analyze, we divided our scheme into two step, Balanced Coloring Algorithm (BCA) and Multiple Coloring Algorithm (MCA).

Figure 4 shows the Cumulative Distribution Function (CDF) of UEs SINR when there are 100 femetocells and 150 UEs in one cell. Compared BCA with Dsatur, about 80% UEs SINR is improved by employing the proposed coloring algorithm scheme. This is because in conversion allocation scheme, a few bands are reused by majority of BSs and the sum of large-scale interference source degrades the channel condition, while others shared by a little number of BSs as we discussed earlier. The BCA scheme releases a part of BSs from heavy bands to the rest on idle condition and eases the burden so that most of UEs SINR are improved. In the MCA scheme, some RBs are allocated to femetocells with greater requirements. Consequently, the increment of reuse brings more tolerated interference and lower performance of SINR.

Figure 5 shows the CDF of throughput of UEs. BCA failed to translate the better SINR into higher capacity without increasing resource while MBA selects more RBs with slight interference for unsatisfied BSs. The conversional scheme perform worst due to its poor SINR. We can conclude that average throughput is approximately improved by 20% by using our proposed resource allocation algorithm. In the case of femetocell dense deployment, the performance by utilizing MCA is much better than the one using BCA because MCA allocates RBs to each cell depending on its data rate requirement.

6   CONCLUSION

In this paper, we propose a novel graph coloring based resource allocation algorithm in hierarchical network. The interference graph is constructed based on the potential co-tier interference while taking the interference toleration into consideration. In our proposed scheme, the BSs with least interference are grouped in clusters so that they can reuse the same frequency bands and the spectrum reusability is improved because the spectrum can be reused appropriately by applying the cost function. In terms of resource allocation, the RBs can be assigned adaptively and flexibly to each BS based on its requirements. The simulation results show that the proposed algorithm achieves a better performance than the conversional. In the future, we can consider the optimization approach to improve the performance further.

ACKNOWLEDGEMENT

This work is supported by project NSFC 61471066 and the open funding project of State Key Lab of Virtual Reality Technology and Systems at Beihang University under Grant No. BUAA-VR-15 KF-19.

REFERENCES

3GPP TR 36.814 V9.0.0. 2010. Evolved Universal Terrestrial Radio Access (E-UTRA): Further advancements for E-UTRA physical layer aspects (Release 9).

Brelaz, D. 1979. New Methods to Color the Vertices of A Graph, *Communications of the ACM 22(4)*: 251256.

Carlaw, S. 2008. IPR and the potential effect on femtocell markets, *FemtoCells Europe, ABIresearch, London, U.K.*

Chang, R.Y. et al. 2009. Multicell OFDMA Downlink Resource Allocation Using a Graphic Framework, *IEEE Trans. Vehicular Technology* 58(7): 3494–3507.

Knisely, D., T. Yoshizawa & F. Favichia. 2009. Standardization of femtocells in 3GPP, *IEEE Commun. Mag.* 47(9): 6875.

Lpez-Prez, D. Alvaro Valcarce & Guillaume de la Roche. 2009. OFDMA Femtocells: A Roadmap on Interference Avoidance, *IEEE Communications Magazine*, 47(9): 41–48.

Mansfield, G. 2008. Femtocells in the US market — Business drivers and consumer propositions, *FemtoCells Europe, London, U.K.*

Sadr, S. & R. Adve. 2012. Hierarchical Resource Allocation in Femtocell Networks using Graph Algorithms, *in Proc. IEEE ICC*: 4416–4420.

Uygungelen, S. G. Auer & Z. Bharucha. 2012. Based Dynamic Frequency Reuse in Femtocell Networks, *in Proc. IEEE 73rd VTC Spring*: 1–6.

Wei Tan, C. 2011. Optimal power control in Rayleigh-fading heterogeneous networks, *INFOCOM, 2011 Proceedings IEEE*: 2552–2560.

Xiang, J. et al. 2010. Downlink Spectrum Sharing for Cognitive Radio Femtocell Networks, *IEEE Systems Journal*, 4(4): 524–534.

Yi, W. et al. 2009. A Novel Spectrum Arrangement Scheme for Femtocell Deployment in LTE Macrocells, *Proc. IEEE 20th Symposium on Personal, Indoor and Mobile Radio Communications*: 6–11.

Zahir, T. et al. 2012. Interference Management in Femtocells, *Communic-ations Surveys & Tutorials, IEEE(99)*: 1–19.

Zhang, Q. et al. 2013. A Coloring-based Resource Allocation for OFDMA Femtocell Networks, *in Proc. IEEE WCNC:* 673–678.

Signal and Information Processing, Networking and Computers – Chen & Huang (Eds)
© 2016 Taylor & Francis Group, London, ISBN 978-1-138-02881-4

Time drift detection in process mining

Haiying Che
Beijing Institute of Technology, Beijing, China

Quentin Machu
Polytech'Tours, Tours, France

Yangguang Zhou
Beijing Institute of Technology, Beijing, China

ABSTRACT: Currently, most of the information systems can record the tracking information and logs, this helps people to know the performance of the process execution. Process Mining techniques allow knowledge extractions such as model discovery, conformance checks and process improvements to take place. Processes are subject to various changes during their execution, for instance, a change in structure may occur when a new regulation comes into force and imposes some change, or may happen under the influence of seasonal effects, natural disasters etc. For many industries, time is a crucial factor in most cases equal to efficiency and profitability. Thus, this research paper presents an approach for detecting time-related changes. Our method extracts time-related characteristics from processes and then compares all of them together by using statistical hypothesis tests in different successive populations. Such a method could not only allow accurate detection when some parts of the processes started to have abnormal behavior: longer or shorter but also enable identification of which parts are involved. Based on the proposed approach in this paper, a ProM6 plug-in is implemented and tested. Further, synthetic data is used to do the experiment, finally, the results are explained and discussed.

1 INTRODUCTION

Business processes are a collection of related and structured activities which use resources in order to serve a specific goal. There are several ways to analyse a business process such as control-flow, data-flow, resources, time and even social outlooks. Currently, more and more business processes store an amount of event logs over a thousand of instances.

While some work has been done on the concept of drift detection to detect when a process changes and to localize the changed parts of the process, in addition to time prediction based on process mining, few has been realized yet regarding the subject of time drift detection, here time drift means time features of the event change. One may want to detect when a process or part of the process goes slower or faster than usual and where the modification is located. In some organizations, this information could be crucial, especially in processes that are extremely complicated. Specifically, it could be used to maintain a process speed, improve the speed or even to understand how different parameters such as re-source organization, workload etc., affects the process run time.

The objective of this paper is to propose an efficient approach to deal with time drifts, then test the approach using CPN Tools and a new ProM frame work plugin.

2 RELATED WORK

2.1 *Discover drifts using features extraction and hypothesis testing*

Over the last decade, several articles have been published on concept drift (both on data mining and process mining). The most significant work on the topic of the concept drift was published by R.P. Jagadeesh Chandra Bose et al. (2011) and then improved in R.P. Jagadeesh Chandra Bose et al. (2014). They first introduce what concept drift in process mining by underlying the fact that assuming a process is in a steady state, it is unrealistic, because a process may change to adapt to changing circumstances. In the introductory part, different perspectives of change and their nature are described. Then the article proposes feature sets and techniques to effectively detect the changes in event logs and identify the regions of change in a process.

Their approach is based on data distribution comparison over two sliding windows. It relies on the principle that characteristics of traces in events logs differ before and after a concept drift. The proposed algorithm computes, in a chronological order, a feature vector for each trace representing the characteristics of the process at the time of the trace. Multiple features could be extracted from event logs such as Relation Type Count and Relation Entropy, and Window Count and J-Measure as described in R.P. Jagadeesh Chandra Bose et al. (2011). Once the feature vectors are computed from the event logs on selected pairs of activities for each trace, two populations (called sliding windows) are initialized and used to perform a statistical hypothesis test (such as the Kolmogorov-Smirnov test and Mann-Whitney test) between each activity pair. They then obtain a vector of p-values (one value per activity-pair). The average p-value is plotted versus w, which is the last index of the left population.

This algorithm has been implemented in ProM Framework and tested on synthetic logs as well as real-life logs. The approach yields good and promising results.

2.2 *Time-based process mining research*

The Process Mining Manifesto states that the scope is not limited to control flow, the organizational, case and time perspectives also play an important role. It also expands extensively on time-based support because process mining techniques are not limited to the past, but also function in the present and the future: some work has been done on prediction, especially on time prediction (Wil. M.P. van der Aalst et al. 2010, Wil. M.P. van der Aalst et al. 2011). The time-based operational support has been first discussed in Wil. M.P. van der Aalst et al. (2010) in which a generic approach in ProM framework has been proposed. Petri nets discovered by ProM can been enriched with time and resource information in A. Rozinat, R.S. Mans et al. (2009). Elder also worked on time in workflow systems in Johann Eder et al. (1999), however, the focus of their work is more on scheduling.

2.3 *This paper's contribution*

The objective of this paper differs slightly from the cited articles on multiple points. First of all, unlike the concept drift discovering (R.P. Jagadeesh Chandra Bose et al. 2011, R.P. Jagadeesh Chandra Bose et al. 2014), we do not use structural features based on precedes and follow relations but only on time aspect. There-fore, we do not intend to find structural changes in processes, we aim at discovering modifications in processes concerning time. Thus, only past event-logs with finished traces are used and event logs with incomplete traces will be filtered out. Inspired by concept drift discovering approaches and time perspective of process mining, the approach to discover time drifts in process event logs is planned.

In actual production, our algorithm can be applied to the process of the production workflow to detect or predict abnormal production process, giving feedback to managers in order to avoid losses timely. In addition, we can also according to the frequency and the type of abnormality that occurs, improve the workflow of production, to improve productivity and production quality.

3 APPROACH PROPOSED

3.1 *Definitions and background*

In this section some basics concepts about the overall structure of process mining and meaning of the terms used are explained.

3.1.1 *What is process mining?*

Process mining relies on event logs for extracting information from event logs. For example, the audit trails of a workflow management system or the transaction logs of an enterprise resource planning system can be used to discover models describing processes, organizations, and products. Moreover, it is possible to use process mining to monitor deviations (e.g. comparing the observed events with predefined models or business rules in the context of SOX).

Process mining techniques work from an event log: a recording of a set of events which occurred or are occurring in a business process. Events logs are defined in standard format like XES or before MXML.

Once we have an event log, we can use it in many different process mining tools such as ProM frameworks plug-in to extract knowledge and insights about the underlying process.

3.2 *Notation*

To simplify this paper, here we define some notations for the objects we will often use.

– L is an event log, a set of n traces for a particular process.
– A is the set of activities over L. |A| is the number of unique activities
– $T \in L$ is a trace of length $|T|$ where $T(k)$ is the kth event of trace T

3.3 *Approach*

In this section, we explain process mining method to detect time drifts in processes. Figure 1 quickly introduces, in a diagram, our approach to give an initial global sight. First of all, there are some preconditions:

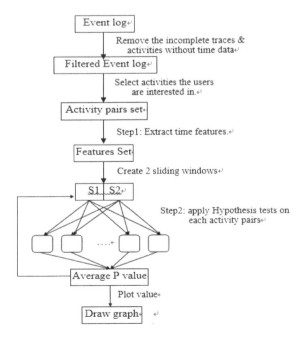

Figure 1.   Approach overview.

101

The input of our approach is an event log L, containing at least for each event in traces; a name (concept-name) and a time stamp.

As we want to work over the past: every incomplete (not finished) trace should be filtered out. The timestamp of the first event for each trace is considered as the start time of the trace. Traces should be ordered in chronological order.

First step is feature extraction. Our approach is based on the fact that if there is a time drift at time t, characteristics of traces before and after differ slightly. We call these characteristics; features. Several features could be extracted from event logs, such as the ones analysed (R.P. Jagadeesh Chandra Bose et al. 2011, R.P. Jagadeesh Chandra Bose et al. 2014). In this paper, we mainly focus on the feature which is defined by the elapsed time between the start time of an activity A and the start time of an activity B where B follows (directly or indirectly) A in trace t, let us call this time Tt(a, b). This feature is mathematically defined below. Dimensionally, we have at most values.

Activities which are used for the time drift analysis could be selected (see Fig. 2) but at least 2 activities must be chosen (for instance the first activity of the process and the last) and at most |A|. The selected activities compose the set.

This process comprises nine activities. We selected activities A, B, H, I. Thus, we will consider the following activity pairs: (A,B);(A,H);(A,I);(B,H);(B,I) in the A* set.

The second step is to analyse these extracted features. As explained earlier, we expect to discover significant characteristics difference in traces before and after a drift point. In or-der to detect that, we create two sliding windows S1 and S2 (see Fig. 3), each of size w where $2w \leq n$ as mathematically defined below.

The first window starts at the first trace and then contains the characteristics of the trace from 1 to w while the second window includes the characteristics of traces w+1 to 2w (see Fig. 4). Instinctively, we understand that if a drift point is contained somewhere in S1 or S2, there is some difference between the traces characteristics contained in S1 and S2. Furthermore, the closer the drift point is to the boundary between S1 and S2, the more distinguishable is the difference between S1 traces and S2 traces. We propose to measure the differences between S1 and S2 using hypothesis tests which are mainly used for comparing groups of data. The w parameter is crucial and should be set wisely according to the staff's experience about their process. A too-small w value will generate noisy results while a too-high w may lead to loss of information (such as missing drift points). To reduce the dependency of w, it is possible to adapt the algorithm to use adaptive windows sizes using ADWIN method in A. Bifet & R. Gavaldà (2007).

$$\forall (a,b) \in A * A^*, S1 = \{t1(a,b) \cdots tw(a,b)\}$$
$$\forall (a,b) \in A * A^*, S2 = \{t(a,b) \cdots t(a,b)\}$$

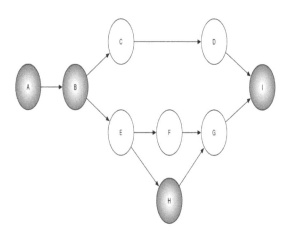

Figure 2.   Activities selection in a process.

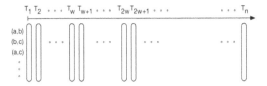

Figure 3.   Features set.

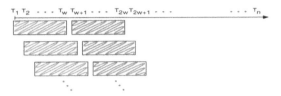

Figure 4.   Populations over features set.

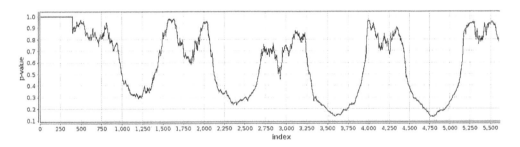

Figure 5.   Example of resulting graph.

There are n vectors of values which correspond to the time between two activities on each trace.

There are several hypothesis tests available, depending on the objectives and on the assumptions we have. Hypothesis tests can be divided into two categories; parametric tests which make assumptions about the probability distributions of the variables being assessed and non-parametric tests. Because we cannot make any premise on the feature distributions, we have to choose non-parametric tests only. Given the two sliding windows, for each activity pairs (a, b), we then have two populations of features; one population from S1, another from S2. For a given activity pair, the two populations are two line vectors containing times between the two activities over the traces of S1 and S2, respectively. We should choose a two-sample test which compares the two populations. There are tests which compare one given population against an a priori distribution (one-sample test) or tests which can compare more than two given populations together. We choose Kolmogorov-Smirnov test (KS) to compare our two populations of features. Given the two sliding windows, for each activity pairs (a, b), we thus have two populations (one from S1, another from S2) of features. We apply statistical tests on each activity pairs and then we compute the average p-value of all these tests for the two current sliding windows. The average p-value is plotted against won a graph and we shift the two sliding windows to the right by one trace. The tests/plot/ shift cycle is repeated until 2w = n. At the end of the algorithm, we obtain a plot with an x-axis corresponding to the n traces and a y-axis with the probability that there is no drift. We should observe gaps around drifts and noticeable troughs on traces corresponding to drifts as shown in Figure 5, in which we can notice drifts around traces 1200, 2400, 3600 and 4800.

We work on two populations of features corresponding to traces 1 to w and w+1 to 2w and then we shift the population to the right.

To summarize our approach, we use event logs in which incomplete traces and activities without time data are filtered out. We chronologically order the traces and select some activity pairs to work on. We then start the first step of our approach; features extraction on the selected set of activities. The second step starts from creating two sliding windows over the extracted features. We apply hypothesis tests on each activity pairs between the two populations (from the two sliding windows), we compute p-values average and we plot that average. Finally, we shift the two populations if we have not reached the end of the features set. By the end of the algorithm, we obtain a readable graph which demonstrates if there are time drifts or not.

4 FIGURE CAPTIONS

4.1 *Generate synthetic logs*

There are mainly two ways to experiment and validate an approach. The first one is to test the approach on synthetic generated data in which we introduced time drift in the process while the second one is obviously to test it with real data, extracted from real-life processes. In this part, we describe how synthetic logs could be generated and used.

We use CPN Tools software which is a tool for editing, simulating, and analysing coloured Petri nets. It also supports timed coloured Petri nets. First we create two (or more) timed models corresponding to a process before and after a drift. Because we are working on time drift, we should modify the time took by an activity to simulate a re-source outage or generate an abnormal time consuming concept drift by adding activities between two important activities in the process. Three simple examples are shown in Figure 6, Figure 7 and Figure 8. These examples are three timed-coloured Petri networks in which i is the coloured-set corresponding to the trace number of the process and in which transitions are used to describe activities (or splits/joins). Notice that after each activity, the time (in seconds) is incremented by a certain amount using a normal random function (be-cause in real life, activities may not always take the same time).

To generate an event log with a drift, we simulate the first model for a certain amount of traces and then the second one. To do that, we have to create another Petri net which will choose between the three models for each trace. We use the hierarchy functionality of CPN Tools to be able to link our three previous Petri nets from this fourth one. Our goal is to generate 6000 traces with two drifts at index 2000 and 4000.

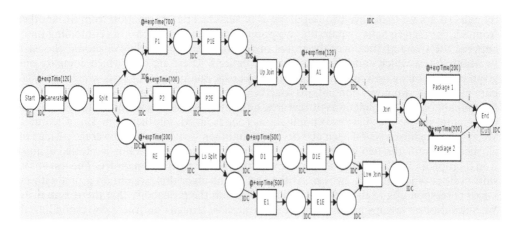

Figure 6.   CPN Tools-synthetic process model 1.

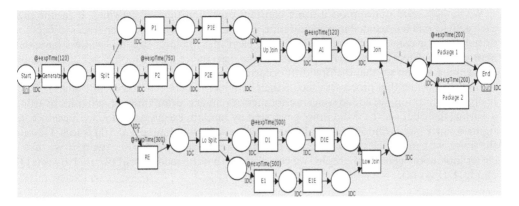

Figure 7.   CPN Tools-synthetic process model 2.

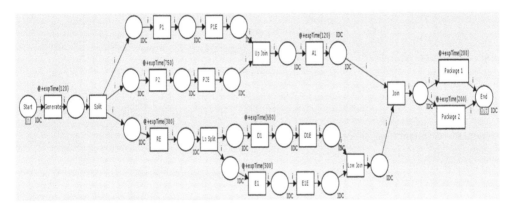

Figure 8.   CPN Tools-synthetic process model 3.

Note that at those indices, we generate two sudden drifts such that there are two abrupt changes in time. We could also generate gradual drifts in which the time would increase slowly.

To generate event logs from CPN Tools, the Simulation function should be used. The general options keys "Save Report" and "Save key bindings" have to be checked. We simulate enough transitions to have all our traces executed. It generates a txt file in a sub folder of the model's folder.

– Trace identifier: CPN variable: i
– Concept: name (Concept): Transition
– Time: timestamp (Time): Time

Finally, we should run the "Filter log using Simple Heuristics" action to filter auxiliary events (orchestrator's ones for example), so we only keep our process activities. Now, we have a synthetic event log which includes two drifts in standard format, ready to be used with the approach plug-in.

4.2   Synthetic logs results

Using the synthetic logs generated by CPN Tools in the previous section (Fig. 6, Fig. 7, Fig. 8), we demonstrate how our approach works. The generated process contains 6000 traces over 13 activities and two time drifts at index 2000 and 4000.

We initially tried our approach using constant execution time of activities. It means that each activity will be executed in constant time, without any variation between traces (except for drifts). In the average result graph (Fig. 9), it is clear that whatever size the windows have, our approach can detect the two drifts perfectly at the right indices. If we look closer at the generated graphs, we can see that the first drift occurred between activities P2 and P2E (Fig. 10).

Because the fact that processes execute their activities in perfect and constant times is not a realistic assumption, we added some randomness on the execution times of activities by using a normal distribution. To avoid noise generated by random execution times, we have to configure a fairly large window size. In Figure 11, we used a window size of 100 traces. Despite the noise, we can still clearly see the two drifts at indices 2000 and 4000. Again, if we take a closer look at the generated graphs, we can observe that the second drift occurs between D1 and D1E (Fig. 12).

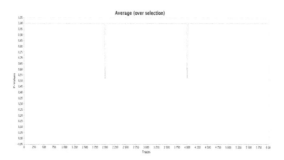

Figure 9.   Constant times-average results.

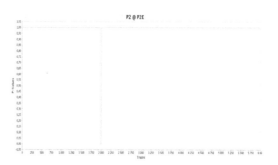

Figure 10.   Constant times-P2@P2E.

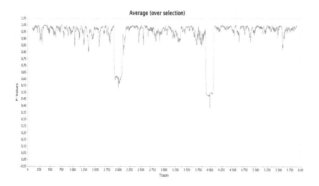

Figure 11.   Noisy times-average results.

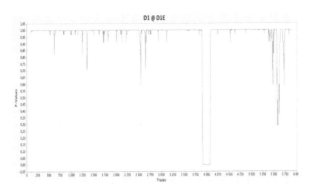

Figure 12. Noisy times-D1@D1E.

However, we observe that our features generate results which are not really smooth. This may lead in certain cases to false positives when activities execution times have too high a variance

5 CONCLUSIONS

In this paper, we proposed an approach to detect changes in processes over time perspective; to detect when parts of processes ran with abnormal durations and in which regions these drifts happened. The approach has been implemented in ProM 6 framework as a plugin and our initial results on synthetic logs show that our method can detect changes even with noise.

We have only considered sudden-drift changes. Our approach has not been tested on gradual drifts (when changes appear slowly on the process) on which further work is required. However, we hope that the gap parameter allows one to deal with these drifts by transforming gradual drifts into several small sudden-drifts.

Time is closely related to resources and social networks. Therefore, this paper may be a starting point for extended work which should take into consideration both a resource and social networks perspective.

REFERENCES

Bifet, A., & Gavaldà, R. 2007. Learning from time-changing data with adaptive windowing.

Jagadeesh Chandra Bose, R.P., Wil. M.P. van der Aalst, Indre Zliobaite & Mykola Pechenizkiy. 2011. Handling concept drift in process mining. 23rd International Conference, CAiSE2011, London, UK, June 20–24, 2011. Pro 391–405.

Jagadeesh Chandra Bose, R.P., Wil. M.P. van der Aalst, Indre Zliobaite & Mykola Pechenizkiy. Dealing with concept drifts in process mining. Neural Networks and Learning Systems 25(1): 154–171.

Johann Eder, Euthimios Panagos & Michael Rabinovich. 1999. Time constraints in workflow systems. Advanced Information Systems Engineering, 11th International Conference, CaiSE "99 Heidelberg, Germany, June 14–18, 1999. Proceedings: 286–300.

Rozinat, A., Mans, R.S. Song, M. & Wil. M.P. van der Aalst. 2009. Discovering simulation models. Information Systems 34(3): 305–327.

Wil. M.P. van der Aalst, Maja Pesic & Minseok Song. 2010. Beyond process mining: From the past to present and future. Advanced Information Systems Engineering, 22nd International Conference, CAiSE 2010, Hammamet, Tunisia, June 7–9, 2010. Proceedings: 38–52.

Wil. M.P. van der Aalst, Schonenberg M.H. & Song, M. 2011. Time prediction based on process mining. Information Systems 36(2): 450–475.

Phil Weber, Behzad Bordbar & Peter Tino. 2011. Real-time detection of process change using process mining.

Signal and Information Processing, Networking and Computers – Chen & Huang (Eds)
© 2016 Taylor & Francis Group, London, ISBN 978-1-138-02881-4

Spectrum sensing for cognitive radio systems with unknown non-zero-mean noise

Mengwei Sun
Key Laboratory of Universal Wireless Communication, Beijing University of Posts and Telecommunications, Beijing, China

Tiehong Tian
China Unicom System Integration Limited Corporation, China

Chenglin Zhao & Bin Li
Key Laboratory of Universal Wireless Communication, Beijing University of Posts and Telecommunications, Beijing, China

ABSTRACT: Prior knowledge of noise distribution is crucial for spectrum sensing. However, the properties of the noise process are often unknown in most of the practical applications. In this paper, a blind spectrum sensing scheme is proposed, which could recover the noise distribution and detect the occupancy of primary frequency band simultaneously. For a given sampling sequence of observation, the spectrum sensing is reformulated into a Bayesian sequence estimation problem and will be solved based on marginalized particle filtering technology. Experimental simulations show that the proposed method could improve the sensing performance and estimate the statistical parameters of noise accurately.

Keywords: spectrum sensing; Bayesian sequence estimation; marginalized particle filtering

1 INTRODUCTION

Cognitive Radio (CR) is an effective technique to increase the utilization rate of the spectrum [1] and has enjoyed high favor in commercial communications recently. Spectrum sensing technology is an important component in CR to enable Secondary Users (SUs) to access the unused licensed spectrum, without causing interference to Primary Users (PUs). Common sensing methods include Energy Detection (ED), Matched Filtering Detection (MFD) and cyclostationary feature detection [1]. However, most existing sensing methods are designed assuming that the properties of the noise process are known and will become ineffective to the practical application in which noise uncertainty is common. There have been several approaches proposed to overcome the impact caused by noise uncertainty, such as a multi-antenna based spectrum sensing method, which is premised on the Generalized Likelihood Ratio Test (GLRT) paradigm [2] and a cooperative sensing method with adaptive thresholds [3]. The use of multi-antenna or cooperative sensing at the receiver poses strict requirements on receivers and increases the complexity.

In this investigation, we focus on the CR system with noise uncertainty and propose a single—antenna single-node spectrum sensing method based on the Bayesian inference framework. The noise follows a Gaussian distribution, but the mean and variance are unknown. We first formulate a Dynamic State-space Model (DSM) to depict the spectrum sensing system in which the PU state and the noise statistical parameters are considered as hidden states to be estimated. Then, a sequential estimation scheme is proposed to estimate the PU state, the noise mean and variance jointly, by utilizing Marginalized Particle Filtering (MPF) technology [4]. The estimated noise parameters will provide important information for the subsequent allocation of cognitive resources to maximize the system throughout.

This rest of this paper is organized as follows. In Section 2, we provide the dynamic state-space model of spectrum sensing. The joint blind sensing algorithm is introduced in Section 3. In Section 4, numerical simulations and performance analyses are provided. Finally, conclusions are drawn in Section 5.

2 SYSTEM MODEL

2.1 Dynamic state-space model

In this paper we consider the following discrete time dynamic state-space model.

$$S_n = \Phi(S_{n-1}) \tag{1}$$

$$y_n = \Psi(\mathbf{x}_n, \mathbf{v}_n) \tag{2}$$

Here, (1) is referred to as a state equation, while (2) is an observation equation. S_n denotes the PU state and $\mathbf{x}_n$ represents the corresponding transmitted signal. y_n is the observation and $\mathbf{v}_n$ represents the noise sequence.

2.2 PU state

Generally, PU state comes into two forms: active and dormant. When S_n is dormant, $\mathbf{x}_n = \mathbf{0}$, otherwise, $\mathbf{x}_n = \mathbf{s}_c$ while $\mathbf{s}_c$ denotes the PU transmitted signal sequence. Different states transfer to each other with specified probability and we use a two-state Markov model to depict the evolution of PU state [5].

2.3 Observation

The function of SU comprises two parts, sensing and signal transmission. What needs to be stressed is that the PU state is assumed to remain unchanged in one sensing-transmission slot. In the sensing time, the sampling of PU signal is taken for decision making; the sampling size is assumed to be M. Then, we adopt a coherent receiver at the receiving end and the observation can be represented by:

$$y_n = \begin{cases} \mathbf{s}_c \otimes \mathbf{v}_n & \text{when the PU is dormant } [H_0] \\ \mathbf{s}_c \otimes (\mathbf{x}_n + \mathbf{v}_n) & \text{when the PU is active } [H_1] \end{cases} \tag{3}$$

Here, '$\otimes$' denotes the convolution operation. $\mathbf{v}_n = [v_{n,0}, v_{n,1}, \ldots, v_{n,M-1}]$ is an i.i.d sequence following the Gaussian distribution with unknown mean μ and variance σ^2, i.e. $v_{n,m} \sim N(\mu, \sigma^2)$, $m = 1, 2, \ldots, M - 1$.

3 SPECTRUM SENSING ALGORITHM

3.1 Problem definition

Based on the formulated DSM, we propose a joint sensing algorithm. The purpose of this algorithm is to detect the hidden PU state, together with unknown noise parameters. We address this problem by concerning the joint posterior probability $p(\mu, \sigma^2, \mathbf{x}_{0:n}|y_{0:n})$, and from the Bayesian perspective, the joint estimation could be achieved by MAP criterion.

$$(\hat{\mu}, \hat{\sigma}^2, \hat{\mathbf{x}}_{0:n})^{\text{MAP}} = \arg \max_{\mathbf{x}_{0:n} \in \{0, \mathbf{s}_c\}} \left[p(\mu, \sigma^2, \mathbf{x}_{0:n} \mid y_{0:n}) \right] \tag{4}$$

110

3.2 Joint estimation

The joint posterior probability in (4) could be decomposed into conditional densities.

$$p(\mu, \sigma^2, \mathbf{x}_{0:n} \mid y_{0:n}) = p(\mu, \sigma^2 \mid \mathbf{x}_{0:n}, y_{0:n}) p(\mathbf{x}_{0:n} \mid y_{0:n}) \tag{5}$$

Following the concept of MPF, $p(\mathbf{x}_{0:n}|y_{0:n})$ will be approximated based on Particle Filtering (PF) technology and the conditional density $p(\mu, \sigma^2 \mid \mathbf{x}_{0:n}, y_{0:n})$ will be computed by a measurement update of the noise distribution. In other words, the joint sensing method proposed consists of two parts: the PU state is detected utilizing PF technology and the noise parameters are updated based on the marginalization concept.

3.2.1 PU state detection

Based on PF, we approximate $p(\mathbf{x}_{0:n}|y_{0:n})$ with particle trajectories and associated weights. However, due to computational complexity, it is desirable to solve this approximation problem sequentially and recursively [6]. The marginal transmitted signal estimation at n-th sensing-transmission slot could be achieved by:

$$\hat{\mathbf{x}}_n = \arg\max_{\mathbf{x}_{0:n} \in \{0, s_c\}} \left[p(\mathbf{x}_n \mid \hat{\mathbf{x}}_{0:n-1}, y_{0:n}) \right] \approx \arg\max_{\mathbf{x}_n \in \{0, s_c\}} \left[\sum_{i=1}^{P} \omega_n^{(i)} \delta(\mathbf{x}_n - \mathbf{x}_n^{(i)}) \right] \tag{6}$$

Specifically, the PU detection is achieved by four steps [5]. First, particles are generated from an important distribution, i.e. $\mathbf{x}_n^{(i)} \sim \pi(\mathbf{x}_n \mid \mathbf{x}_{0:n-1}^{(i)}, y_{0:n})$. Second, the associated importancew weights $\omega_n^{(i)}$ are computed recursively [7]. Third, we normalize the importance weights and conduct a re-sampling procedure when weight degeneracy happens. Finally, the MAP estimation of PU state could be achieved by (6), based on the calculated particles and weights.

3.2.2 Estimation of noise parameters

The noise follows the Gaussian distribution with unknown μ mean and variance σ^2. From (6), we can see that the posterior probability of observation y_n also follows the Gaussian distribution, i.e., $(y_n \mid \mathbf{x}_n^{(i)} \mu, \sigma^2) \sim N(\mathbf{x}_n^{(i)} \mathbf{s}_c^T + \|\mathbf{s}_c\|_1 \mu, \|\mathbf{s}_c\|_2 \sigma^2)$, in other words, $[(y_n - \mathbf{x}_n^{(i)} \mathbf{s}_c^T)|\mu, \sigma^2 \sim] \sim N(\|\mathbf{s}_c\|_1 \mu, \|\mathbf{s}_c\|_2 \sigma^2) = N(U, \Sigma)$. U and Σ are unknown parameters and will be estimated based on the application of conjugate prior.

Without the loss of generality, a Normal-inverse-Gamma distribution is utilized to define the conjugate prior of the Gaussian distribution with unknown mean U and variance Σ [4], and the hierarchical Bayesian model is given below:

$$U \mid \Sigma \sim N(U_0, \Sigma/\kappa_0) \tag{7}$$

$$\Sigma \sim iG(\nu_0/2, \Lambda_0/2) \tag{8}$$

Here, $iG(\cdot)$ denotes the Inverse Gamma distribution, the hyper-parameters $\hat{\mu}_n$, κ_n, ν_n, and Λ_n of the posterior distribution are updated recursively as follows [4]:

$$\kappa_n = \kappa_0 + n \tag{9}$$

$$U_n^{(i)} = \frac{\kappa_0}{\kappa_0 + n} U_0 + \frac{n}{\kappa_0 + n} \bar{e}_n^{(i)} \tag{10}$$

$$\nu_n = \nu_0 + n \tag{11}$$

$$\Lambda_n^{(i)} = \Lambda_0 + \sum_{n=1}^{n} \left(e_n^{(i)} - \bar{e}_n^{(i)} \right)^2 + \frac{\kappa_0 n}{\kappa_0 + n} \left(\mu_0 - \bar{e}_n^{(i)} \right)^2 \tag{12}$$

Here, $e_n^{(i)} = y_n - \mathbf{x}_n^{(i)} \mathbf{s}_c^T$ and $\bar{e}_n^{(i)}$ represents the mean of $e_{1:n}^{(i)}$.

111

Finally, the estimation of U_n and Σ_n could be achieved by relying on the marginalization concept and unbiased estimation. Then, based on the mathematical relation between μ and U, σ^2, and Σ, which has been mentioned above, the estimation of noise mean and variance will be calculated.

$$\hat{\mu}_n = \frac{\hat{U}_n}{\|s_c\|_1} = \frac{1}{\|s_c\|_1} E\left(U \mid y_{0:n}\right) \approx \frac{1}{\|s_c\|_1} \sum_{i=1}^{P} E\left(U \mid \mathbf{x}_{0:n}^{(i)}, y_{0:n}\right) \omega_n^{(i)} = \frac{1}{\|s_c\|_1} \sum_{i=1}^{P} U_n^{(i)} \omega_n^{(i)} \qquad (13)$$

$$\hat{\sigma}_n^2 = \frac{\Sigma_n}{\|s_c\|_2} = \frac{1}{\|s_c\|_2} E\left(\Sigma_n \mid y_{0:n}\right) \approx \frac{1}{\|s_c\|_2} \sum_{i=1}^{P} E\left(\Sigma_n \mid \mathbf{x}_{0:n}^{(i)}, y_{0:n}\right) \omega_n^{(i)} = \frac{1}{\|s_c\|_2} \sum_{i=1}^{P} \frac{\Lambda_n^{(i)}}{v_n - 1} \omega_n^{(i)} \qquad (14)$$

4 SIMULATIONS

In this section, we illustrate the sensing performance of the proposed sensing method, compared with the MFD, when the noise statistical parameters are unknown. Then, we evaluate the accuracy of the noise parameters estimation.

The actual value and known value of the Signal power to Noise Ratio (SNR) are denoted by SNR and SNR_0, and $SNR_0 = SNR + snr$. The random number snr follows a uniform distribution and the float rang could be written as $[-\vartheta, \vartheta]$. We take $\vartheta = 15$ dB. The detection probability [5] results are shown in Figure 1. The performance of the proposed algorithm is close to the sensing performance achieved by MFD with given noise parameters, and well above the MFD with unknown noise parameters. The estimated noise mean and variance are shown in Figure 2.

5 CONCLUSION

In order to address the inevitable challenge engendered by unknown noise statistical parameters in realistic applications, a novel DSM is formulated in this paper, and on this basis, an iterative joint sensing scheme is presented, which could detect the PU state as well as estimate the noise mean and variance. The recovered unknown noise parameters could promote the sensing performance and provide important information for the SUs to modify the transmitted power, in order to maximize the system throughout. The simulation results have demonstrated an excellent performance of the proposed sensing algorithm. To conclude, the designed joint estimation scheme provides a promising solution in enhancing the sensing performance in realistic CR networks.

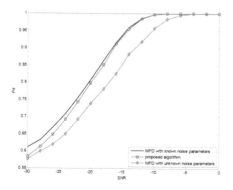

Figure 1. Comparison of detection performance.

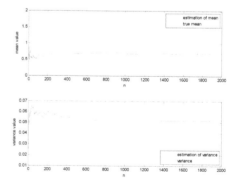

Figure 2. Estimated mean and variance.

REFERENCES

Li, B., Sun, M.W., Li, X.F., Nallanathan, A. & Zhao, C.L. 2014. Energy Detection based Spectrum Sensing for Cognitive Radios over Time-Frequency Doubly Selective Fading Channels. *IEEE Transactions on Signal Processing* 63(2): 402–417.

Li, B., Zhao, C.L., Sun, M.W. & Nallanathan, A. 2014. Spectrum Sensing for Cognitive Radios in Time-Variant Flat Fading Channels: A Joint Estimation Approach. *IEEE Transactions on Communications* 62(8): 2665–2680.

Lu, L., Zhou, W., Onunkwo, U. & Li, G.Y. 2012. Ten Years of Cognitive Radio Technology. *EURASIP Journal on Wireless Communications and Networking*, 28: 1–16.

Saha, S., Ozkan, E., Gustafsson, F. & Šmídl, V. 2010. Marginalized particle filters for Bayesian estimation of Gaussian noise parameters. In *Proc. 2010 13th Conference on Information Fusion (FUSION)*, *Edinburgh, July 2010.*

Song, C., Alemseged, Y.D & Tran, H.N., et al. 2010. Adaptive two thresholds based energy detection for cooperative spectrum sensing. In *Proc. of Consumer Communications and Networking, Las Vegas, January 2010.*

Sun, M.W., Li, B., Song, Q., Zhao, L. & Zhao, C.L. 2014. Joint detection scheme for spectrum sensing over time-variant flat fading channels. *IET Communications* 8(12): 2064–2073.

Zhang, R., Lim, T.J., Liang, Y.C., Zeng, Y. 2010. Multi-antenna based spectrum sensing for cognitive radios: A GLRT approach. *IEEE Transactions on Communications* 58(1): 84–88.

Signal and Information Processing, Networking and Computers – Chen & Huang (Eds)
© 2016 Taylor & Francis Group, London, ISBN 978-1-138-02881-4

A secure fuzzy-based cluster head election algorithm for Wireless Sensor Networks

J. Shi
China Mobile (Tianjin) Co. Ltd., Tianjin, China

X.N. Han
School of Instrumentation Science and Opto-Electronics Engineering, Beihang University, Beijing, China

ABSTRACT: Cluster-based topology is frequently utilized to improve the performance of wireless sensor networks. Cluster heads election is important in designing reliable clustering networks as the existence of malicious cluster heads is a serious security threat to the entire network. To solve this problem, we propose a secure fuzzy-based cluster head election scheme (SFCE) for wireless sensor networks. The detection of malicious sensor nodes takes advantage of the proposed Bayesian-based trust management scheme, and fuzzy logic is adopted to select nodes with the maximum priority as cluster heads. To prove the efficiency of SFCE, we conduct simulations and results that demonstrate that compared with LEACH, SFCE can effectively deter these compromised sensor nodes from being chosen as cluster heads and reduce energy consumption.

Keywords: cluster head election; trust; fuzzy logic; wireless sensor networks

1 INTRODUCTION

Wireless Sensor Networks (WSNs) are a collection of tiny sensors, which have the capability of data sensing, information processing as well as short distance wireless communication. In large scale networks, cluster-based architecture is widely used to improve the scalability and prolong network lifetime. Clustering decreases energy consumption of sensor nodes and provides an easy way to maintain topology structure (Masdari et al. 2013).

In many applications, wireless sensor networks are usually deployed in unwatched and very hostile environments. The inherent characteristics of wireless communications make clustering networks vulnerable to various malicious attacks (Khalid et al. 2012). In clustering networks, Cluster Heads (CHs) collect data from Cluster Members (CMs) at the first place and then forward the aggregated messages to the Base Station (BS). If a malicious node declares itself as a CH to take part in data transmission, the whole network will be under serious security threats. Therefore, security of CH election is extremely important (Feng et al. 2012). However, the traditional security schemes with cryptographic protection are invalid in dealing with internal attacks. For another thing, encryption algorithms call for more resources and this makes them unfit for sensors with limited energy and resources. Trust management schemes help sensors to predict uncertainty about the future behaviors and compute the reliability of another node (Lopez et al. 2010). Consequently, a trust-based CH election algorithm is a better solution to improve the secure attribute of clustering networks.

Low-energy adaptive clustering hierarchy, or LEACH (Heinzelman et al. 2000) is one of the most classic cluster routing protocols. Although, LEACH saves nodes energy greatly; it is easy to be attacked without security support. Crosby et al. puts forward a distributed CH election scheme based on trust (Crosby et al. 2006). In this method, all CMs voted for their neighbors whom they trusted most and the nodes that got the most votes were chosen as CHs. Chatterjee et al. proposed a trust aware clustering algorithm. The CHs were the most qualified and trustworthy nodes elected by an authenticated voting scheme using parallel multiple signatures (Chatterjee et al. 2014). In addition, the proposed trust scheme used self and recommendation evidences to build nodes' trust so as to eliminate network wide flooding. Mao and Zhao proposed an unequal clustering scheme on the basis of fuzzy logic (Mao & Zhao 2011). They considered energy level, distance to BS when selecting CHs. Fuzzy logic was used to compute each CHs' competition radius. Bajaber and Awan proposed Adaptive Decentralized Re-Clustering Protocol (ADRP) (Bajaber & Awan 2011). In ADRP, the CHs were selected considering each node's remaining energy and each clusters' average energy. Nevertheless, these CH election approaches suffer from the following disadvantages: (1) nodes trust value is not considered when electing CHs; (2) CHs may locate at areas where nodes are sparse; and (3) nodes' residual energy is ignored.

Aiming at solving the mentioned problems, we propose a secure fuzzy-based cluster head election algorithm for WSNs. A trust management scheme based on Bayesian theory is built to conduct trust evaluation and detect malicious nodes. The proposed algorithm adopts fuzzy logic to compute nodes' priority to be chosen as CHs. Three fuzzy inputs are nodes trust degree, energy parameter and density, and nodes with the highest priority, which is the fuzzy output, are elected as CHs.

2 BAYESIAN-BASED TRUST MANAGEMENT SCHEME

The Beta distribution has been proven to be useful in the description of probability distribution of binary events and reputation estimation for entities (Jøsang & Ismail 2002). In this paper, we build up the trust relationship among nodes according to Bayesian Theory. We utilize the watchdog mechanism (Ganeriwal et al. 2008) to supervise both data forwarding behavior and data outlier.

Suppose that node i is the evaluating node and node j is the evaluated node. The direct trust calculation is on the basis of node i's first-hand observation on node j. Suppose α_{ij} is node j's normal action number that node i has observed and β_{ij} is node j's abnormal action number that node i has observed. It's proven that the direct trust degree node i has on node j, signified by DT_{ij}, can be expressed by the mathematical expectation of Beta distribution (Jøsang & Ismail 2002):

$$DT_{ij} = \frac{\alpha_{ij} + 1}{\alpha_{ij} + \beta_{ij} + 2} \qquad (1)$$

As for indirect trust computation, we collect the recommendation from common neighbors of node i and node j, which are referred to as recommend nodes. At first, we compute the direct trust DT_{ik} that node i has about the recommended node k according to Equation 1. Let γ be the threshold. If $DT_{ik} < \gamma$, which indicates that node i doesn't trust node k, node i will totally reject node k's recommendation. If $DT_{ik} \geq \gamma$, which means that node k is trustworthy, we compute the weight of different recommender, ω_k as follows:

$$\omega_k = \frac{DT_{ik}}{\sum_{m=1}^{n} DT_{im}} \qquad k = 1, 2, \ldots, n \qquad (2)$$

where n is the number of reliable recommenders. As expected, t4he more reliable the recommenders, the more weights they are assigned. Suppose α_{ij}^r and β_{ij}^r are the recommendation information about node j. The indirect trust value IT_{ij} can be expressed via the following way:

$$IT_{ij} = \frac{\sum\limits_{k=1}^{n} \omega_k * \alpha_{kj} + 1}{\sum\limits_{k=1}^{n} \omega_k * \alpha_{kj} + \sum\limits_{k=1}^{n} \omega_k * \beta_{kj} + 2} \qquad (3)$$

In Equation 3, α_{kj} and β_{kj} are the number of normal actions and the number of abnormal actions that recommend k monitors about node j, respectively.

Finally, the overall trust T_{ij} that node i has about node j can be computed as:

$$T_{ij} = \mu * DT_{ij} + (1 - \mu) * IT_{ij} \qquad (4)$$

where μ is the weighing factor reflecting the significance of direct trust and its value varying in different application scenes.

In order to improve the adaptability, we update the trust values at the period of τ. Because the trust value is related to interaction records, the evaluating node updates the interaction records at time $(t + \tau)$ via Equation 5:

$$\begin{cases} \alpha^{t+\tau} = (1 - \theta) * \alpha^t + \theta * \Delta\alpha \\ \beta^{t+\tau} = (1 - \theta) * \beta^t + \theta * \Delta\beta \end{cases} \qquad (5)$$

where $\Delta\alpha$, $\Delta\beta$ are the behavior record during period τ and the parameter θ is called forgetting factor, which reflects the weight of history experience. The forgetting factor θ is dynamic during the updating process:

$$\theta = \begin{cases} \theta_h, & \text{if } \Delta\alpha < \Delta\beta \\ \theta_l, & \text{if } \Delta\alpha \geq \Delta\beta \end{cases} \qquad (6)$$

where $0 < \theta_l < \theta_h < 1$. That means when the evaluated node behaves badly during period τ, its abnormal actions are strictly punished. While it behaves well during period τ, its normal actions have less effect on current trust evaluation and this prevents the evaluated node from improving its credibility by masquerade.

3 CLUSTER HEAD ELECTION BASED ON FUZZY LOGIC

Unlike traditional control systems, which depend on exact environment representation, fuzzy logic systems don't need accurate mathematical model systems and can make decisions in real time even when the information is inadequate. On the other hand, the computational complexity of fuzzy logic system is low. Hence, fuzzy logic system has been widely used for WSNs (Singh et al. 2012).

There are three input variables taken into consideration when electing CHs for the fuzzy logic system: trust value denoted by T, energy parameter denoted by E, and density denoted by D. Here, energy parameter indicates nodes' residual energy and density denotes the intensive degree of nodes' neighbors. E and D can be computed as:

$$E = \frac{E_r}{E_o} \qquad (7)$$

$$D = \frac{n}{N} \qquad (8)$$

where E_r is nodes' residual energy and E_o represents the initial energy. n is the quantity of the neighbors of the node. N represents the total number of sensor nodes in the networks.

When the current CH has served for a preset period of time or its residual energy is lower than a predefined value, it broadcasts a new election message. Once receiving a new election message, each node generates a random number in the range of 0–1. Once the random number is less than the threshold $T(n)$, this node becomes a candidate cluster head. And the threshold $T(n)$ can be computed in the following way:

$$T(n) = \begin{cases} \dfrac{P}{1 - P \times [r \bmod (1/P)]} & n \in G \\ 0 & otherwise \end{cases} \tag{9}$$

where P is the proportion of cluster heads and r is the number of current round. G represents the set of nodes, which have not been selected as CHs in the last $r mod(1/P)$ rounds.

If a node becomes a candidate CH, it obtains the number of its neighbors through broadcast inquiry. Then the candidate CH broadcasts a *Candidate_Message* with its residual energy and density to its neighbors. For a common node, it may receive multiple *Candidate_Messages*. Upon receiving these *Candidate_Messages*, the common node excludes untrustworthy candidates whose trust value is lower than the threshold Tr according to its trust record. Then, the node computes the priority of those credible candidates to be chosen as CHs according to fuzzy logic. Due to its simplicity, the most used Mamdani Method is adopted for fuzzy inference. The computation process is as follows.

At first, the three fuzzy input variables: trust value, energy parameter, and density are fuzzified into different fuzzy sets and allocates each fuzzy set a membership degree. The membership degree describes the level of membership of an element to this fuzzy set. The most used membership functions are triangular membership function, Gaussian membership function and trapezoidal function. In this proposal, we use triangular membership function and trapezoidal function. Table 1 and Figure 1 show the fuzzy sets and membership functions for fuzzy input and output variables.

Afterwards, the fuzzified variables are processed to infer the output according to the fuzzy if-then rules. In this phase, the if-then rules are used to infer the output with fuzzy inputs and we define 27 rules shown in Table 2. As shown in Table 2, if the trust value is low, the priority is low, which means trust value has more effects than the other two input variables. Thus, nodes with higher trust value, more energy and higher density are more likely to be chosen as CHs.

According, to the if-then rules, the fuzzy output variable *priority* can be obtained. This fuzzy variable has to be transformed to an exact number that can be used in reality. We call this process defuzzification. Here, we adopt the most widely used COA (center of area) method, which computes the output as follows:

$$Priority = \frac{\int x \cdot \mu(x)\, dx}{\int x\, dx} \tag{10}$$

where $\mu(x)$ represents the membership function of fuzzy set *priority*.

Then the common nodes choose the credible candidates with the max priority as their CHs and send *join-in Message*. The CHs receive the members' *join-in Message* and reject the requests from those whose trust value is low so as to isolate the malicious nodes.

Table 1. Fuzzy sets for fuzzy input and output variables.

Variables	Fuzzy sets
Trust value (T)	High, medium, low
Energy parameter (E)	High, medium, low
Density (D)	High, medium, low
Priority (P)	High, little high, medium, little low, low

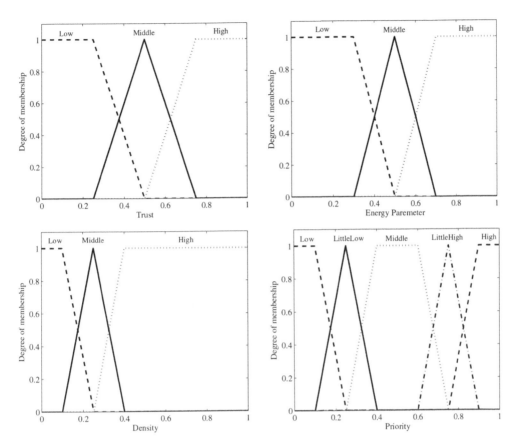

Figure 1. Degree of membership function for fuzzy variables.

Table 2. Fuzzy if-then inference rules.

Rules	Trust value	Energy parameter	Density	Priority
1	Low	Low	Low	Low
2	Low	Medium	Low	Low
3	Low	High	Low	Low
4	Low	Medium	Low	Low
...	...	...	...	...
26	High	High	Medium	Slightly high
27	High	High	High	High

4 SIMULATION RESULTS AND ANALYSIS

In this part, we simulate our SFCE model on Matlab platform and compare its performance with LEACH protocol. Hundred nodes are randomly deployed in the 100 m × 100 m area and BS is located at (50,150). The communication radius of nodes is 20 m.

4.1 *Analysis of trust computation*

In this subsection, we do research on the validity of our proposed trust mechanism. The simulation scenario is that a malicious node behaves well in the first 30 rounds and behaves

119

poorly in the following 30 rounds. After that, it cooperates abidingly. Here, we set γ, θ_h, and θ_l to be 0.7, 0.7, and 0.3 respectively. As shown in Figure 2, the trust value of this malicious node tracks was its current status. The trust value declines rapidly when the node launches attacks and recovers when it behaves well. However, the recovery needs much longer time and continuous good actions. This indicates that it takes much longer time for trust collapse than trust accumulation. This results from the use of adaptive forgetting factor. When the evaluated node behaves badly, its abnormal actions are strictly punished. When it behaves well, its normal actions have less effect on current trust evaluation and this prevents the evaluated node from improving its credibility by masquerade.

4.2 Analysis of CH distribution

Figure 3 shows the CH distribution generated by SFCE and LEACH. Here, *Tr* is set to be 0.7. We can see that SFCE can exclude malicious nodes from the network and prevent them from being elected as CHs, while LEACH protocol may select malicious nodes as CHs. Moreover, the CHs in SFCE won't be distributed in areas where nodes are sparse, whereas in LEACH, nodes located in sparse areas may be selected as CHs. The reason is that LEACH doesn't consider trust value and density.

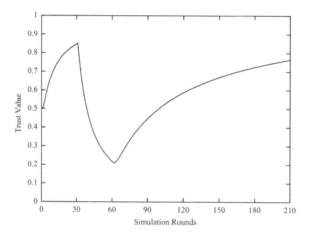

Figure 2.   Simulation result of trust computation.

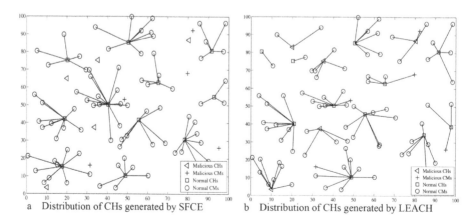

a   Distribution of CHs generated by SFCE   b   Distribution of CHs generated by LEACH

Figure 3.   The distribution of CHs generated by SFCE and LEACH protocol.

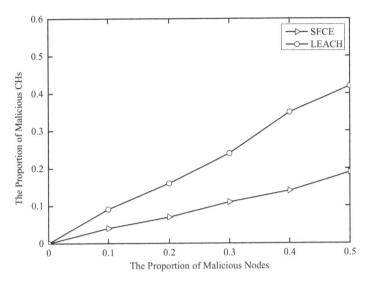

Figure 4. The proportion of malicious CHs with the increase of malicious nodes.

4.3 *Analysis of CH security*

In this subsection, we investigate the proportion of malicious CHs generated by SFCE and LEACH. In Figure 4 that no matter how many malicious nodes there are, the proportion of malicious CHs generated by SFCE is always lower than that which is generated by LEACH. This is because in LEACH, nodes elect CHs without any security mechanism, while in SFCE, only trustworthy nodes can be chosen as CHs. Trust value is set as the decisive premise when we build the fuzzy if-then rules. If the trust value is low, regardless of how much the energy and density is, the priority is low. Only when the trust value increases gradually, the priority increases with the rise of energy and density. In reality, malicious nodes may provide false residual energy and density information to improve the priority. By adopting such if-then rules, our proposal can effectively resist the deceptive behaviors.

4.4 *Analysis of energy efficiency*

In this subsection, we analyze the energy efficiency of SFCE that adopts the energy consumption model proposed by Heinzelman (Heinzelman et al. 2000). Here, the original energy of each node is 1 J and the length of data packets is 3kbit. The other parameters are: $E_{ele} = 50nJ/bit$, $\varepsilon_{amp} = 100nJ/bit/m^2$. At that time they used the first node die (FND) to represent the network lifetime. Figure 5 gives the results of 50 rounds simulations. As can be seen, SFCE prolongs the network lifetime. This can be explained by the fact that SFCE considers residual energy and density of nodes so that nodes with more energy and higher density are more likely to be elected as CHs. It's intuitive that CHs with more energy are able to support a large number of nodes within the cluster so that CHs will not die too early. Also, CHs with higher density reduce the energy consumption of intra-cluster communication.

5 CONCLUSION

In order to achieve network security, we proposed a secure fuzzy-based cluster head election algorithm (SFCE). In our proposal, a trust management based on beta distribution is utilized to implement trust computation and identify malicious nodes. Considering nodes' trust values, residual energy and density, we use fuzzy logic to select nodes with the highest priority as

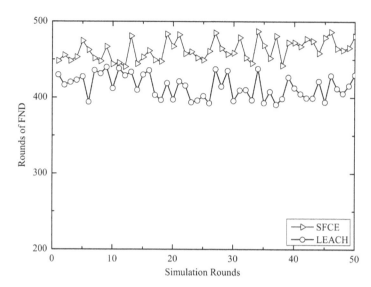

Figure 5. Simulation result of energy consumption.

CHs. Experimental results demonstrate that SFCE is capable of preventing malicious nodes from being elected as CHs and optimizing CHs' distribution. Moreover, this method is also energy efficient.

REFERENCES

Bajaber, F. & Awan, I. 2011. Adaptive decentralized re-clustering protocol for wireless sensor networks. *Journal of Computer and System Sciences.* 77(2): 282–292.

Chatterjee, P, Ghosh, U, Sengupta, I. & Ghosh, S.K. 2014. A trust enhanced secure clustering framework for wireless ad hoc networks. *Wireless Network* 20(7): 1669–1684.

Crosby, G.V, Pissinou, N. & Gadze, J. 2006. A framework for trust based cluster head election in wireless sensor networks. *DSSNS 2006: 2nd IEEE Workshop on Dependability and Security in Sensor Networks and Systems. 24–28 Aprial 2006.* Columbia: Institute of Electrical and Electronics Engineers Computer Society: 13–22.

Feng, R.J, Che, S.Y. & Wang, X. 2012. A credible cluster-head election algorithm based on fuzzy logic in wireless sensor networks. *Journal of Computational Information Systems* 8(15): 6241–6248.

Ganeriwal, S, Balzano, L.K. & Srivastava, M.B. 2008. Reputation-based framework for high integrity sensor networks. *ACM Transactions on Sensor Network* 4(3): article 15.

Heinzelman, W.R, Chandrakasan, A. & Balakrishnan, H. 2000. Energy-efficient communication protocol for wireless sensor networks. *The 33rd Hawaii International Conference on System Sciences. 4–7 January 2000,* Maui: IEEE: 223–232.

Jøsang, A. & Ismail, R. 2002. The beta reputation system. *15th Bled Electronic Commerce Conference e-Reality: Constructing the e-Economy.* 17–19 June 2002, Bled: 41–54.

Khalid, O, Khan, S.U, Madani, S.A, Hayat, K, Khan, M.I, Min-Allah, N, Kolodziej, J, Wang, L.Z, Zeadally, S. & Chen, D. 2013. Comparative study of trust and reputation systems for wireless sensor networks. *Security and Communication Networks* 6(6):669–688.

Lopez, J, Roman, R, Agudo, I. & Gago, C.F. 2010. Trust management systems for wireless sensor networks: Best practices. *Computer Communications* 33(9): 1086–1093.

Masdari, M, Bazarchi, S.M & Bidaki, M. 2013. Analysis of secure LEACH-Based clustering protocols in wireless sensor networks. *Journal of Network and Computer Applications* 36(4): 1243–1260.

Singh, A.K, Purohit, N. & Varma, S. 2014. Fuzzy logic based clustering in wireless sensor networks: A survey. *International Journal of Electronics* 100(1): 126–141.

Song, M., & Zhao, C.L. 2011. Unequal clustering algorithm for WSN based on fuzzy logic and improved ACO. *The Journal of China Universities of Posts and Telecommunications* 18(6): 89–97.

Signal and Information Processing, Networking and Computers – Chen & Huang (Eds)
© 2016 Taylor & Francis Group, London, ISBN 978-1-138-02881-4

Using game theory to optimize GOP level rate control in HEVC

Jiahui Zhu, Songlin Sun & Yaoyao Guo
Key Laboratory of Trustworthy Distributed Computing and Service (BUPT), Ministry of Education,
Beijing University of Posts and Telecommunications, Beijing, China

ABSTRACT: HEVC is the latest video coding standard and has a great efficiency improvement over the preceding standards. This paper proposes an algorithm to optimize GoP level rate control in HEVC, which is based on game theory. We model the bit allocation problem as a non-cooperative bargaining game. The NBS of this model is the best strategy that can meet the highest quality of the sequence. Compared with the standard rate control algorithm, the simulation results show that the algorithm could lower the bit rate error and get better performance of the visual quality, though the coding efficiency is at a little loss.

Keywords: game theory; rate control; HEVC

1 INTRODUCTION

High Efficiency Video Coding (HEVC) (Gary. J. Sullivan et al, 2013) is the latest video compression standard, which succeeds the previous standard called H.264/MPEG-4 AVC (Advanced Video Coding) (Wiegand. T. et al, 2003). Compared with H.264, HEVC can reduce the bit rates consumption with 50%, without lowering the image quality.

As a crucial module in HEVC, rate control helps to adjust the coding parameters so that the quality of the video can adapt to the current bandwidth, and thus utilize the bandwidth more efficiently. Generally, rate control algorithm can be divided into two parts. The first part is bit allocation, and the second part is adjusting encoding parameters to achieve the allocated bits. In the first part, appropriate bits are allocated to each level, which can be categorized into Group Of Pictures (GOP) level, picture level and Coding Unit (CU) level. Once the bits are allocated appropriately, the next step is to adjust the coding parameters so that the actual amount of bits consumed is close to the pre-allocated target bits.

In the previous research process of video coding, several rate control models have been investigated in different video coding standards. Since H.264, the rate quantization (R-Q) model has been proposed. T. Chiang & Y. Zhang (2002) investigated a quadratic R-Q model, which assumed that the predicted residues were related to the Laplacian Distribution and used the Mean of Absolute Difference (MAD) to estimate the complexity of basic coding units. After that, Choi et al. (2012) proposed the quadratic pixel based Unified Rate Quantization (URQ) model. Since the Lagrange multiplier, which is used in the Rate-Distortion Optimization (RDO) process, is more important in determining the bit rate, Li Bin et al. (2012) proposed a new rate control algorithm for HEVC via R-λ model to control the bit rate more accurately. In order to control the bit rate allocation in intra-coded pictures better, Karczewicz. M. & Wang. X. (2013) further introduced a SATD (Sum of Absolute Transformed Differences) based bit allocation method.

In R-λ model, a sliding window is designed to allocate bits according to the characteristics of the current sequence (Li Bin et al. 2012). The target bits for each GoP is allocated by:

$$T_{AvgPic} = \frac{R_{PicAvg} \cdot (N_{coded} + SW) - R_{coded}}{SW} \tag{1}$$

$$T_{GOP} = T_{AvgPic} \cdot N_{GOP}$$

(2)

where SW is the size of a smooth window, which is used to make the bitrate change smoother. Generally, this model allows the former GoPs to cost more bits to guarantee the visual quality of preference frames. However, under some circumstances, especially in the screen content sequence, it does not work very well. Different from the traditional videos captured on camera, screen content sequence, which consists of both natural content and computer generated content, has many unique characteristics. The key preference frames of the screen content sequence may occur in the middle or latter GoPs. Figure 1 shows the comparison of absolute differences within the camera captured sequence and the screen content sequence. It is easy to see that the screen content sequences, like SlideEditing, have much more abrupt pictures than BasketDrill, which is captured on camera and almost has no screen content. For the video sequence, which contains many abrupt frames if too many bitrates are consumed in the former GoPs, the visual quality of the frames, after abrupt pictures, will descend quickly. In this case, we try to find a more proper way of bitrates allocation.

This paper proposes an algorithm based on the multi-player game theory to optimize GoP level bitrates allocation. First, the bit allocation among GoPs is modeled as a bargaining game. All the GoPs are regarded as players working under an overall constraint, aiming to share the bandwidth resources effectively and fairly. Second, the linear distortion model

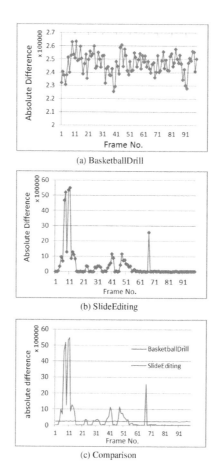

(a) BasketballDrill

(b) SlideEditing

(c) Comparison

Figure 1. The comparison of absolute difference within camera captured sequence and the screen content sequence.

124

and dependency relationship among GoPs are employed to define the utility function in the bargaining game. Finally, the Nash bargaining solution of this model is used to allocate the bandwidth for each GoP.

Next is the structure of this paper. Section 2 is a short introduction to game theory, including a brief review of bargaining game and Nash Bargaining Solution (NBS) (Nash John, 1953). In Section 3, the bargaining model of bitrate allocation among GoPs is proposed. In Section 4, experimental results are provided to demonstrate the efficiency and fairness of the proposed RC schemes. Finally, the conclusion is presented in Section 5.

2 BRIEF INTRODUCTION TO GAME THEORY

Game theory, which mainly studies strategic decision making, arises in almost every facet of human interaction (and inhuman interaction as well). Modern game theory started with Von Neumann, who proved the existence of mixed-strategy equilibria in two-person zero-sum game first. The theorem used to prove this problem by Von Neumann is called Brouwer fixed-point theorem, which became a standard method in game theory and mathematical economics.

According to the difference between the formalization of interdependence among players, game theory can be divided into two branches, called the cooperative and non-cooperative branches. If the players in a game are willing to form binding commitments to each other, and try to maximize the overall gain instead of the individual gain, then this game is cooperative. By contrast, in the non-cooperative game, the participators have conflicting interests. Maximizing individual utility is the target of players in a non-cooperative game, which means the players must make decisions independently.

Nash equilibrium is a basic concept in non-cooperative game theory, and widely used to predict the outcome of a strategic interaction. Nash equilibrium is a strategy meaning that, for every player in the game, there is no better strategy than playing according to this strategy. In other words, if each player has one strictly dominant strategy, then the strategy profile where all play their dominant strategies is the only Nash Equilibrium. In addition, the solution of the bargaining game is called the Nash bargaining solution, for it maximizes the product of each player's utility on the bargaining set.

As the bandwidth is limited in video coding, and each GoP competes for a share of resources to maximize its own utility, the non-cooperative Nash game theory is deployed in this paper.

3 PROPOSED METHOD

The problem of identifying the optimal quantization scale is equivalent to finding an optimal allocation of the GoP target bits to maximize the perceptual quality, which is a resource optimization problem. In the proposed algorithm, we suppose the aim of the bargaining game is allocating the constrained bandwidth to different rational players. We can configure that the players are the GoPs within the video sequence, and the strategy that decides the utility of each player is denoted by R_i, meaning the bits it requested. Since the target number of bits for a frame is constrained, the sum of the bits requested by the remaining GoPs should be no more than the remaining bits for the sequence, i.e.

$$\sum_{i=1}^{N} R_i \leq R_c \tag{3}$$

where R_c is the total of bitrates of the sequence.

Visual quality is an important measurement for the performance of video coding, and we use it as the utility of each GoP in this bargaining game. The definition of the utility function of the whole video sequence is $u = (u_1, u_2, \ldots, u_N)$. Given a combination of strategies carried out by all the GoPs $R = (R_1, R_2, ..., R_N)$. Since the visual quality is related only to the bitrates it consumes, the utility of the game could be presented by $u = (u_1(R_1), u_2(R_2), ..., u_N(R_N))$.

Suppose D_{ref} is the sum of Mean Squared Error (MSE) in a GoP, and is the corresponding distortion, the approximately relation is

$$\Delta D = D_{ref} \cdot \phi \tag{4}$$

The factor ϕ is related to the video sequence.

Define $u_i = \frac{1}{D_i}$, where D_i means the MSE of the ith GoP. Consider u_i^0 as the initial utility of the ith GoP, which is under the distortion D_i. Denote the initial quality of the game $u^0 = (u_1^0, u_2^0, ..., u_N^0)$, and the number of bits achieving u_i^0 as R_i^0. Since $u > u_0$, we have $R > R^0$, where $R^0 = (R_1^0, R_2^0, ..., R_N^0)$, which means

$$R_i > R_i^0 \tag{5}$$

Define U the set of achievable utilities, then $\langle U, u^0 \rangle$ represents the bargaining game. In order to solve the problem, we set strategy $R^* = (R_1^*, R_2^*, ..., R_N^*)$ the NBS of this game, and $u^*(\langle U, u^0 \rangle)$ is the Nash bargaining point.

As we know, the number of bits in a frame is mainly influenced by its own QP. In GoP level, we take the average QP as the parameter that influences its bit consumption. Besides, because of the motion-compensated prediction mechanism, the visual quality for each frame located in a GoP depends on the quality change of its reference frame located in former GoPs. The MSE also plays an important role in bit allocation. Therefore, the R-D relationship could be represented by:

$$R_i(\bar{Q}_{step}^i) = \frac{\xi_i \cdot d_i}{\bar{Q}_{step}^i} \tag{6}$$

$$D_i(\bar{Q}_{step}^i) = \eta_i \cdot \bar{Q}_{step}^i \tag{7}$$

where ξ_i is related to index, η_i is the system parameter and d_i is the sum of MSE for the ith GoP. Therefore, we define the utility function as:

$$u_i = \frac{1}{D_i} = \frac{R_i}{\xi_i \cdot d_i \cdot \eta_i} \tag{8}$$

Since the Nash bargaining solution R^* satisfied $\prod_{i=1}^{N} (u_i(R_i) - u_i^0) \geq \prod_{i=1}^{N} (u_i(R_i^*) - u_i^0)$, u_i is concave and injective, then R^* is also satisfied with

$$\prod_{i=1}^{N} \ln(u_i(R_i^*) - u_i^0) \geq \prod_{i=1}^{N} \ln(u_i(R_i) - u_i^0)$$

$$s.t. R_i > R_i^0, \sum_{i=1}^{N} R_i \leq R_c \tag{9}$$

The above inequality constrained optimization problem can be solved by maximizing the following Lagrangean, using the theorem of Kuhn and Tucker:

$$J = \sum_{i=1}^{N} \ln(u_i - u_i^0) + \lambda \left(R_c - \sum_{i=1}^{N} R_i \right) + \sum_{i=1}^{N} \theta_i(R_i - R_i^0) \tag{10}$$

where λ and θ_i are the Lagrange multiplier. The optimized solution can be obtained by solving

126

$$\begin{cases} \dfrac{\partial J}{\partial R_i} = \dfrac{\partial \ln(u_i - d_i)}{\partial R_i} - \lambda + \theta_i = 0 \\[3mm] \dfrac{\partial J}{\partial \lambda} = R_c - \displaystyle\sum_{i=1}^{N} R_i \geq 0 \\[3mm] \lambda \dfrac{\partial J}{\partial \theta_i} = \lambda\left(R_c - \displaystyle\sum_{i=1}^{N} R_i\right) = 0 \\[3mm] \dfrac{\partial J}{\partial \theta_i} = R_i - R_i^0 \geq 0 \\[3mm] \theta_i \dfrac{\partial J}{\partial \theta_i} = \theta_i\left(R_i - R_i^0\right) = 0 \end{cases} \tag{11}$$

where $i \in (1,2,...,N)$.

Therefore, the NBS for the game is given by:

$$R_i = \varphi_i \cdot \frac{R_c - \displaystyle\sum_{i=1}^{N} u_i^0 \xi_i d_i \eta_i}{\displaystyle\sum_{i=1}^{N} \varphi_i} + u_i^0 \xi_i d_i \eta_i \tag{12}$$

where $\varphi_i = \sum_{k=i}^{N} \frac{D_i - D_i^2 - u_i^0}{D_k - D_k^2 - u_k^0} \cdot \phi^{k-i}$, k and i are the index of GoPs.

From Eq.(12), the bit allocation scheme could be divided into two parts. One part is $u_i^0 \xi_i d_i \eta_i$, which are the minimum bits to guarantee the minimum utility (visual quality) of the GoP. Another part is influenced by the importance of the corresponding frame, which can be calculated by $\varphi_i / \sum_{i=1}^{N} \varphi_i$.

4 EXPERIMENTAL RESULTS

To verify the algorithm proposed in this paper, HM-12.1+RExt-5.0(HEVC model) is used as simulation platform. The bit rate of HM-12.1+RExt-5.0 without rate control is used as the target bit rate of the proposed methods. HM-12.1+RExt-5.0 is also used as the anchor of all the comparisons. Class F is used for testing. The common test conditions (Rosewarne. C., 2014) are followed, and the proposed algorithm is compared with the HM anchor and the R- model, which is the currently recommended rate control algorithm in HM. The rate estimation accuracy is measured by:

$$M\% = \frac{R_{target} - R_{actual}}{R_{target}} * 100\% \tag{13}$$

where R_{actual} and R_{target} denote the actual and target bit rates when encoding a video sequence, respectively.

YUV color space is widely used in video coding, especially YUV 4:2:0. In this color space, luma component Y has more samples than chroma samples U and V. In other words, the brightness carries more information than the color components. Figure 2 shows the different pictures generated by only Y components, U components, or V components. It is obvious

| (a) original picture | (b) Y component | (c) U component | (d) V component |

Figure 2.   YUV components comparison.

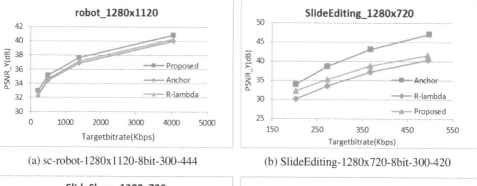

(a) sc-robot-1280x1120-8bit-300-444

(b) SlideEditing-1280x720-8bit-300-420

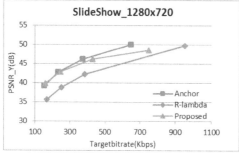

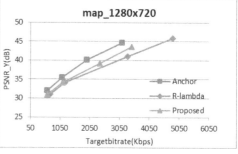

(c) sc-SlideShow-1280x720-8bit-500-444

(d) map-1280x720-8bit-300-444

Figure 3.   RD curves of the performance comparison under hierarchical B configuration.

Table 1.   Performance comparison for the proposed algorithm with R-λ model under IBBB configuration.

Sequence	Bitrate Error		BD-Rate Y
	R-lambda	Proposed	
BasketballDrill	0.2416%	0.1322%	–0.4258%
ChinaSpeed	0.0213%	–0.0043%	0.5209%
SlideEditing	4.1785%	–0.2813%	–4.6130%
SlideShow	7.9810%	1.0808%	–6.6629%
Map	9.4180%	2.4076%	–2.9150%
Programming	6.4869%	1.3739%	–0.7847%
Robot	0.1593%	0.0536%	–0.7638%
SlideShow 444	1.8912%	–1.3433%	–3.9627%
WebBrowsing	0.0310%	–0.0132%	–4.6500%
WordEditing	7.4463%	1.1517%	–0.6165%
All other	3.7855%	0.4558%	–2.4875%

that Figure 2(b), which only contains luma components, resembles the original picture most. In this case, we take PSNR-Y as the measurement of video quality.

The experimental results regarding RD performance are measured in the form of BD-Rate. Figure 3 shows the RD curves of the performance comparisons of Hierarchical B. Table 1 illustrates the experimental results of CLASS F and screen content sequences under Hierarchical B configurations, which are obtained by using the current rate control algorithms in HM and in this paper, respectively. It can be seen from the tables that the average BD-Rate is reduced by 2.49% under Hierarchical B configuration, by utilizing the proposed mechanism. In addition, the bitrate error is reduced with 3.3% with regard to the R-lambda model.

5 CONCLUSION

In this paper, we investigated a bargaining game-based GOP level bit allocation in HEVC. To meet the requirements efficiently and fairly under the constraint of bandwidth, we modeled the bit allocation problem as a bargaining game. The linear R-D model is employed to define the utility function of multiple GOPs. After that, we built the dependency relationship between distortion and the features of GOPs. Finally, the bitrate allocation between GOPs was proposed as the Nash equilibrium of the bargaining game. The experimental results showed that, compared with standard RC algorithm, the proposed algorithm can deliver higher performance on bit rate accuracy.

REFERENCES

Chiang, T. & Y. Zhang. 2002. A new rate control scheme using quadratic rate distortion model. *IEEE Transactions on Circuits and Systems for Video Technology* 7:287–311. Feb. 1997.

Choi, Hyomin et al. 2012. Rate control based on unified RQ model for HEVC. *ITU-T SG16 Contribution, JCTVC-H0213, in San Jos:*1–13. April. 2012.

Gary. J. Sullivan et al. 2013. Overview of the High Efficiency Video Coding (HEVC) Standard. *IEEE Transactions on Circuits and Systems for Video Technology* 22:1649–1668. Dec. 2012.

High Efficiency Video Coding (HEVC) Model [Online]. Available: https://hevc.hhi.fraunhofer.de/trac/hevc/ browser/tags.

Karczewicz. M. & Wang. X. 2013. Intra Frame Rate Control Based on SATD. *Jt. Collab. Team Video Coding (JCT-VC) of ITU-T SG* 16:18–26. Apr. 2013.

Li Bin et al. 2012. Rate control by R-lambda model for HEVC.*Jt. Collab. Team Video Coding (JCT-VC) of ITU-T SG, in Shanghai, China* 16: 10–19. Oct. 2012.

Nash, John. 1953. Two-person cooperative games. *Econometrica: Journal of the Econometric Society*: 128–140. Jan. 1953.

Rosewarne. C. 2014. Common Test Conditions and Software Reference Configurations. *JCTVCL10 06, 12th JCT-VC Meeting, in U.S.* Jan. 2014.

Wiegand. T. et al. 2003. Overview of the H.264/AVC video coding standard. *IEEE Transactions on Circuits and Systems for Video Technology* 13:560–576. July. 2003.

Signal and Information Processing, Networking and Computers – Chen & Huang (Eds)
© 2016 Taylor & Francis Group, London, ISBN 978-1-138-02881-4

Reinforcement learning based cooperative sensing policy in Cognitive Radio network

Xingrui Ye, Xiao Jun Jing & Songlin Sun
School of Information and Communication Engineering, Beijing University of Posts and Telecommunications, Beijing, China

Yan Li
International School, Beijing University of Posts and Telecommunications, Beijing, China

Xiaohan Wang & Dongmei Cheng
305 Hospital of PLA, Beijing, China

ABSTRACT: In this paper a cooperative spectrum sensing policy for cognitive radio based on reinforcement learning is presented. The goal of the proposed strategy is to allow the secondary users to find the idle spectrum, and meanwhile the secondary users will not affect the primary users. Since the radio spectrum varies temporally and spatially, combining the policy with reinforcement learning is a suitable solution to this situation. The proposed ε-greedy method is a tradeoff between computational complexity and sensing performance. Simulation results have shown the miss detection probability decreased and the overall secondary network throughput improved a lot at the same time because of the proposed policy.

Keywords: cognitive radio; sensing policy; cooperative sensing; reinforcement learning

1 INTRODUCTION

The usable radio spectrum has become a scarce and expensive resource since the demand for wireless services is increasing. In D. Cabric, S.M. Mishra & R.W. Brodersen (2004), the measurement results can be seen that most of the spectrum has not been fully utilized, which means that the licensed spectrum is sporadically used. We can learn from X. Wang, H. Li & H. Lin (2011) that only 15% of the licensed spectrum has been used at any location and at any time, which is very low. Cognitive Radio (CR) is a technology that enables Second User (SU) to dynamically utilize unused licensed spectrum without interference with the primary or licensed user. Cognitive radio is an intelligent system which can dynamically perceive the environment around. Cognitive radio management techniques include spectrum sensing, spectrum analysis and spectrum decision. Spectrum sensing is a very important step, moreover the spectrum sensing provides the basis for accurate dynamic spectrum access (S. Haykin 2005).

Radio spectrum resources are divided into frequency bands. Only the operators have the license of using the spectrum. The policy exists to reduce the reflection between the users when they are using the same frequency band. Now, the spectrum is divided into two major categories: the licensed spectrum and unlicensed spectrum. For the unlicensed frequency band, users can freely access to the spectrum to transmit data. Of course the users may produce a certain interference. Especially when the number of users accessing to the same band is large, the interference is very serious. In the licensed spectrum, the user's use of the spectrum is authorized by the government. The user will not interfere with each other when transmitting data, because the user's frequency band and the usage are fixed. This fixed spectrum allocation is very effective in the past, but now the number of mobile devices is growing fiercely. It means that there are too many devices accessing to the unlicensed spectrum, which will lead to severe interference. New spectrum accessing policy need to be developed. There are some difficulties in the reassignment of licensed spectrum. In order to avoid the disadvantages of the fixed access policy, the emergence of cognitive radio allows users to access the spectrum dynamically. Of course, these frequency bands they accessed are not occupied by the licensed primary user in the moment, and when the primary user need to access the spectrum, secondary users need exit the spectrum timely to avoid the interference. CR can intelligently perceive the surrounding environment, and the system can adapt to it when the environment changes. There are two main objectives of CR: one is to ensure the secondary user can get effective and reliable transmission anytime and anywhere, and the other one is to make fully use of the spectrum resource.

A CR network is considered to include N_s secondary users which are spatially distributed wireless terminals. The wide spectrum have been divided into N_B subbands. In the proposed strategy, the way of cooperative spectrum sensing is considered. Through the introduction of cooperative spectrum sensing, the impact of the fading has been reduced (S. Haykin 2005, J. Lundén et al. 2009, S. Chaudhari et al. 2009). Cooperative spectrum sensing can bring good performance, and it also needs to be optimized to achieve the purpose of energy saving, which is a very important factor (H. Huang et al. 2009). Cooperative spectrum sensing is usually divided into three categories: centralized, distributed and relay assisted cooperative sensing. In this paper, centralized cooperative sensing policy has been used. In centralized cooperative sensing, every single SU senses the spectrum independently first, then the SUs send the result to the Fusion Center (FC). The FC will broadcast the final information of sensing result to each SU. There is no fusion center in the distributed cooperative sensing. In this policy, SUs make their own local decision firstly, then they will share the decision with each other. It should be known that usually the final decision is made after several iterations, which is the main disadvantage of this policy. The relay-assisted cooperative sensing is suitable for the situation where the sensing channels or the reporting channels are weak. According to the cooperative way of sensing, the accuracy of sensing has been improved. The settings of cognitive radio in this paper has been shown in Figure 1.

Reinforcement learning is an important kind of machine learning method which has many applications in the field of intelligent control robot, analysis and prediction. Reinforcement learning is the learning mapping from the environment to the behavior, in order to make the reward signal function value biggest. Reinforcement Learning System (RLS) must learn from their own experience. Through this way, RLS obtains knowledge in action—evaluation of the environment and improves the action plan to adapt to the environment.

This paper will focus on the spectrum sensing policy, which is one of the important function performed by the fusion center. The procedure of spectrum sensing strategy can be divided into two parts. It needs to tell SUs which time and which frequency band to sense, in addition, which SU will sense the frequency band need to be decided considering of the accuracy and energy saving. There are generally two indicators to evaluate if policy is good or not. The first one is to ensure that the sensing error probability is low, and the wrong decision will have a big bad impact on the whole network. In addition, the throughput needs be improved, which means that the CR system can provide a stable data transmission to more secondary users, which will improve the utilization of spectrum.

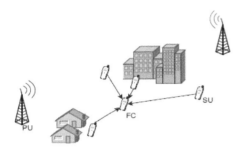

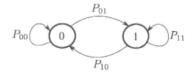

Figure 1. A cognitive radio setting.   Figure 2.   The Gilbert–Elliot channel model.

2   SYSTEM MODEL

Since the interested frequency band is considered to be a wide band, it has been divided into N_B subbands in this paper. As mentioned in the introduction, cooperative spectrum sensing policy is used. It is assumed that there are N_S SU terminals in the secondary network. The SU can not only sense one subband at a time, it can sense K_S subbands at most. Moreover the bandwidth of different subbands can be different.

Centralized cooperative sensing policy is used in this paper. There is a Fusion Center (FC) in the network which will control the procedure of the cooperative sensing. The SU perform its local sensing at first, and then it sends the result to the FC. After collecting the information from different SUs, the FC identifies the idle spectrum hole and broadcasts the information to all the SUs.

A two-state Markov chain is proposed to model the activity of the primary user, which is shown in Figure 2. In the model states 0 and 1 represent idle and busy state respectively. The idle state means that the subband is not used by the primary user so it can be used by the secondary user. The busy state means the opposite.

3   REINFORCEMENT LEARNING BASED SENSING POLICY

The sensing policy in this paper is based on the ε-greedy method, so a brief introduction of this method is given in the first sub part. Then the sensing policy is introduced in the sub part 3.2.

3.1   The ε-greedy method

The ε-greedy method is used for sensing policy in this paper. The method is proposed considering of the balance between exploration and exploitation. Because the use of the spectrum may be different in different places at different time, e.g. the spectrum may be not that busy in the suburbs at night as that in the densely populated areas in the daytime. The sensing performance of exploration may be not that good as exploitation is, but it has the advantage of simple and energy conservation.

$Q_t(a)$ represents the estimated value of action a at time t. The selected action is represented by a_t^* at time t. In the ε-greedy policy, the action with the highest value will be selected considering the tradeoff between exploration and exploitation, i.e. $a_t^* = \arg\max_a Q_t(a)$, with the probability of $1-\varepsilon$.

The Q-value will be updated after the action a is taken, and the $r(a)$ is the reward of the action a (R.S. Sutton, A.G. Barto 1998). The updating method is seen as

$$Q_{t+1}(a) = Q_t(a) + \alpha_t \left[r_{t+1}(a) - Q_t(a) \right] \tag{1}$$

133

where $r_{t+1}(a)$ represents the reward at time $t+1$ for taking action a and $\alpha_t \ (0 < \alpha_t < 1)$ is a step size parameter.

3.2 The details of the policy in this paper

In this sub part, the details about the sensing policy is going to be talked about. The advantages of the cooperative sensing has been discussed before. Fusion center manages the sensing policy in centralized cooperative sensing. Usually, the sensing strategy is managed by the fusion center. The proposed sensing policy has two purposes, one is to determine the subbands to be sensed, and the other to determine which SU will sense the band. Based on this we need two kinds of Q-values: the one of each subband, and the one of SUs. The first Q-value is used to pick out the best subband which can accommodate the largest number of secondary users at the same time. The best subband can provide high data transmission rate to the terminals. The second Q-value is used to select secondary user. The SU which has high Q-value is easier to be chosen to sense the frequency band, because its estimation of the band is more accurate than the others.

When calculating the Q-values of the subbands, we usually need to define a reward. According to the actual meaning, the reward $r_{t+1}(m)$ is defined as the following formula:

$$r_{t+1}(m) = \begin{cases} R_{t+1}(m), & \text{if } m \text{ is accessed and free,} \\ 0, & \text{if } m \text{ is occupied,} \end{cases} \tag{2}$$

where $R_{t+1}(m)$ is the throughput, and it is obtained by SU that accessed to the subband m instantaneously. So this is also called feedback. According to (1), the FC can update the Q-values of each subband using this reward. Then the SU's Q-value is updated according to

$$Q_{t+1}(a) = Q_t(a) + \alpha_k r'_{t+1}(a) \tag{3}$$

As to the reward of SUs, it can be obtained by comparing its local decision with the FC decision, the reward is formulated as:

$$r'_{t+1}(s, m) = \begin{cases} d_{t+1}(s, m), & d_{t+1}(FC, m) = 1, \\ 1 - d_{t+1}(s, m), & d_{t+1}(FC, m) = 0, \end{cases} \tag{4}$$

This formula is obtained according to its practical implication. We will give a little more explanation on it. In the formula, $d_{t+1}(s, m)$ is the decision made by the secondary user, and the $d_{t+1}(FC, m)$ represents the decision made by the fusion center. When the decision made by the FC equals 1, if the decision made by SU is 1 too, that means the SU makes a correct decision. Then the SU is given a reward values 1. On the opposite, if the decision made by SU is o, that means the SU makes a wrong decision. Then the SU is given a reward values 0. The situation when the decision made by the FC equals 0 can be dealt with in the same way.

After all the Q-value updates, based on the ε-greedy method, the policy of sensing include 2 kinds of solution. One is in a very gentle way, the other is in a rude way. The gentle way consists of two stages, it may seem a little complicated, but this way is very smart and accurate. The rude way of sensing assignment is very simple, but its accuracy of sensing is lower than the gentle way. The combination of this two way in the proposed policy is a tradeoff between computational complexity and accuracy. The rude way is used with possibility ε, which means the sensing assignment is pseudo-random. With possibility $1 - \varepsilon$ the sensing assignment made by the fusion center is in the gentle way. At the first stage, the subband with the highest Q-value is selected to sense. Than the secondary users with high Q-values will be chosen to sense the frequency band. We should notice that the number of the chosen SUs is not fixed. At last, the FC broadcast the information to all the SUs. The process will be shown in Figure 3.

In J. Oksanen et al. (2010), more details about the sensing policy which is based on the pseudorandom frequency hopping can be seen.

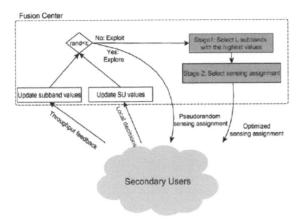

Figure 3.   Flow diagram.

In the first stage of exploitation, the subbands with the highest Q-values which means providing the largest throughput of the secondary network have been chosen to be sensed. Then the selection of SUs to Sense is a sensing Assignment Problem (SAP).The SAP is formulated as

$$\min_{X} \sum_{m \in M} \sum_{s \in S} x_{sm}$$

$$s.t. \ \hat{P}_{miss,FC}^{m}(X) \le P_{miss,target}^{m} \quad \sum_{m \in M} x_{sm} \le K_s \ x_{sm} \in \{0,1\}$$

(5)

X is a sensing assignment matrix which is binary. x_{sm} equals 1 when SU s is assigned to sense subband m and x_{sm} equals 0 when SU s is not assigned.

The OR-rule is used in the sensing assignment, which means that only when all the SUs report the subband is idle, the subband is considered to be idle by the FC. Then the SAP becomes a BIP problem. An integer programming problem is a mathematical optimization or feasibility program in which some or all of the variables are restricted to be integers. In BIP, the integer is binary. The BIP problem is usually solved by the Branch-and-Bound (BB) algorithms. An iterative Hungarian algorithm has also been studied to be suitable for solving BIP. BB algorithm has been used in this paper.

4   SIMULATION EXAMPLES

A great deal of simulation has been done to prove the excellent performance of the sensing policy. The results are obtained in a stationary scenario. There are three figures in this section showing the simulation results. We will demonstrate the excellent performance of the proposed strategy from three aspects. First, we consider the throughput of the whole system. High throughput means the spectrum resource has been fully used, which is one of the main objectives of CR. Then the convergence of the proposed method is verified. If it can converge to a lower value, it means that policy is reliable and can be used to detect the spectrum. Finally, the number of SUs required to be considered. As the number of SUs reduces, the energy efficiency of secondary users will be improved.

The curves are obtained with BB algorithm in Figure 4. The percentages are gotten by comparing with an ideal policy. The curve with $\varepsilon = 1$ means exploration only case. For $\varepsilon < 1$ cases the performances are better. There are four curves in the figure with different s. When the value of s goes down, the curve is on the upside, which means the throughput is higher. Particular for $\varepsilon = 0.1$, the throughout is high up to 87% of the ideal policy.

135

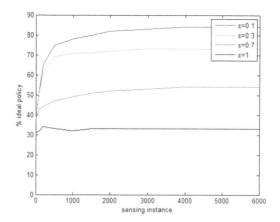

Figure 4.   SU network's throughput.

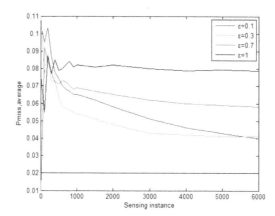

Figure 5.   Convergence of the policy.

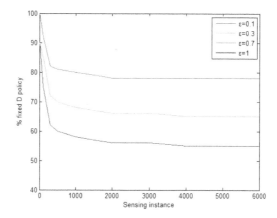

Figure 6.   Number of sensing SUs.

Figure 5 shows the convergence curves. For $\varepsilon < 1$ the missed detection probability finally converges below $P_{miss,target} = 0.1$ as it can be seen. Smaller s results in lower miss detection.

The curves in Figure 6 are obtained comparing with another sensing policy. The number of the sensing secondary user has been fixed to be two in that policy. But the proposed policy in this paper has not been fixed. It can be seen that the number of secondary user is smaller, because the fusion center only selects the secondary users with high Q-value to sense. The number of sensing SUs is variable each time, rather than fixed. This improves the efficiency of the SU network. Particularly it can be seen that when $\varepsilon = 0.1$, the number drops to close to 50% of the fixed number case.

5 CONCLUSIONS

In this paper, we proposed a reinforcement learning based cooperative sensing policy. The policy shows the SUs which subbands to sense and which SUs should sense the subband considering the efficiency problem. The Sensing Assignment Problem (SAP) has been proved to be a BIP problem for the OR-rule. Exact BB method and approximative IH method both can be used to solve this problem. Simulations have shown that the throughput of the network has been improved, the number of the sensing SU decreases and the miss detection probability is low at the same time due to the proposed sensing policy.

REFERENCES

Cabric, D., S.M. Mishra & R.W. Brodersen. 2004. Proceedings of the Asilomar Conference on Signals, Systems and Computers. *Implementation issues in spectrum sensing for cognitive radios* vol.1: pp. 772–776.

Chaudhari, S., V. Koivunen & H.V. Poor. 2009. Autocorrelation-based decentralized sequential detection of OFDM signals in cognitive radios, *IEEE Transactions on Signal Processing* 57(7): 2690–2700.

Haykin, S. 2005. Cognitive radio: brain-empowered wireless communications. *IEEE Journal on Selected Areas in Communications* 23(2): 201–220.

Huang, H., Z. Zhang, P. Cheng & P. Qiu. 2009. Opportunistic spectrum access in cognitive radio system employing cooperative spectrum sensing. *Proceedings of IEEE Vehicular Technology Conference*: pp. 1–5, 26–29.

Lundén, J., V. Koivunen, A. Huttunen & H.V. Poor. 2009. Collaborative cyclostationary spectrum sensing for cognitive radio systems. *IEEE Transactions on Signal Processing* 57(11): 4182–4195.

Oksanen, J., V. Koivunen, J. Lundén & A. Huttunen. 2010. Diversity-based spectrum sensing policy for detecting primary signal over multiple frequency bands. *Proceedings of the ICASSP Conference*, Dallas Texas: pp.3130–3133.

Sutton, R.S., A.G. Barto. 1998. *Reinforcement Learning: An Introduction*. Cambridge, MA: MIT Press.

Wang, X., H. Li & H. Lin. 2011. A new adaptive OFDM system with precoded cyclic prefix for dynamic cognitive radio communications. *IEEE Journal on Selected Areas in Communications* vol. 29 no. 2: pp. 431–442.

Signal and Information Processing, Networking and Computers – Chen & Huang (Eds)
© 2016 Taylor & Francis Group, London, ISBN 978-1-138-02881-4

A combined BD-SO precoding scheme for next generation HetNets

Di Lu, Jingyi Sun & Cheng Qian
International School, Beijing University of Posts and Telecommunications, Beijing, China

ABSTRACT: Heterogeneous Network (HetNet) is pervasively introduced in the future generation of mobile communication system in order to address seamless switch and better Quality of Service (QoS). The key technology of HetNet is Multiple Input Multiple Output (MIMO), which allows multiplexing of spatial resources by increasing the number of antennas. We are supposed to adopt appropriate precoding schemes using Channel State Information (CSI) to solve interference problems resulted from the multi-user objective of massive MIMO. For this purpose, we proposed an algorithm that combines Block Diagonalization (BD) and Successive Optimization (SO) algorithm. We first use the BD method within groups of users, and then adopt inter-group SO method iteratively to reach a balance between system overhead and Co-Channel Interference (CCI). Results can visibly confirm a better performance for this proposed combination technique.

Keywords: MIMO; HetNets; BD; SO; precoding matrix

1 INTRODUCTION

The MIMO (Multiple Input Multiple Output) techniques are based on Heterogeneous Network (HetNet), which consists of two or more different wireless communication systems adopting different access technologies. Network convergence in HetNet is aiming at selecting a suitable network and then providing better Quality of Service (QoS) for users according to user features, transmission requirements or network services. Currently, MIMO technique has become an international priority and research hotspot within the communication field due to its superiorities on transmission rate, usage of system capacity and spatial resources. By directly increasing the number of antennas at the transmitter and receiver, a higher performance can be achieved so that spatial multiplexing gain will be improved. Consequently, data streams with very high rates can be transmitted to multiple users.

Several precoding algorithms have been proposed in order to solve the most obvious encumbrance in MIMO elimination or reduction of Co-Channel Interference (CCI). They are all aimed at precoding at the transmitter using perfect Channel State Information (CSI). Dirty Paper Coding (DPC) is one of the non-linear precoding schemes, which can increase the sum capacity of the system. For instance, the Tomlinson-Harashima Precoding (THP) method is equivalent to a decision feedback equalizer at the receiving side employing QR decomposition to pre-cancel the inter-user interference. However, these kinds of non-linear precoding algorithms are quite impractical and very difficult to implement due to their complexities. Under this situation, another type of precoding schemes with better practicability, referred to as linear precoding, has been proposed. Typically, systems adopt the channel inversion method. One of the common design criteria for channel inversion method is Zero Forcing (ZF), which has a relative poor performance for Signal to Noise Ratio (SNR). To meet the MIMO condition that multiple users in the network have multiple antennas, the Block Diagonalization (BD) algorithm is a better solution. The BD principle is to eliminate the interference by placing all the intended users at null space of all the unintended user channels, so that the maximum information rate for each user can be achieved.

In this paper, we researched and adopted another precoding algorithm Successive Optimization (SO), to relatively reduce the system overhead for the BD method. SO better orders the precoding sequence and uses the previous i-1 users precoding vectors to optimize the design criterion when designing the i_{th} user precoding vectors. To reach the balance between the system overhead for BD to eliminate inter-user interference and a certain requirement for elimination of interference, we proposed a combination algorithm to merge these two methods. We divided all users in the massive MIMO system into three groups, which have 15–20 members each, according to descending values of SNR. For users within the same group, we use the BD method. Furthermore, the SO method is adopted in different groups. Goals of this proposed algorithm is to maximize the extent for elimination of interference and minimize system overhead at the same time. We can come to the conclusion from the simulation results that the proposed algorithm at the transmit side in the MIMO system outperforms the traditional block diagonalization algorithm to a certain degree.

This paper is organized in several components. 2 describes the massive MIMO system model. 3 shows the related work. 4 states and explains the proposed optimization algorithm for precoding. 5 shows the simulation results and comparisons. In section 6, we come up with our conclusions.

2 SYSTEM MODEL

In a MU-MIMO downlink channel, each system has a single base station and K users. M_t transmit antennas are equipped at the base station, and M_t receive antennas are equipped at the user, where i is the natural number from 1 to K. In each of the users, we can get the transmitted signal x which is:

$$x_i = v_i d_i \tag{1}$$

where d_i is the data symbol and v_i is precoding vector. Thus, the signal vector is transmitted from the base station. Additionally, we can get the total received signal:

$$y = Hvx + n \tag{2}$$

H is the combined channel matrix of all users, given by:

$$H = \left[H_1^T, H_2^T \cdots H_K^T \right]^T \tag{3}$$

where $H_i \in C^{M_{R_i} \times M_T}$. Taking co-channel interference and white Gaussian noise into account, the received signal for each of the user can be expressed as:

$$y_i = H_i v_i d_i + H_i \sum_{k \neq i}^{K} v_k d_k + n_i \tag{4}$$

where $H_i \sum_{k \neq i}^{K} v_k d_k$ represented co-channel interference between each of the other user and n_i is the Gaussian noise vector. After the transmission, there is a matched receive filter at each of the users. The channel matrix H_i is just clarified to the intended i_{th} user and not to the unintended users. Thus, we can get the matched receive filter as:

$$R_i = \frac{\left(H_i v_i \right)^H}{\left\| H_i v_i \right\|_F^2} \tag{5}$$

3 CONVENTIONAL PRECODING SCHEMES

In this component, we briefly reviewed and explained two existing fundamental precoding algorithms for mitigating MIMO channel interference, i.e. Successive Optimization (SO) and Block Diagonalization (BD).

3.1 Review of the existing algorithms

3.1.1 SO precoding

SO first sorts the precoding sequence and then uses the previous $i1$ users precoding vectors to optimize the design criterion for the i_{th} users precoding vectors. We assume a precoding vector v_i, and data symbol s_i. All of these are normalized. Sequentially, we assume there is a narrow band channel. So the received signal can be expressed as:

$$y = Hvx + n \tag{6}$$

The matched receive filter at receiver is:

$$R_i = \frac{(H_i v_i)^H}{\|H_i v_i\|_F^2} \tag{7}$$

The signal to interference noise ratio at the receiver is:

$$SINR_i = \frac{\|R_i H_i v_i\|_F^2}{\sum_{k \neq i}^K \|R_i H_i v_k\|_F^2 + \|R_i\|_F^2 \sigma_i^2} \tag{8}$$

Combined with the filter we obtain:

$$SINR_i = \frac{\|H_i v_i\|_F^2}{\frac{\sum_{k \neq i}^K \|v_i^H H_i^H H_i v_k\|_F^2}{\|H_i v_i\|_F^2} + S_i \sigma_i^2} \tag{9}$$

According to the SINR equation, interference caused by the users is:

$$\sigma_{in,i}^2 = \sum_{k=1}^{i-1} \|R_i H_i v_k\|_F^2 \tag{10}$$

Combined that with the receive filter, we obtain:

$$\sigma_{in,i}^2 = \frac{\frac{\sum_{k=1}^{i-1} \|v_i^H H_i^H H_i v_k\|_F^2}{\|H_i v_i\|_F^2}}{\|H_i v_i\|_F^2} \tag{11}$$

From the result we can conclude that this ratio represents the signal power of i_{th} user to partly CCI. SINR at receiver will increase if we decrease the ratio. When building the precoding vector, it can be rewritten as:

$$v_i^o = \max_{v_i \in C^{N_t \times L_i}} \frac{\|H_i v_i\|_F^2}{\|\tilde{H}_i v_i\|_F^2 + S_i \sigma_i^2 + \sum_{k=1}^{i-1} \|v_i^H H_i^H H_i v_k\|_F^2} \tag{12}$$

Therefore, the reasonable strategy is to maximize the signal power and minimize the interference, CCI and noise as much as possible. The CCI is supposed to not be omitted when we build the precoding vector. Obviously, we ought to sort the K users before i because if 1 to i1 i1 users has smaller interference, user i will definitely get a better performance. We can choose to minimize the average symbol error rate to achieve that:

$$P_{avg} = \frac{1}{K} \sum_{i=1}^{K} P_i \qquad (13)$$

We use P_i to represent the symbol error rate of user i. There will be a smaller error rate and P_{avg} related to larger SINR. As P_i affects the average error rate mostly when there is a poor quality of channel, we can obtain the maximum value of E_i the larger $SINR_i$. Users who have a bad quality of channel will be precoded later. By using this kind of iterative logic, the performance can be improved after accessing all the user vectors.

3.1.2 BD precoding

BD is aimed at eliminating the interference by placing all the intended users at null space of all the unintended user channels. As a consequence, it minimizes the total transmit power for a predefined QoS level or maximizes total system throughout under a transmit power constraint. It must be related that the number of transmit antennas F_T is larger than the total number of receive antennas F_R in the network. A precoding matrix can be expressed as:

$$H = \left[H_1 H_2 \cdots H_k \right] \in^{M_T \times r} \qquad (14)$$

where $H_i \in^{F_T \times r}$ is the i_{th} users precoding matrix. And $r \le F_R$ represents the total number of transmitted data stream sequences, $r_i \le F_{R_i}$ means the number of data stream sequences transmitted to the i_{th} user. Our goal is to find the precoding matrix J such that Ji lies in the null space of the other users channel matrices. Let's define $\tilde{H}_i$ as:

$$\tilde{H}_i = \left[H_1^T \cdots H_{i-1}^T H_{i+1}^T \cdots H_K^T \right]^T \qquad (15)$$

From the Singular Value Decomposition we get:

$$\tilde{H}_i = \tilde{Y}_i \tilde{\Sigma}_i \left[\tilde{V}_i^{(1)} \tilde{V}_i^{(0)} \right]^H \qquad (16)$$

for whose rank is $\tilde{L}_i$, we choose the last right $F_T - \tilde{L}_i$ singular vectors $\tilde{V}_i^{(0)} \in^{F_T \times F_T - \tilde{L}_i}$ which construct an orthogonal basis for $\tilde{H}_i$ null space. After eliminating, we can obtain an equivalent channel for user i is $H_i \tilde{V}_i^{(0)}$. Now, let us define the SVD:

$$H_i \tilde{V}_i^{(0)} = Y_i \Sigma_i \left[V_i^{(1)} V_i^{(0)} \right]^H \qquad (17)$$

We multiply the first L_i singular vectors $V_i^{(1)}$ and $V_i^{(0)}$ and use the result as an orthogonal basis of dimension L_i. Also, it's responsible for representing the transmission vectors that can be used to maximize the information rate for user i.

4 OUR PROPOSED SCHEME

We divide all users into three groups, users with low, middle, and high CSI. Like the Figure 1 shows below, the three inner circular region cover all the users in each group and the entire circular region cover all the three user groups. We adopt BD algorithm in each group primarily and then SO algorithm in different groups.

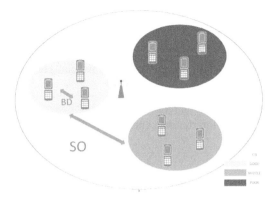

Figure 1. Our proposed scheme.

Suppose we have x users with CSI, y users with middle CSI, and z users with high CSI. First, the x users are the interference user; the other y + z user channel is projected on to its null space. Using Schmidt orthogonalization for each user channel matrix H_k,

$$\bar{H}_k = \left(I - \sum_{i=1}^{k-1} A_i A_i^H \right) H_k \tag{18}$$

by SVD we get:

$$\bar{H}_k = \begin{pmatrix} A_k & B_k \end{pmatrix} \begin{pmatrix} D_k \\ 0 \end{pmatrix} V_k^H = A_k D_k V_k^H \tag{19}$$

where D_k is diagonal matrix and the diagonal elements is non-zero eigen values of $\bar{H}_k$. Define $Q = [A_1 \ A_2 \ ... \ A_K]$, $H = [H_1 \ H_2 \ ... \ H_K]$ and $Q^H Q = I$. What is more,

$$Q^H H = M \tag{20}$$

M is block upper triangular matrices. From the SVD decomposition, we know that A_k satisfies orthogonality and

$$span(\bar{H}_k^H) = span(A_k)$$
$$A_k^H A_j = 0 (k \neq j)$$

Also

$$\bar{H}_k^H \bar{H}_j = 0 (k \neq j)$$

which means

$$H_k = \bar{H}_k + \sum_{i=1}^{k-1} A_i A_i^H H_k \tag{21}$$

that is

$$H_k = [A_1 \ A_2 \ ... \ A_k] \bullet [A_1^H \ H_k \ ... \ A_{k-1}^H \ H_k \ D_k V_k^H \ 0 \ ... \ 0]^T \tag{22}$$

The base station receive signal is:

143

$$Y = \sum_{k=1}^{K} H_k x_k + n = \left[H_1 \ H_2 \ \dots \ H_K \right] \left[x_1 \ x_2 \ \dots \ x_K \right]^T + n \qquad (23)$$

Then the left multiply by Q^H. Following these steps we can get the receive signal for each user:

1. From the K user, use $L_k = V_K D_K^{-1}$ get $\bar{x}_K = L_K \bar{y}_k$
2. According to the $\bar{x}_K$ cancel interference and get

$$\hat{y}_i = \hat{y}_i - \sum_{j=i+1}^{K} A_i^H H_j \bar{x}_j$$

3. $L_i = V_i D_i^{-1} \Rightarrow \bar{x}_i = L_i \bar{y}_i$
4. Repeat (2),(3), until we find $\bar{x}_K$ Secondly, use this method to x + y users and get the channel matrix H1 and put it into the null space.

For each group, we use BD algorithm to eliminate interference within the group. For channel with perfect CSI, $W_u[n]$ is the precoding matrix and $H_u[n]$ is the $N_r \times N_t$ channel matrix from the transmitter to the u_{th}. 4 The design of the precoding matrix for the u_{th} user is based on the SVD of the aggregated channel matrix of the other users. Constrained $W_u^*[n]W_u[n] = I_{N_r}$, in this way the received signal becomes:

$$\begin{aligned} y_u[n] &= H_u[n]W_u[n]x_u[n] + z_u[n] \\ &= H_{eff,u}[n]x_u[n] + z_u[n] \end{aligned} \qquad (24)$$

As $W_u[n]$ is a unitary matrix, which is independent of $H_u[n]$, $H_{eff,u}[n]$ is a Gaussian matrix as $H[n]$. We assume that the number of data streams for user u is equal to the number of receive antennas, and with equal power allocation, the input covariance is $\Phi_{BD,u} = \frac{\gamma}{N_u} I_{N_r}$. Within an $N_r \times N_t$ effective channel, the achievable ergodic rate for the u_{th} is given by:

$$M_{BD,u} = E\left[\log_2 \det\left(I_{N_r} + \frac{\gamma}{N_t} H_{eff,u}[n] H_{eff,u}^*[n] \right) \right] \qquad (25)$$

Therefore, the achievable rate for the BD system with perfect CSI is given as:

$$M_{BD} = \sum_{u=1}^{A} M_{BD,u} \approx A N_r C_{iso}(1, \gamma/\beta) \qquad (26)$$

$$C_{iso} = E\left[\log \ \det\left(I_{N_r} + \frac{\gamma}{N_t} H[n] H^*[n] \right) \right] \qquad (27)$$

For channel with poor CSI or middle CSI, same as the channel with perfect CSI.

5 SIMULATION RESULTS

In this component, assuming that the number of transmit antennas is 32, we divided users into three groups, and each has two users with four antennas. We simulated our proposed BD-SO precoding algorithm and compared performances on bit error rate and spectral efficiency for different groups in terms of distinct values of SNR. The Block Diagonalization precoding scheme was adopted in one group, while the Successive Optimization method was used in different groups. That means group 2 is an optimization of group 1 and group 3 is an optimization of group 2. After the simulation, we obtained result images for these two comparative items, respectively.

Figure 2 shows the BER performance of different groups, or we can say, different optimization iterations according to ascending SNR values at 5 dB. From the simulation results, about a 0.5 BER for the first group and a 0.35 BER for the second group at the 30 dB SNR are observed in this figure. Furthermore, we note that the further optimized group 3 has a marvel improvement compared to group 2 on the BER performance. It reached almost 0 BER when SNR is at 5 dB.

Figure 3 describes the spectral efficiency with the same antenna configuration as in Figure 2. From the simulation results, with the increasing values of SNR at the receiver of each user, the spectral efficiency of group 2 will be marginally improved compared to group 1. The first iteration (group 2) improves the spectral efficiency by almost 2.4 bps/Hz with regard to group 1 at the 30dB SNR. Similarly, the second iteration (group 3) has a dramatic improvement of 25.6 bps/Hz compared to group 2 when SNR is 30 dB on the spectral efficiency performance.

From all of these simulation results, we can see that the advanced BD-SO precoding method improves the system performance dramatically. It gains tremendous strengths on relative low BER and indeed high spectral efficiency.

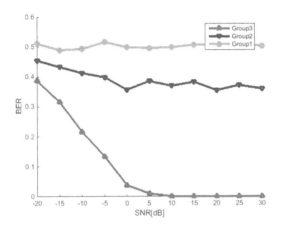

Figure 2.   BER performance comparison for each group.

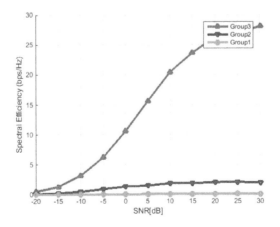

Figure 3.   Spectral efficiency comparison for each group.

6 CONCLUSION

With a view to the advanced **BD-SO** algorithm for precoding schemes in the massive **MIMO** system in this paper, we can find that the modification method addresses a good balance between system overhead and interference coming from multiple users. Under HetNet condition, the system performance can be further improved by adopting our algorithm. It is based on Block Diagonalization with users in one group and Successive Optimization method in different groups. From the simulation results, it is obvious that our proposed **BD-SO** algorithm overcomes the drawbacks of the original **BD** method and obtains a better system performance on bit error rate and spectral efficiency.

REFERENCES

Gao Xiangchuan, Fei Xiong & Lei Song. 2010. A successive iterative optimization precoding method for downlink multi-user MIMO system. *Wireless Communications and Signal Processing (WCSP), 2010 International Conference, Suzhou, China*: 1–5.

Jun Zhang, Andrews J.G. & Heath R.W. 2009. Block Diagonalization in the MIMO Broadcast Channel with Delayed CSIT. *Global Telecommunications Conference, 2009. GLOBECOM 2009. IEEE, Honolulu, United States of America*: 1–6.

Kudo R., Ishihara K., Murakami T., Abeysekera B.A.H.S., Asai Y. & Mizoguchi M. 2012. Successive Optimization Transmission for High and Low SNR Stations in Wireless LAN Systems. *Vehicular Technology Conference (VTC Fall), 2012 IEEE, Quebec City, Canada*: 1–5.

Stankovic V. & Haardt M. 2005. Successive optimization Tomlinson-Harashima precoding (SO THP) for multi-user MIMO systems. *Acoustics, Speech, and Signal Processing, 2005. Proceedings. (ICASSP '05). IEEE International Conference, Philadelphia, United States of America*: 3–4.

Signal and Information Processing, Networking and Computers – Chen & Huang (Eds)
© 2016 Taylor & Francis Group, London, ISBN 978-1-138-02881-4

Secure transmission with Artificial Noise in heterogeneous massive MIMO network

Chang Li & Songlin Sun

School of Information and Communication Engineering, Key Laboratory of Trustworthy Distributed Computing and Service, Ministry of Education, Beijing University of Posts and Telecommunications, Beijing, China

Wei Liu

Department of Network and Platform, M2M & Internet of Things Institute, China Mobile Research Institute, Beijing, P.R. China

Hai Huang

School of Information and Communication Engineering, Key Laboratory of Trustworthy Distributed Computing and Service, Ministry of Education, Beijing University of Posts and Telecommunications, Beijing, China

ABSTRACT: Heterogeneous Network (HetNet) is a promising technology to improve the capacity of future generations of cellular network, in which a mobile station can be served by multiple Base Stations (BSs) with different scales of coverage range, including short range Low Power Nodes (LPNs). In this paper, we study security provisioning in heterogeneous massive Multiple-Input–Multiple-Output (MIMO) systems. Specifically, we consider secure downlink transmission in a massive MIMO system with Matched-Filter (MF) precoding and Artificial Noise (AN) generation at the BS, and transmit the signal to the MT through the LPN in the presence of a passive multi-antenna eavesdropper. Thereby, we consider an AN shaping matrix, where the AN is transmitted in the null-space of the matrix formed by all MT channels. Our analytical and numerical results reveal that, in massive MIMO systems employing MF precoding and AN generation, the Signal-to-Interference-and Noise Ratio (SINR) of intended MTs in LPNs is higher than eavesdropper, we can also obtain the positive secrecy capacity which prove this method can guarantee the secure transmission.

Keywords: Heterogeneous Network (HetNet); massive Multiple-Input–Multiple-Output (MIMO) systems; Artificial Noise (AN); secure transmission

1 INTRODUCTION

The architecture of Heterogeneous Network (HetNet) has received much attention as a means of meeting the requirements defined by LTE-Advanced (LTE-A) mobile system (R.Q. Hu & Y. Qian. 2013). A typical cell of HetNet consists of eNodeBs and Low Power Nodes (LPNs), and the two types of nodes are categorized by different transmission powers so as to cover different sizes or layers (S. Sun et al. 2015). The core concept is to density the topology to enable very high spatial and spectral reuse, which allows the spectral efficiency to increase significantly. Massive Multiple-Input–Multiple-Output (MIMO), namely, deploying large scale antenna arrays at existing macro Base Stations (BSs) is expected to dominate the implementation of next generation HetNet (F. Rusek et al. 2013).

Security is an important issue for cellular networks due to the broadcast nature of the medium. Traditionally, security has been ensured through cryptographic encryption implemented at the application layer. This scheme is based on certain assumptions regarding computational complexity, and is thereby potentially vulnerable (A. Mukherjee et al. 2014). Recently physical layer security as a complement has drawn significant research based on cryptographic methods. The pioneering work based on physical layer security in A. D. Wyne (1975) considered the classical three-terminal network which consists of a transmitter (Alice), an intended receiver (Bob), and an eavesdropper (Eve). It was shown that a source-destination pair can exchange secure messages with a positive rate as long as the desired receiver enjoys better channel conditions than the eavesdroppers. Meanwhile, more recent studies have considered physical layer security provisioning in multi-antenna multi-MT networks (A. Khisti & G. Wornell. 2010a, A. Khisti & G. Wornell. 2010b, E. Ekrem & S. Ulukus. 2011, F. Oggier & B. Hassibi. 2011). For certain practical transmission strategies, it is interesting to investigate the achievable secrecy rates of such networks though the secrecy capacity region for multi-MT networks is still an open problem. Eavesdroppers are typically passive so as to hide their existence, and thus their Channel State Information (CSI) cannot be obtained by Alice. In this case, multiple transmit antennas can be exploited to enhance secrecy by simultaneously transmitting both the information signal and Artificial Noise (AN) (S. Goel & R. Negi. 2008). Specifically, precoding is used to make the AN invisible to Bob while degrading the decoding performance of possibly present Eves (S. Goel & R. Negi. 2008, X. Zhou & M.R. McKay. 2010, A. Mukherjee & A. L. Swindlehurst. 2011).

Recently, a new promising design approach in wireless networks, known as massive or large-scale MIMO, has been proposed (H. Q. Ngo et al. 2013), where BS antenna arrays are equipped with an order of magnitude more elements than what is used in current systems, namely a hundred antennas or more. Massive MIMO enjoys all the benefits of conventional multi-MT MIMO, such as improved data rate, reliability and reduced interference. In fact, massive MIMO employing simple matched filter precoding/combining enables large gains in bandwidth and power efficiency compared to conventional MIMO systems. Furthermore, in Time-Division Duplex (TDD) systems, channel reciprocity can be exploited to estimate the downlink channels via uplink training such that the resulting overhead scales linearly with the number of MTs but is independent of the number of BS antennas.

Massive MIMO systems offer an abundance of BS antennas, while multiple transmit antennas can be exploited for secrecy enhancement. Therefore, the combination of both concepts seems natural and promising, which is the main motivation for the present work. However, in Hetnet, the intended MTs not only receive the signals transmitted by the BSs, but also receive the signals transmitted by the LPNs, meantime the eavesdroppers seek to decode the signals in these transmission process. So the security cannot be neglected. To solve this issue, in this paper, we study secure downlink transmission in heterogeneous massive MIMO systems in the presence of a multi-antenna eavesdropper, which attempts to intercept the signal intended for one of the MTs. We consider MF precoding and AN shaping methods in BS, and AN is transmitted in the null-space of the channel matrix.

The rest of the paper is organized as follows. The applied system model is introduced in section II. Section III analyzes the downlink data transmission and the design of AN shaping matrix. Then derive the secrecy capacity. The simulation and numerical results are described in section IV. Finally, the conclusions are drawn in Section V.

2 SYSTEM MODEL

In this paper, we consider a flat-fading heterogeneous system, as depicted in Figure 1. A two-tier macro cell consisting of a BS and 4 LPNs with 4 eavesdroppers, where the BS is equipped with N_t antennas serving K single-antenna Mobile Terminals (MTs), and $L(L>0)$ LPNs are deployed arbitrarily in the macro cell forming an overlay layer. Each LPN is equipped with $M(1 \leq M \leq 8)$ antennas.

Note that the coverage area (depicted in dashed circles) of the LPNs is limited due to power constraints. The antenna number at BS, N_t, can scale from 8 to hundreds, i.e. massive MIMO. A MT with a single-antenna equipment can either be served by the BS or one of

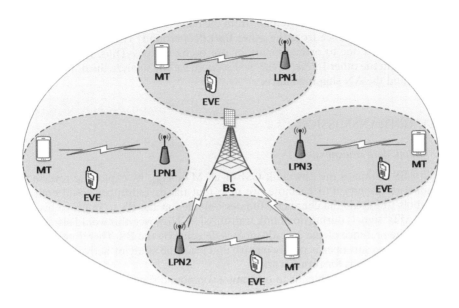

Figure 1. The layout of the system model.

the LPNs, though they can be affected by the BS or neighboring LPNs. Every eavesdropper equipped with N_e antennas (equivalent to N_e cooperative single-antenna eavesdroppers) is located in the coverage area of every LPN. The eavesdroppers are passive and seek to recover the information transmitted to the MT which belong to the same cell with them.

Let $H_B \in \mathbb{C}^{1 \times N_t}$ and $H_B^{eve} \in \mathbb{C}^{N_e \times N_t}$ denote the stationary channel between macro BS and the MTs it serves, and the channel between macro BS and the eavesdropper, respectively. $H_B = \sqrt{l_B} \tilde{H}_B$ comprises the path-loss, l_B, and the small-scale fading vector, $\tilde{H}_B \sim \mathbb{CN}\left(0_{N_t}^T, I_{N_t}\right)$. Similarly, we model the eavesdropper channel as $H_B^{eve} = \sqrt{l_B^{eve}} \tilde{H}_B^{eve}$, where l_B^{eve} and $\tilde{H}_B^{eve}$ denote the path-loss and small-scale fading components, respectively. The elements of $\tilde{H}_B^{eve}$ are modeled as independent and identically distributed (i.i.d.) Gaussian random variables (r.v.s) with zero mean and unit variance. We define that H_j denotes the stationary channel between a certain LPN j and the MTs it serves, where

$$
H_j = \begin{bmatrix} H_{j,1} \\ H_{j,2} \\ \dots \\ H_{j,K_j} \\ H_{j,K_{j+1}} \\ \dots \\ H_{j,K_j+L_j} \end{bmatrix}, H_j \in \mathbb{C}^{(K_j+L_j) \times M}, H_{j,k} \in \mathbb{C}^{1 \times M}. \tag{1}
$$

The k th row vector $H_{j,k}$ is the channel between the j th LPN and the k th MT. $H_{B,k}$ is the channel between the macro BS and the k th MT. Let $K_j + L_j \leq M$, K_j denotes the number of MTs of the j th LPN, L_j denotes the number of MTs interfered by the j th LPN and served by other nodes.

Since the BS performs massive MIMO and LPNs perform MIMO, the transmitted signal can be recognized as Gaussian variance by LPN MTs. To improve the capacity, interference between MTs should be effectively avoided. The interference to a MT can be categorized by

149

two different types, one is the interference from the remaining MTs in the same LPN, and the other one is the interference from the other BS/LPNs around. The interference of the first type can be ignored due to orthogonality of precoding vectors. Thus, we will focus on the interference from the other LPNs. In the next section, we will analyze the downlink transmission format and the AN shaping matrix.

3 SECURE TRANSMISSION

3.1 *Downlink data transmission*

The macro BS estimates the downlink CSI of all MTs, $H_{B,k}$, by exploiting reverse training and channel reciprocity. Furthermore, we note that the eavesdropper could emit his own pilot symbols to impair the channel estimates obtained at the BS to improve his ability to decode the MTs' signals during downlink transmission. However, this would also increase the chance that the presence of the eavesdropper is detected by the BS. Therefore, in this paper, we assume the eavesdropper is purely passive and leave the study of active eavesdroppers in massive MIMO systems for future work.

In Hetnet, the macro BS intends to transmit a compound signal $x_k\,(k=1,\ldots,K)$ consisting of a confidential signal $s_k\,(k=1,\ldots,K)$ and the AN signal $z_i\,(i=1,\ldots,N_t-K)$ to the j th LPN, and the j th LPN transmits the signal to k th MT. The signal vector for the K MTs is denoted by $s=[s_1,\ldots,s_K]^T\in\mathbb{C}^{K\times1}$ with $\mathbb{E}[ss^H]=I_K$. Each signal vector s is multiplied by a transmit beamforming matrix, $T=[t_1,\ldots,t_k,\ldots,t_K]\in\mathbb{C}^{N_t\times K}$, before transmission. As typical for massive MIMO systems, we adopt simple matched-filter precoding, i.e. $t_k=H_{Bk}^H/\|H_{Bk}\|$. Furthermore, we assume that the CSI of eavesdropper is not available at the macro BS. Hence, assuming that there are $K(K<N_t)$ MTs, the BS may use the remaining N_t-K degrees of freedom offered by the N_t transmit antennas for emission of AN to degrade the eavesdropper's ability to decode the data intended for the MTs. The AN vector, $z=[z_1,\ldots,z_{N_t-K}]^T\sim\mathbb{CN}(0_{N_t-K},I_{N_t-K})$, is multiplied by an AN shaping matrix $V=[V_1,\ldots,V_i,\ldots,V_{N_t-K}]\in\mathbb{C}^{N_t\times(N_t-K)}$ with $\|V_{ni}\|=1,i=1,\ldots,N_t-K$. The considered choices for the AN shaping matrix will be discussed in the next section. The signal vector transmitted by the macro BS is given by

$$x=\sqrt{p}Ts+\sqrt{q}Vz=\sum_{k=1}^{K}\sqrt{p}t_ks_k+\sum_{i=1}^{N_t-K}\sqrt{q}v_iz_i,\qquad(2)$$

where p and q denote the transmit power allocated to each MT and each AN signal, respectively, i.e. for simplicity, we assume uniform power allocation across MTs and AN signals, respectively. Let the total transmit power be denoted by P. Then, p and q can be represented as $p=\phi P/K$ and $q=(1-\phi)P/(N_t-K)$, respectively, where the power allocation factor $\phi,0\leq\phi\leq1$, strikes a power balance between the information bearing signal and the AN.

The $L-1$ small-cells of LPNs adjacent to the current small-cell of LPN transmit their own signals. In this work, in order to gain some fundamental insights, we assume that all small-cells of LPN employ identical values for p and q as well as ϕ. Accordingly, the received signals at the k th MT served by j th LPN, y_{jk}, and at the eavesdropper, y_{eve}, are given by

$$y_{jk}=\sqrt{p}H_{Bk}t_ks_k+H_{jk}\left(\sqrt{p}H_{Bk}t_ks_k+\sqrt{q}H_{Bk}v_kz_k\right)+\sqrt{p}H_{jk}\sum_{i\neq j}t_is_i+\sqrt{q}H_{Bk}v_kz_k+n_k,\qquad(3)$$

$$y_{eve}=\sqrt{p}\sum_{l=1}^{L}H_B^{eve}T_ls_l+\sqrt{q}\sum_{l=1}^{L}H_B^{eve}V_lz_l+n_{eve},\qquad(4)$$

where $n_k\sim\mathbb{CN}(0,\sigma_k^2)$ and $n_{eve}\sim\mathbb{CN}(0_{N_e},\sigma_{eve}^2I_{N_e})$ are the Gaussian noises at the k th MT and at the eavesdropper, respectively. The first term on the right hand side of (3) is the signal intended for the k th MT in the macro BS with effective channel gain $\sqrt{p}H_{Bk}t_k$,

150

which is assumed to be perfectly known at the k th MT in the macro BS. The second term on the right hand side of (3) is the signal intended for the k th MT in the j th LPN. The third and forth terms on the right hand side of (3) represent the interference from other LPNs and AN leakage, respectively. On the other hand, the eavesdropper observes an $M\,N_t \times N_e$ MIMO channel comprising K local MT signals, $(L-1)K$ out-of-cell MT signals, N_t-K local cell AN signals, and $(N_t-K)(L-1)$ out-of-cell AN signals. In order to obtain a lower bound on the achievable secrecy rate, we assume that the eavesdropper can acquire perfect knowledge of the effective channels of all MTs, i.e. $H_B^{eve}t_k, \forall k$. We note however that this is a quite pessimistic assumption because the uplink training performed in massive MIMO makes it difficult for the eavesdropper to perform accurate channel estimation.

In this paper, our aim is to guarantee secure transmission, which can be considered as the achievable channel secrecy capacity for each MT. The SINR of the k th MT in the small-cell of the j th LPN is given by

$$SINR_{jk} = \frac{\left|\sqrt{p}H_{Bk}t_k\right|^2 + \left|\sqrt{p}H_{jk}H_{Bk}t_k\right|^2}{\sigma_k^2 + \sum_{i=1}^{N_t-K}\left|\sqrt{q}H_B^{eve}v_i\right|^2 + \sum_{j'=1, j'\neq j}^{L}\left|\sqrt{p}H_{j'k}t_{j'}\right|^2}. \qquad (5)$$

Here σ_k equals to the variance of the sum of white Gaussian noise and received signal from BS. Similarly, the SINR of the eavesdropper in the small-cell of the j th LPN is given by

$$SINR_{eve} = \frac{\left|\sqrt{p}H_B^{eve}t_k\right|^2}{\sigma_{eve}^2 + \sum_{i=1}^{N_t-K}\left|\sqrt{q}H_B^{eve}v_i\right|^2 + \sum_{j'=1, j'\neq j}^{L}\left|\sqrt{p}H_{j'k}t_{j'}\right|^2 + w_{eve}^2}. \qquad (6)$$

Here σ_{eve} equals to the variance of the sum of white Gaussian noise at eavesdropper, and W_{eve} is the redundancy caused by the unknown AN shaping matrix.

3.2 Design of AN shaping matrix

In this paper, we consider two different designs for the AN shaping matrix V_n.

Null-Space Method: For massive MIMO, V_n is usually chosen to lie in the null-space of the estimated channel, H_{Bk}, i.e. $H_{Bk}V = 0_{N_t-K}^T, k=1,...,K$, which is possible as long as $N_t > K$ holds (A. Mukherjee et al. 2014). We refer to this method as N in the following. If perfect CSI is available, the N-method prevents impairment of the MTs by AN generated by the macro BS. However, in case of pilot contamination, AN leakage to the MTs in the local cell is unavoidable. The process of N-method as shown in Figure 2. More importantly, for the large values of N_t and K typical for massive MIMO systems, computation of the null-space of H_{Bk}, is computationally expensive. This motivates the introduction of a simpler method for generation of the AN shaping matrix.

Random Method: In this case, the columns of V_n are mutually independent random vectors. We refer to this method as R in the following. Here, we construct the columns of V_n as $v_i = \tilde{v}_i / \|\tilde{v}_i\|$, where the $\tilde{v}_i, i=1,...,N_t-K$, are mutually independent Gaussian random vectors. Note that the R-method does not even attempt to avoid AN leakage to the MTs in the local cell. However, it may still improve the ergodic secrecy rate as the precoding vector for the desired MT signal, t_k, is correlated with the MT channel, H_{Bk}, whereas the columns of the AN shaping matrix are not correlated with the MT channel.

Although the N-method always achieves a better performance than the R-method, if pilot contamination and inter-cell interference are significant, the performance differences between both schemes are small. This makes the R-method an attractive alternative for massive MIMO systems due to its simplicity.

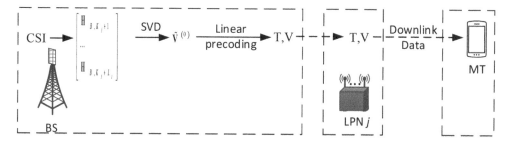

Figure 2. The process of N-method.

3.3 *The secrecy capacity*

The secrecy capacity, namely the achievable ergodic secrecy rate of the k th MT, can be expressed as the difference between the achievable ergodic rate of the k th MT and the ergodic capacity of the eavesdropper.

The ergodic secrecy rate is an appropriate performance measure if delays can be afforded and coding over many independent channel realizations is possible. Considering the k th MT in the small-cell of the j th LPN, the considered channel is an instance of a Multiple-Input, Single Output, Multiple Eavesdropper (MISOME) wiretap channel (A.D. Wyne. 1975). We provide an expression for an achievable ergodic secrecy rate of the k th MT in the small-cell of the j th LPN as

$$C_s = \left[R_{jk} - C_{jk}^{eve} \right]^+,$$ (7)

where $[x]^+ = \max\{0, x\}$, R_{jk} is an achievable ergodic rate of the k th MT in the small-cell of the j th LPN, C_{jk}^{eve} is the ergodic capacity between the macro BS and the eavesdropper seeking to decode the information of the k th MT in the small-cell of the j th LPN. According to the Shannon information theory, we can easily obtain the following formulas

$$R_{jk} = B \log\left(1 + SINR_{jk}\right),$$ (8)

$$C_{jk}^{eve} = B \log\left(1 + SINR_{eve}\right).$$ (9)

4 SIMULATION RESULTS

In this section, we present the simulation results to demonstrate the performance of our scheme. Here we combines massive MIMO and small cells by implementing our algorithm to achieve the target SINR with low complexity and high security. We take the property of null-space to enhance the security to the MTs. At the same time, we evaluate the system performance over channel realizations and different simulation parameters. The main simulation parameters that characterize the macro cell and the LPNs can be found in Table 1 and the main channel model parameters are summarized in Table 2.

Figure 3 shows the SINR of MTs and EVEs in two conditions, they are the systems with AN and without AN. In the condition with AN, EVEs get worse SINR than the condition without AN due to the unknown AN shaping matrix, and it makes difficult for the eavesdroppers to perform accurate channel estimation and decode the original signal. In comparision, the MTs which know the AN shaping matrix have an improved SINR, which results in a better system performance and guarantee the secure transmission. Moreover we can also see in the condition with AN, the difference between the SINR of MTs and EVEs is higher than the condition without AN.

Table 1. Simulation parameters.

Parameters	Values
Number of the macro cell MTs	20, 28, 36, ..., 60
Number of per LPN MTs	4
Macro cell radius	500 m
LPNs radius	50 m
Antennas of the macro BS	$N_t \in \{52,56,60,...,112\}$
Antennas of the LPNs	$M \in \{4,5,6,8\}$
Antennas of the EVEs	$N_e \in \{4,8,12,...,32\}$
Total bandwidth	100 Mhz
Requirement of the rate (per MT)	$\{2, 3, 4, ..., 9\}$ Mbps

Table 2. Channel parameters in the numerical evaluation.

Parameters	Values
Small-scale fading distribution	$\tilde{H}_B \sim \mathbb{CN}\left(0^T_{N_t}, I_{N_t}\right)$
Standard deviation of log-normal shadowing	7 dB
Path and penetration loss at distance d (km)	$148.1 + 37.6\log_{10}(d)\,dB$
Normal cases for the BS and LPN	$128.1 + 37.6\log_{10}(d)\,dB$
Special cases for the LPN: Within 40 m from LPN	$127 + 30\log_{10}(d)\,dB$

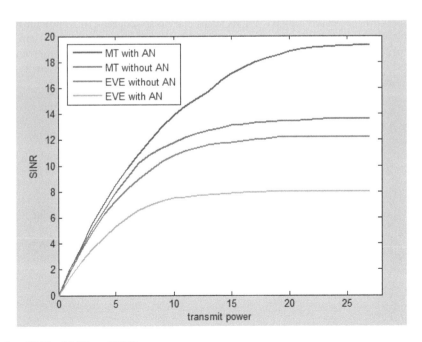

Figure 3. SINR of MTs and EVEs.

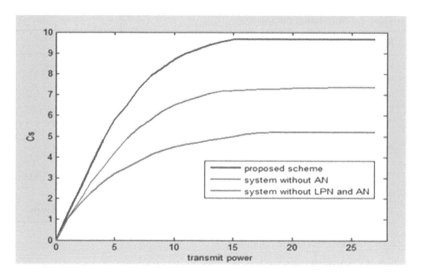

Figure 4.   Secrecy capacity in different conditions.

Figure 4 demonstrates the secrecy capacity in different conditions. The lowest secrecy capacity appeared in the condition without LPN, the BS transmits the signals to MTs directly and the eavesdroppers can decode the original signal more easily. It can indicate that the LPN can increase security to some extent. The condition with AN has the highest secrecy capacity among the three conditions. The BS transmit a compound signal consisting of a confidential signal and the AN signal to MTs through LPNs. The MTs can decode the original signal because they know the AN shaping matrix and can achieve high SINR. On the contrary, EVEs cannot decode accurately and generate the redundancy caused by the unknown AN shaping matrix. Therefore the proposed scheme can achieve higher secrecy capacity.

In general, the proposed massive MIMO artificial noise scheme can significantly alleviate the interference and enhance the security, meanwhile achieve a desirable system performance for next generation HetNet. The application of AN guarantees the secure transmission. Compared with the conventional design, the new system has a satisfying security performance and low complexity by taking advantage of the AN based null-space.

5   CONCLUSION

In this paper, we considered a massive MIMO system in HetNet with matched-filter precoding and artificial noise generation at the BS and LPN for secure downlink transmission in the presence of a multi-antenna passive eavesdropper. We considered both the null-space based AN shaping matrix design and a random AN shaping matrix design. We derived the secrecy capacity through the SINR of intended MTs and eavesdropper. In particular, our results reveal that for the heterogeneous massive MIMO system with MF precoding and AN, the SINR of intended MTs in LPNs covered with small cells is higher than eavesdropper. We can also obtain the positive secrecy capacity which prove this method can guarantee the secure transmission.

ACKNOWLEDGEMENTS

This work is supported by project NSFC 61471066 and the open funding project of State Key Lab of Virtual Reality Technology and Systems at Beihang University under Grant No. BUAA-VR-15 KF-19.

REFERENCES

Ekrem, E. & Ulukus, S. 2011. The secrecy capacity region of the Gaussian MIMO multi-receiver wiretap channel. *IEEE Trans. Information Theory.* 57(4): 2083–2114.

Goel, S. & Negi, R. 2008. Guaranteeing secrecy using artificial noise. *IEEE Trans. Wireless Communication.* 7(6): 2180–2189.

Hu, R.Q. & Qian, Y. 2013. Heterogeneous cellular networks. *John Wiley and Sons.*

Khisti, A. & Wornell, G. 2010. Secure transmission with multiple antennas I: The MISOME wiretap channel. *IEEE Trans. Information Theory.* 56(7): 3088–3104.

Khisti, A. & Wornell, G. 2010. Secure transmission with multiple antennas II: The MIMOME wiretap channel. *IEEE Trans. Information Theory.* 56(11): 5515–5532.

Mukherjee, A. & Swindlehurst, A.L. 2011. Robust beamforming for security in MIMO wiretap channels with imperfect CSI. *IEEE Trans. Signal Process.* 59(1): 351–361.

Mukherjee, A., Fakoorian, S.A.A., Huang, J. & Swindlehurst, A.L. 2014. Principles of physical-layer security in multi MT wireless networks: A survey. *IEEE Communication. Surveys Tuts.*

Ngo, H.Q., Larsson, E.G. & Marzetta, T.L. 2013. Energy and spectral efficiency of very large multi MT MIMO systems. *IEEE Trans. Communication.* 61(4): 1436–1449.

Oggier, F. & Hassibi, B. 2011. The secrecy capacity of the MIMO wiretap channel. *IEEE Trans. Information Theory.* 57(8): 4961–4972.

Rusek, F., Persson, D., Lau, B., Larsson, E., Marzetta, T., Edfors, O. & Tufvesson, F. 2013. Scaling up MIMO: Opportunities and challenges with very large arrays. *IEEE Signal Process. Magazine.* 30(1): 40–60.

Sun, S., Kaoch, M. & Ran, T. 2015. Adaptive SON and cognitive smart LPN for 5G heterogeneous networks. *Journal of Mobile Networks and Applications—Special Issue on Networking 5G Mobile Communications Systems: Key Technologies and Challenges.*

Wyne, A.D. 1975. The wire-tap channel. *Bell Syst. Tech. J.* 54(8): 1355–1387.

Zhou, X. & McKay, M.R. 2010. Secure transmission with artificial noise over fading channels: Achievable rate and optimal power allocation. *IEEE Trans. Vehicle Technology.* 59(8): 3831–3842.

Signal and Information Processing, Networking and Computers – Chen & Huang (Eds)
© 2016 Taylor & Francis Group, London, ISBN 978-1-138-02881-4

User scheduling algorithm based on Null-Space precoding scheme for heterogeneous networks

Qing Zhou, Chen Fu & Min Lu
International School, Beijing University of Posts and Telecommunications, Beijing, China

ABSTRACT: Heterogeneous cellular network (HetNet) has great potential to bring radical changes to wireless communication system on future since it offers the great probability to enlarge the channel capacity by allowing one mobile station to be served by multiple base stations with different coverage areas. In view of HetNet, the null-space based precoding scheme, an emerging precoding scheme applied on massive Multiple-Input Multiple-Output (MIMO) appears. This scheme takes advantages of the multi-antenna feature of base stations to eliminate the interference caused by Small Cell Network (SCN) and warrants the Quality-of-Service (QoS). Nonetheless, the confined numbers of LPN antennas poses a dilemma for the null-space based precoding scheme. We proposed a user scheduling algorithm incorporated into the original precoding scheme to alleviate the impacts which arise from this issue. Numerical simulation results manifest the increment on data throughput of overall system after the proposed algorithm applied.

Keywords: HetNet; Null-Space precoding scheme; MIMO; user scheduling; throughput

1 INTRODUCTION

Two acknowledged truths of wireless proposed in (Marzetta 2015) are the mounting demand of throughput and the constant quantity of accessible electromagnetic spectrum. Under the condition of finite user numbers, conventional macro-cell network topology can meet almost all the requirements for both throughput and spectrum reuse. The mounting user numbers, nonetheless, makes it more and more difficult to serve all users with ideal QoS on the same network topology. This situation results in the imminent emerge of new technologies to ease or even solve the problems from the above two aspects.

Heterogeneous network (HetNet) as one of the focus on wireless system researches, has been acknowledged as a crucial component on LTE-Advanced mobile system (LTE-A) (Sun et al. 2013). The fundamental structure of HetNet is a cell constituted by one eNodeBS and several Low Power Nodes (LPNs), which are distinguished by power consumption and coverage areas. Despite the advantages given by the compressed multi-tier structure, there are still a lot further to go with some technologies which are valuable as well.

Large-scale antenna arrays is considered to be one of the most feasible and effective solution to increase the throughput of a system and inhibit the impact of thermal noise (Rusek et al. 2013). MIMO with augmented amounts of antennas can be applied on both transmitters and receivers to raise the upper limitation of accessible channel capacity and improve the system performance from both data rate and link liability (Tse & Viswanath 2005). Also compared with the traditional mode, the progress on propagation performance of MIMO is obvious, especially MU-MIMO and Massive MIMO (Marzetta 2015, Hoydis 2013). Practically, MU-MIMO has already found its place on the construction of wireless communication system. For example, the BS equipped with up to eight antennas has been used on LTE (Dahlman et al. 2008). A collection of classical linear precoding algorithms including ZF, MF, and MMSE.etc are suitable to very large antenna arrays (Hoydis et al. 2012) and other precoding algorithms always regard them as prototypes to further optimize the system performance.

(Zhang et al. 2015) proposes the null-space based precoding scheme, a novel massive MIMO precoding scheme which is applied on LTE-A HetNet to optimize QoS of systems through the multi-antenna feature of MIMO without extra power consumption. It takes the conventional precoding algorithms as foundations and exploits excess antennas on both BS and LPNs (Hoydis et al. 2013) to mitigate corresponding interference caused by small-cell network. Under realistic assumption, BS can be equipped with hundreds or thousands of antennas, while LPNs only have several antennas owing to the restriction of its own condition. That is to say, the excess antennas cannot satisfy the demand on increasing user numbers and the null-space based precoding scheme cannot work on the ideal circumstance to eliminate the interference completely.

In this paper, we introduce the concept of user schedule to the null-space based precoding scheme within a real scenario to optimize the performance of system and increase the overall data throughput under the high requirements on QoS. The proposed user scheduling algorithm on this paper mainly focuses on the optimal selection of victim users under the condition of limited excess antennas of LPNs with a concern on data throughput.

The rest of this paper is organized as follows. Section 2 introduces the system model of a practical massive MIMO scenario enabled HetNets with corresponding functions. Section 3 is an overview of the null-space based algorithm with comparison of classical precoding algorithms. In Section 4, we propose our user scheduling algorithm and demonstrate exhaustive operating circumstance, whereas the simulation results are discussed on Section 5. At last, a brief conclusion of this paper is presented on Section 6.

2 SYSTEM MODEL

One of the major concern of LTE-A is to optimize indoor and hotspot scenario. The placement of LPN (Low Power Node) is a solution to such scenarios. Introducing LPNs can decrease the distance from end user to station, which improves the signal quality user receive and reduce traffic load on BS (Base Station). Therefore it enables the great cover depth and optimizes the throughput of a system. Our proposed model is a two-tier micro cell consists of one BS and several LPNs. These LPNs are deployed under one umbrella macro BS coverage, as shown in Fig. 1. The BS is equipped with large amount number of antennas which is $\mathbf{N}_{BS}$ (MIMO). The number of LPNs is N and each LPN is equipped with M antennas. One user with single antenna in the micro cell can be served by either BS or any one of LPNs. Within the range of LPN j's coverage, the number of the proper LPN user is K_j and there are also L_j victim users which served by any other power nodes. $K_j + L_j > M$.

$$\mathbf{H}_j = \begin{bmatrix} \mathbf{H}_{j,1} \\ \mathbf{H}_{j,2} \\ \cdots \\ \mathbf{H}_{j,K_j} \\ \mathbf{H}_{j,K_j+1} \\ \cdots \\ \mathbf{H}_{j,K_j+L_j} \end{bmatrix}, \mathbf{H}_j \in C^{(K_j+L_j)\times M}, \mathbf{H}_{j,k} \in C^{1\times M}. \tag{1}$$

$\mathbf{H}_j$ represents the channel matrix between LPN j and all users within its covering range. Specifically, $\mathbf{H}_0$ represents the stationary channel matrix of eNB, which is the channel matrix between base station and its users. The k-th row vector $\mathbf{H}_{j,k}$ denotes the channel between LPN j and user k. $\mathbf{H}_{0,k}$ stands for the channel between eNB and the k-th user. L_j is the number of all victim users. For $i \leq K_j$, the precoding vector of i-th user is $\mathbf{T}_{j,i}$, and $\mathbf{S}_{j,i}$ denotes the information sent to user i. The received signal of the k-th user of LPN or BS is given by

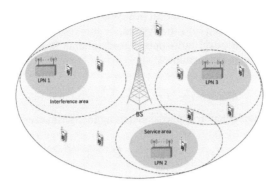

Figure 1. The layout of heterogeneous networks.

$$y_{j,k} = \mathbf{H}_{j,k} \sum_{i=1}^{K_j} \mathbf{T}_{j,i} s_{j,i} + \mathbf{H}_{0,k} \mathbf{T}_{0,k} s_0 + \omega_{j,k}, \tag{2}$$

where $\omega_{j,k}$ is the additive white Gaussian noise of zero mean and variance σ^2 [Zhang et al. 2014].

In order to optimize the throughput of the system, it's equal to maximize the channel capacity of each user, aka the maximum information rate per user. According to Shannon equation, $C = B log_2 (1 + SINR)$ (Signal-to-Interference plus Noise Ratio), users with higher SINR, the system experiences greater throughput. The SINR of user k from LPN j is given by

$$SINR_{j,k} = \frac{\left| \mathbf{H}_{j,k} \mathbf{T}_{j,k} \right|^2}{\sigma_{j,k^2} + \sum_{k'=1, k' \neq k}^{K_j} \left| \mathbf{H}_{j,k} \mathbf{T}_{j,k'} \right|^2 + \left| H_{0,k} T_{0,k} \right|^2}. \tag{3}$$

SINR decreases significantly due to the interference. Hence in order to improve the SINR, we should decrease the interference. If the k-th user is a BS user, which can be greatly affected by the nearby LPN. This kind of interference is cross-tier interference which we mainly focus on. The precoding scheme we adopt is able to alleviate cross-tier interference, and thus minimize the interference caused by victim users.

3   CONVENTIONAL PRECODING SCHEMES

3.1   *Review of the existing algorithms*

In downlink of MIMO system, the fundamental linear precoding techniques is considered to be used, which includes ZF, MMSE and MRT. In this section, we mainly take ZF precoding as an example.

ZF: Zero-Forcing precoding is an algorithm that eliminates the interference takes no account of the noise by implementing a pseudo-inverse of the channel status matrix. The according precoding matrix can be viewed as

$$\mathbf{T}_{ZF} = \frac{1}{\beta} \mathbf{H}^H \left(\mathbf{H} \mathbf{H}^H \right)^{-1}, \tag{4}$$

where $\beta = \sqrt{\frac{tr(\mathbf{B}\mathbf{B}^H)}{P_{tr}}}$, and $\mathbf{B} = \mathbf{H}^H (\mathbf{H}\mathbf{H}^H)^{-1}$.

Some drawbacks are shared by all those algorithms. For those algorithms, the main impairment is that the CSI must be precisely accurate which means that all of them know perfect

CSI of the users, which is impossible in practice. Without an accurate CSI, more imprecision and distortion can be introduced into the system.

To solve the problems, a novel null-space based scheme has been considered as its low complexity and considerable performance.

3.2 Our foundation precoding scheme: Precoding based on null-space

For the j-th LPN, the complementary space concerning the victim users can be viewed as follow:

$$\tilde{\mathbf{H}}_{j,v} = \left[\mathbf{H}_{j,K_j+1}^{H} \cdots \mathbf{H}_{j,K_j+L_j}^{H}\right]^{H}, \tag{5}$$

In which $\tilde{\mathbf{H}}_{j,v}$ is combined with all the channel matrix of all the victim users interfered by the j-th LPN. In order to avoid the interference caused by victim users, $\mathbf{T}_{j,i}$ should satisfy the following condition:

$$\tilde{\mathbf{H}}_{j,v}\mathbf{T}_{j,i} = 0^{L_j \times 1} \tag{6}$$

To be more accurate, $\tilde{\mathbf{H}}_{j,v}$ can be written as

$$\tilde{\mathbf{H}}_{j,v} = \tilde{\mathbf{U}}_{j,v}\tilde{\mathbf{\Lambda}}_{j,v}\tilde{\mathbf{V}}_{j,v}^{H} = \tilde{\mathbf{U}}_{j,v}\left[\sum_{j,v} 0\right]_{L_j \times M}\left[\tilde{\mathbf{v}}_{j,v,1}\tilde{\mathbf{v}}_{j,v,2}\cdots\tilde{\mathbf{v}}_{j,v,M_j}\right]^{H}. \tag{7}$$

Consider the use of singular value decomposition. We define $\tilde{\mathbf{V}}_{j,i}^{(0)} = [\tilde{\mathbf{V}}_{j,i,L_{i+1}} \cdots \tilde{\mathbf{V}}_{j,i,M_j}]$. As all column vectors in $\tilde{\mathbf{V}}_{j,i}^{(0)}$ locate in the null-space of all victim users. We have $\tilde{\mathbf{H}}_{j,v}\tilde{\mathbf{V}}_{j,i}^{(0)} = 0$. For the k-th user in the j-th LPN, define the projection matrix $\mathbf{P}_{j,k}$ on to the null-space of victim users by

$$\mathbf{P}_{j,k} = \mathbf{H}_{j,k}\tilde{\mathbf{V}}_{j,v}^{(0)}\left(\tilde{\mathbf{V}}_{j,v}^{(0)}\right)^{H}, \tag{8}$$

So that the projection matrix for all K_j users served by the j-th LPN can be written as $\mathbf{P}_j$. With $\mathbf{P}_j$ the precoding matrix of the corresponding ZF algorithm can be achieved as follow

$$\mathbf{T}_{j,ZF} = \frac{1}{\beta_j}\mathbf{P}_j^{H}\left(\mathbf{P}_j\mathbf{P}_j^{H}\right)^{-1}, \tag{9}$$

where $\mathbf{B}_j = \mathbf{P}_j^{H}\left(\mathbf{P}_j\mathbf{P}_j^{H}\right)^{-1}$, and $\beta_j = \sqrt{\frac{tr(\mathbf{B}_j\mathbf{B}_j^{H})}{P_{tr,j}}}$.

3.2.1 The advantages of null-space based precoding algorithm
Accurate CSI acquirement: In traditional HetNets system and corresponding algorithms, the user terminals end devices report their CSI to the access point it attaches, but not to the other interference nodes. As a result, the LPN does not know the CSI of the victim users within its coverage. But in the null-space based algorithm, both the BSs and LPNs know the information of all the users to achieve an accurate precoding vector so that the interference to the victim users can be avoided. As the victim users' channel matrix is orthogonal to the precoding matrix, the interference to the victim user is prevented so that SINR is must similar to the practice.

Reduced complexity: By combining the Block-Diagonal (BD) algorithm and linear prediction method, the proposed scheme is of lower complexity and better performance. As all the users receive signals from the LPNs, which is impossible in high complexity computing, so the system defines the serve area and interference area within each LPNs. The very LPNs that cause significant interference to a particular victim user will be picked out and the unnecessary LPNs are ignored.

4 OUR PROPOSED ALGORITHM

4.1 Applied situation

Under ideal condition, for LPN j,

$$\tilde{\mathbf{V}}_{j,i}^{(0)} = [\tilde{\mathbf{V}}_{j,i,L_j+1} \cdots \tilde{\mathbf{V}}_{j,i,M_j}]. \tag{10}$$

The null space vector of victim users $\tilde{\mathbf{V}}_{j,i}^{(0)}$ contains $M_j - L_j$ number of elements, which represents the null-space precoding scheme is only feasible for eliminating the interference from limited number of victim users. However, for the general case $1 < M < 8$, which in actual scenario is far less than the number of all its users plus the additional victim BS users. In this context we ignore the victim LPN user due to the interference experienced by which is minor. Therefore, under the circumstances of deficient amount of antennas LPN j equipped with, two explicit schemes that achieve the elimination the interference come from certain $M_j - L_j$ number of victim users are outlined in the rest of this section.

4.2 User scheduling algorithm based on SINR

The algorithm we propose is applied in BS. We suppose the precoding scheme BS adopted is null—space precoding in order to alleviate the interference to other LPN users.

Step 1:

$$\tilde{\mathbf{H}}_{0,v} = \left[\mathbf{H}_{1,1}^H \cdots \mathbf{H}_{1,K_1}^H \mathbf{H}_{n,1}^H \cdots \mathbf{H}_{n,K_n}^H \mathbf{H}_{N,1}^H \cdots \mathbf{H}_{N,K_N}^H \right]^H. \tag{11}$$

From (12), the expression of complementary space $\tilde{\mathbf{H}}_{0,v}$ consists of channel matrix of all LPN users. Owing to the fact that the served user equipment will only have its CSI reported to the access point it attached to, the channel matrix of all LPN users can't be obtained by BS directly. In this case each LPN should transmit all its user's channel matrix to BS in order to calculate the null space, as indicated in Fig. 2.

Step 2: The same as null space precoding scheme used for LPN mentioned above, the expression for $\tilde{\mathbf{V}}_{0,i}^{(0)}$ and $\mathbf{P}_0$ are given after obtaining $\tilde{\mathbf{H}}_{0,v}$. using these quantities, the precoding vector $\mathbf{T}_0$ can be calculated. Note that CSI of BS users which are interfered by LPN j also can be obtained by BS, which is $\mathbf{H}_{j,k_j+n}$, $1 < n \le L_j$. Using the quantities acquired, SINR of all victim BS users interference by LPN j can be calculated by the given equation

$$SINR_{0,k} = \frac{\left| \mathbf{H}_{j,k_j+n} \mathbf{T}_{0,k} \right|^2}{\sigma_{0,k}^2 + \sum_{k'=1, k' \neq k}^{K_0} \left| \mathbf{H}_{j,k_j+n} \mathbf{T}_{0,k'} \right|^2 + \sum_{n=1}^{N} \left| \mathbf{H}_{n,k} \mathbf{T}_{n,k} \right|^2}. \tag{12}$$

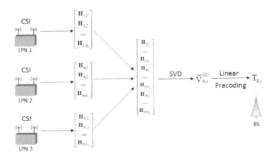

Figure 2. BS performs null space precoding scheme.

161

where $\mathbf{SINR}_{0,k}$ denotes the SINR of the k-th BS user which is interfered by LPN j, $\mathbf{T}_{0,k}$ stands for the precoding matrix regarding to the k-th user.

Step 3: After the acquisition of all SINR of victim BS users interfered by LPN j, we apply bubble sort, ranking from the highest SINR to the lowest. For the $M_j - L_j$ number of victim BS users with the highest SINR, the complimentary space $\tilde{\mathbf{H}}_{j,v}$ can be formed by these chosen users, which is given by

$$\tilde{\mathbf{H}}_{j,v} = \left[\mathbf{H}_{j,K_j+1}^H \cdots \mathbf{H}_{j,M}^H \right]^H. \tag{13}$$

Step 4: Similarly, with $\tilde{\mathbf{H}}_{j,v}$, the null space precoding scheme conducted in LPN j can be achieved.

4.3 User scheduling algorithm based on channel matrix

Another victim user choosing scheme is to make decisions only base on channel matrix of each victim BS user. For the reason that all the channel matrix of BS users interfered by LPN j are accessible for BS, then for each $\mathbf{H}_{j,K_j+n}, n \leq L_j$, the value of $| \mathbf{H}_{j,k_j+n} |^2$ can be calculated. Selecting the users with the highest $M_j - K_j$ number of value, the according $\mathbf{H}_{j,k_j+n}$ are chosen to form the matrix $[\mathbf{H}_{j,K_j+1} \cdots \mathbf{H}_{j,M}]$.

4.4 Advantages of our proposed scheme

Optimize the overall throughput of the whole system:

Regarding the scheduling algorithm based on SINR, by Shannon equation, $C = Blog_2(1 + SINR)$, channel capacity will increment with the increasing SINR. Bigger channel capacity implies higher maximal information rate. Applying greedy algorithm, the higher information rate of a user can lead to a more satisfying system throughput. Hence the optimize throughput problem can be simplified to the interference mitigation problem. During the precoding procedure in LPN j, by greedily placing those BS users which has higher SINR into null space, we are able to eliminate the LPN interference to these $M_j - K_j$ number of BS users, which guarantee the service experience and maximal information rate. As a consequence, it leads to the optimized overall system throughput. For the scheduling algorithm based on $\mathbf{H}$, BS users with the high $|\mathbf{H}|^2$ are protected from the interference from LPN j. Reserving the BS users with the great channel situation and service experience, the optimized system throughput can be achieved.

Low complexity in user management:

The algorithm whose selection principal is SINR will surly increase the computational complexity in BS. Because on one hand, the calculation of SINR requires multiply precoding vector with channel matrix, on the other hand, the variance of white Gaussian noise also needed to be considered. However, the incremental calculation complexity for BS stays within a small scale. Comparing to algorithm regarding SINR, our proposed scheme which base on channel matrix is processed with much lower complexity. Recall that the algorithm depends on merely the channel matrix, which can be obtained by BS without any further calculation.

5 SIMULATION RESULT

Our proposed user scheduling algorithm has been tested by means of system level simulation, whose parameters are shown in Table 1. To evaluate our algorithm, we mainly compare it with null-space precoding scheme which viewed as the foundation as. A two-tier HetNet topology which made up of one BS and four LPNs presented as Fig. 3 is considered.

Table 1. Simulation parameters.

Parameters	Values
Number of the macro cell users	$N_{BS} \in \{42, 46, 50, 54, 58, 62\}, 42$
Number of per LPN users	$N_{LPN} \in \{1, 2, 3, 4\}, 2$
Macro cell radius	500 m
LPNs radius	50 m
Antennas of the macro cell	100
Antennas of the LPNs	6
Total bandwidth	100 Mhz
Requirements of the rate (per user)	20 Mbps
Parameters	Values
Small scale fading distribution	$h_{k,j} \sim CN(0, R_{k,j})$
Standard deviation of log-normal shadowing	7 dB
Path and penetration loss at distance d(km)	$148.1 + 37.6log_{10}(d)$ dB
Normal cases for the BS and LPN	$128.1 + 37.6log_{10}(d)$ dB
Special cases for the LPN: Within 40 m from LPN	$127 + 30log_{10}(d)$ dB
Noise Variance σ_k^2 (5dB noise figure)	−127dBm

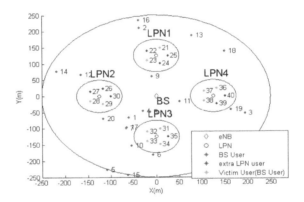

Figure 3. The topology of the macro cell.

Where the BS stands in the center of the micro-cell. Within the coverage of each LPN, victim BS users which denoted by green mark experience the interference from the proper LPN. Especially, for a single LPN, the sum of its user which marked by blue and victim BS users surpasses the number of LPN antennas. While the marked red represents BS users which are free from high interference. Under this circumstance, we assess the system perform-ance over the overall power which measures the optimized system throughput. The following two figures can show the merits of the user scheduling algorithm we introduced in this paper by comparing to conventional ZF scheme and the null-space precoding scheme.

In general, of the equal basis of power, the higher SINR a signal has, the more effective transmission information it possesses, thus leading to a higher throughput. Our proposed user scheduling algorithm aims at optimize the overall throughput, hence in order to achieve this goal, we should satisfy the condition that minimize the power when transmitting the same quantity of information. As shown in (Fig. 4), we consider the relationship between user number and power within one particular LPN. When the user number represented by X-axis no more than 4, they are all considered as LPN users, thus our proposed user schedul-ing algorithm shows no quite difference to the null-space precoding scheme. However with the increasing number of overall users within the LPN covering range, 4 BS users are added. When the the number of additional BS users is smaller than two, which means that when the LPN antennas are sufficient, the interference from other BS/LPNs can be totally eliminated

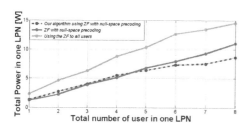

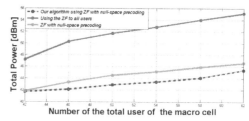

Figure 4.   Impact of different number of users.   Figure 5.   Impact of the total users.

(Zhang et al. 2015) as aforementioned. When there are more than two BS users, the extra antennas of proper LPN is inadequate, and thus the advantages of our algorithm has been embodied by the rather low power consumption.

To evaluate the overall system performance, the simulation is conducted of which the result is shown in (Figs. 4–5). Increasing the number of overall users represents more victim BS users are involved in the macro cell. As can be seen in (Fig. 5), with more victim BS users, and when the number of victim BS users surpasses the number of available extra LPN antennas, our algorithm is significantly better than the null-space precoding scheme and ZF precoding scheme, which the power can reach. Especially under the circumstance of large number of victim BS users, after the ranking procedure in our algorithm, victim users with high SINR receives better service for the reason that the interference is alleviated, thus result in the decreased power and improved throughput.

Consequently, in the scenario we assumed for the simulation results as shown in (Fig. 5), our proposed user scheduling algorithm has been proved to be an effective technique for improving the throughput no matter within one LPN or for the overall system in HetNets. As a result, through mitigating the interference to the selected victim BS users, the system performance can be significantly improved under the condition of inadequate LPN antennas to eliminate all the interference to victim BS users.

6   CONCLUSIONS

In this paper we have tackled the problem of mitigating the cross-tier interference in heterogeneous cellular network environment. In particular, We mainly concentrate on the BS user which interfered by the nearby LPN. Our proposed algorithm based on a novel precoding scheme called null-space precoding scheme. The precoding scheme can significantly alleviate the certain amount of interference by victim BS user, which the amount is restricted by the extra number of LPN antennas. Our proposed scheme is conducted under the circumstance of victim BS users are far more than the extra LPN antennas. The selecting principal our algorithm adopted lies in either SINR or channel matrix of these BS users. By applying our algorithm, LPN is able to mitigate interference from certain number of victim BS users in order to achieve the optimized system throughput and at the same time, no sharp rise in calculation complexity or acquired power in BS exists. Simulation result demonstrates that the victim BS user selecting algorithm we proposed is practically feasible, which is shown by the improved information rate. The analytical result is supported by the simulation outcome.

REFERENCES

Dahlman, E., Parkvall, S., Skold, J. & Beming, P. 2007. 3 g evolution: hspa and lte for mobile broadband.:22.

Feng, W., Wang, Y., Ge, N., Lu, J. & Zhang, J. 2013. Virtual mimo in multi-cell distributed antenna systems: coordinated transmissions with large-scale csit. *IEEE Journal on Selected Areas in Communications* 31(10):2067–2081.

Hoydis, J., Ten Brink, S. & Debbah, M. 2012. Comparison of linear precoding schemes for downlink massive MIMO. *Communications (ICC), 2012 IEEE International Conference*:2135–2139.

Hoydis, J., Hosseini, K. Brink, S.T. & Debbah, M. 2013. Making smart use of excess antennas: massive mimo, small cells, and tdd. *Bell Labs Technical Journal* 18(2):5–21.

Hoydis, J., Ten Brink, S. & Debbah, M. 2013. Massive mimo in the ul/dl of cellular networks: how many antennas do we need? *Selected Areas in Communications IEEE Journal* 31(2): 160–171.

Marzetta, T.L. 2015. Massive mimo: an introduction. *Bell Labs Technical Journal* 20: 11–22.

Parkvall, S., E. Dahlman, G.J. Ongren, S. Landstrom & L. Lindbom 2011 Heterogeneous network deployments in LTE *The soft-cell approach Ericsson Review* (2).

Stefan, I., Burchardt, H. & Haas, H. 2013. Area Spectral Efficiency Performance Comparison between VLC and RF Femtocell Networks. *Communications (ICC), 2013 IEEE International Conference*: 3825–3829.

Sun, S., Ju, Y. & Yamao, Y. 2013. Overlay cognitive radio of dm system for 4 g cellular networks. *Wireless Communications IEEE* 20(2):68–73.

Sun, S., Kadoch, M. & Ran, T. 2015. Adaptive son and cognitive smart lpn for 5 g heterogeneous networks. *Mobile Networks Applications*.

Tse, D. & Viswanath, P. 2004. Fundamentals of wireless communications. *Eth Z"rich Lecture Script* 3(5): B6-1–B6-5.

Yin, H., Gesbert, D., Filippou, M. & Liu, Y. 2012. A coordinated approach to channel estimation in large-scale multiple-antenna systems. *IEEE Journal on Selected Areas in Communications* 31(2): 264–273.

Zhang, F., S. Song, B. Rong, F.R. Yu & K. Lu A Novel Massive MIMO Precoding Scheme for Next Generation Heterogeneous Network.

Signal and Information Processing, Networking and Computers – Chen & Huang (Eds)
© 2016 Taylor & Francis Group, London, ISBN 978-1-138-02881-4

Risk assessment method of multidimensional AHP based on SoS architecture

Yun Huang, Huaizhi Yan, Yi Zheng & Chong Zhao
Beijing Institute of Technology, Beijing, China

ABSTRACT: This paper researches the characteristics of a cloud information system in combination with the theories of information security risk assessment, aiming to put forward a feasible and effective risk assessment method for large, complex cloud information systems. Based on the AHP method as the foundation, the assessment method introduces the idea of multidimensional modeling of the SoS (System of Systems) architecture in level classification, and presents the multidimensional AHP analysis method which is used to analyze the complex cloud information system risks in order to reach a more comprehensive and objective conclusions. In combination with the actual cloud information system, it is verified that the proposed theory is feasible and effective.

Keywords: risk assessment; cloud information system; AHP; SoS

1 INTRODUCTION

As computer and network information technologies develop quickly, cloud technology has been rapidly popularized to change people's access to software and hardware, and also greatly reduced the construction cost of the information system. The information system based on cloud computing consists of many interconnected and interdependent subsystems, which takes the place of a traditional client server architecture. However, along with the convenience, it also brings new threat and risk to us.

In May 2014, NetIQ released cloud security survey results, which show that 45% of the respondents doubt safety processing of cloud service providers, whereas 70% of the respondents said cloud services threaten sensitive data. As the cloud computing technology becomes more and more mature, risk management and control of cloud computing have become the key for development of cloud computing. To implement effective risk assessment, it can be very beneficial to improve the industry recognition.

This paper is based on complex and open characteristics of the cloud information system, and adopts the hierarchical analysis theory. By virtue of multidimensional modeling of the SoS architecture basis and based on the complex relationship among the subsystems, a holographic and multi-dimensional comprehensive evaluation model was established. In combination with knowledge of information security risk assessment, risk assessment of information system based on cloud computing was carried out.

2 CLOUD INFORMATION SYSTEM RISK ANALYSIS

Cloud information system refers to an interdependent network system with a complex structure and large scale network connection which are constructed through virtualization and distributed computing technology in order to realize the large-scale software system infrastructure. Cloud computing is featured by virtualization, remote network access, shared resource pool and scalability, etc. Because of these characteristics, the information system based on cloud computing becomes more vulnerable to malicious attacks. In addition, once the security problem happens, it will affect a large number of users, resulting in more serious consequences. Therefore, the information system based on cloud computing requires higher security, and thus it is necessary to carry on risk management.

Cloud information system is a special case of the information system, which complies with the same security principles of the traditional information system, and encounters the same security risk. However, the introduction of new technologies and service patterns will surely bring new security risks. Cloud computing under different service types of technical support will encounter different technical risks. Iaas's virtualization technology, Paas's distributed processing technology and Saas's application virtualization technology are crucial for construction of cloud information system, and will also generate main risks to be faced by the cloud information system.

Inheritance of cloud service ability determines the continuity of service mode risk (Chi C. & Jing Y. 2014). Upper cloud services will encounter new risks besides the above technical risks. In addition, ownership and management of the cloud information system are separated, so that the system will encounter unsustainable cloud services, identity management, and more management risks. Because of the characteristics of cloud information system, many new laws and regulations as well as industry application risks are taken into account.

3 MULTIDIMENSIONAL ANALYTIC HIERARCHY MODEL

3.1 Analytic hierarchy process

Risk assessment is an essential part in information security risk management. Risk analysis generally involves the qualitative, quantitative and comprehensive analyses of the system in three aspects including assets, threats and vulnerability. Analytic Hierarchy Process (AHP) is a comprehensive analysis of pairwise comparison and weight of the calculation. It is very suitable for the complex cloud computing information system to realize the risk assessment. Due to the strong competence of AHP in dealing with multi criteria problems, it can effectively make the complex problem into an ordered hierarchy, and can quantify the decision maker's experience judgment in the absence of quantitative data. However, in traditional AHP hierarchical modeling process, the human subjective factors used to establish the hierarchical structure model greatly influence the given pairwise comparison matrix, making the results difficult to be accepted by all the decision makers.

3.2 SoS architecture modeling

Any interaction or mutual dependence of a set of entities in the form of a unity and purpose of Systems is called as SoS (System of Systems). SoS is used in system analysis, design and implementation of system structure in the system. In system analysis, the SoS is featured by loose coupling, coincidence with changing dynamic, complexity and interactive difficulties (Jingjing L., Jing L., Zhiguo W. 2014). Referring to V2.0 DoDAF, the framework of the DoD (Department of Defense of the United States) Architecture Framework, a risk assessment method is proposed for the application of SoS architecture modeling in the analytic hierarchy process.

Cloud information system is a complex system comprising multiple subsystems, wherein all the subsystems are interdependent, the internal structure is very complex, and there are many threats which appear across regions. If it fails to make comprehensive modeling analysis of cloud information system, the criterion layer and layer scheme obtained by the hierarchy analysis will not involve all the risk factors, which will inevitably and greatly reduce validity and rationality of risk analysis results. The system is abstracted by SoS, and the cloud information system is regarded as a whole system. According to the theory of system behaviors, the multi-dimensional modeling of cloud information system is carried out.

4 MULTIDIMENSIONAL ANALYSIS AND EVALUATION

Based on the theory of previous chapters, we've verified that the multidimensional and hierarchical analysis method is feasible and effective to analyze a cloud services platform built by a company.

4.1 *Project background*

A company in Beijing wants to make use of cloud services technologies and solutions in combination with openstack, zookeeper and other open source software as the foundation to

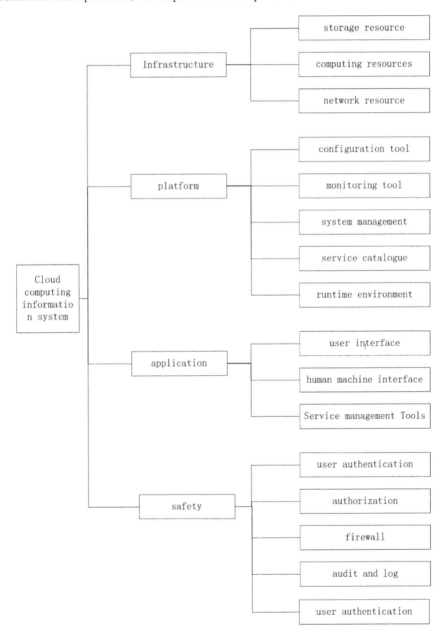

Figure 1. SoS function hierarchical model.

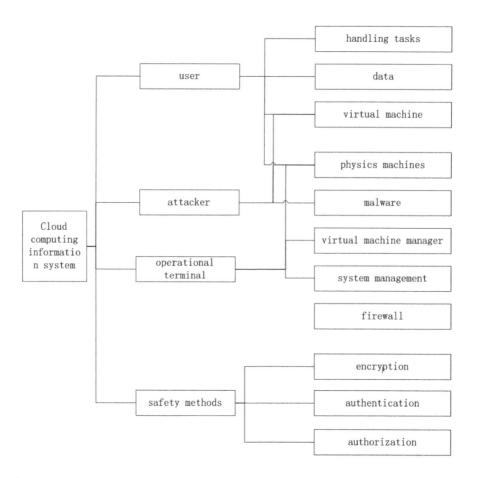

Figure 2.   SoS stakeholder hierarchical model.

build a high performance computing platform and realize unified management of software and hardware resources. To realize the deployment of resources in the form of virtualization and automation, and provide services to the unity, the system should be developed well in order to provide support for related IT services of the company.

4.2   *Multidimensional modeling*

Firstly, hierarchy structure model of system shall be established. Cloud computing information system is a complex system which is interconnected and interdependent, because the complex SoS architecture cannot be modeled by a single model. On the contrary, from different points of view, systems in different subsystem models are established respectively. We call them sub models. We integrate the relationship among these sub models through the shared state variables which are integrated into a meta model. Coordination and integration of the meta model is to establish interconnected and interdependent relationship through direct and indirect analysis model. Through the SoS architecture for multi model modeling and coordinated and unified modeling method, more efficient hierarchical models can be obtained to reveal and represent different aspects of risk factors.

The hierarchical analysis model is divided into three layers: target layer, criterion layer, and project layer. Application of the multidimensional modeling highlights analysis of relations among subsystems and dependency relations from multiple perspectives (Yacov y. H., Barry M.H., Zhenyu G., et al., 2014), so that above SoS architecture is decomposed from

170

multiple perspectives. For example, stakeholders analysis, obtains the hierarchy models shown in Figures 1–2 are obtained according to analysis of stakeholders.

4.3 Analysis and verification

4.3.1 Multidimensional modeling
In the first section of this chapter, the risk factors of the system were studied, while the twenty point risks are the risk factors of the system, corresponding to R1-R20. Combined with the above analysis, the function of multi-dimensional modeling of the cloud computing information system and the two angles of the stakeholders are taken as the criteria. The hierarchical model was obtained as Table 1 and Table 2:

4.3.2 Constructs the multidimensional correlation matrix
Based on hierarchical modeling analysis, the risk factors are evaluated, and the structure of the comparison matrix is as follows:

Functional hierarchy model:

$$A = \begin{bmatrix} 1 & 1/5 & 2 & 1/4 \\ 5 & 1 & 3 & 2 \\ 1/2 & 1/3 & 1 & 1/2 \\ 4 & 1/2 & 2 & 1 \end{bmatrix}$$

$$B_1 = \begin{bmatrix} 1 & 8 & 4 \\ 1/8 & 1 & 2 \\ 1/4 & 1/2 & 1 \end{bmatrix}$$

$$B_2 = \begin{bmatrix} 1 & 3 & 1/2 & 1/3 & 1/2 & 1/4 \\ 1/3 & 1 & 4 & 5 & 4 & 9 \\ 2 & 1/4 & 1 & 2 & 1 & 2 \\ 3 & 1/5 & 1/2 & 1 & 1/2 & 1 \\ 2 & 1/4 & 1 & 2 & 1 & 2 \\ 4 & 1/9 & 1/2 & 1 & 1/2 & 1 \end{bmatrix}$$

Table 1. Functional hierarchy model.

Target layer	Standard layer	Scheme layer
Cloud computing information system	Infrastructure	R,R6,R14
	Platform	R2, R3, R5, R7, R8, R10
	Application	R4, R11, R12, R13, R17
	Security	R9, R15, R16, R18, R19, R20

Table 2. Stakeholder hierarchy model.

Target layer	Standard layer	Scheme layer
Cloud computing information system	General user	R7, R9, R13, R15, R16, R19
	Attacker	R3, R4, R6, R8
	Operable terminal	R1, R11, R14
	Security means	R2, R5, R10, R12, R17, R18, R20

$$B_3 = \begin{bmatrix} 1 & 1/3 & 1/2 & 1 & 1/3 \\ 3 & 1 & 2 & 3 & 1 \\ 2 & 1/2 & 1 & 2 & 1/2 \\ 1 & 1/3 & 1/2 & 1 & 1/3 \\ 3 & 1 & 2 & 3 & 1 \end{bmatrix}$$

$$B_4 = \begin{bmatrix} 1 & 2 & 1/2 & 1 & 1/2 & 2 \\ 1/2 & 1 & 1/4 & 1/2 & 1/3 & 1 \\ 2 & 4 & 1 & 2 & 2 & 4 \\ 1 & 2 & 1/2 & 1 & 1/2 & 2 \\ 2 & 3 & 1/2 & 2 & 1 & 3 \\ 1/2 & 1 & 1/4 & 1/2 & 1/3 & 1 \end{bmatrix}$$

Stakeholder hierarchy model:

$$A = \begin{bmatrix} 1 & 2 & 3 & 1/2 \\ 1/2 & 1 & 2 & 1/3 \\ 1/3 & 1/2 & 1 & 1/4 \\ 2 & 3 & 4 & 1 \end{bmatrix}$$

$$B_1 = \begin{bmatrix} 1 & 1/2 & 1/2 & 1/6 & 1/3 & 1/5 \\ 2 & 1 & 1 & 1/3 & 1/2 & 1/3 \\ 2 & 1 & 1 & 1/3 & 1/2 & 1/3 \\ 6 & 3 & 3 & 1 & 2 & 1 \\ 3 & 2 & 2 & 1/2 & 1 & 1/2 \\ 5 & 3 & 3 & 1 & 2 & 1 \end{bmatrix}$$

$$B_2 = \begin{bmatrix} 1 & 2 & 1/2 & 4 & 2 \\ 1/2 & 1 & 1/3 & 2 & 1/2 \\ 2 & 3 & 1 & 6 & 2 \\ 1/4 & 1/2 & 1/6 & 1 & 1/3 \\ 1/2 & 2 & 1/2 & 3 & 1 \end{bmatrix}$$

$$B_3 = \begin{bmatrix} 1 & 2 & 8 \\ 1/2 & 1 & 4 \\ 1/8 & 1/4 & 1 \end{bmatrix}$$

$$B_4 = \begin{bmatrix} 1 & 2 & 4 & 3 & 3 & 2 & 4 \\ 1/2 & 1 & 3 & 2 & 2 & 1 & 3 \\ 1/4 & 1/3 & 1 & 1/2 & 1/2 & 1/3 & 1 \\ 1/3 & 1/2 & 2 & 1 & 1 & 1/2 & 2 \\ 1/3 & 1/2 & 2 & 1 & 1 & 1/2 & 2 \\ 1/2 & 1 & 3 & 2 & 2 & 1 & 3 \\ 1/4 & 1/3 & 1 & 1/2 & 1/2 & 1/3 & 1 \end{bmatrix}$$

172

4.3.3 *Weight vector*

The comparison matrix is calculated, while the maximum eigenvalues and eigenvectors are obtained, as shown in Table 3:

4.3.4 *Check consistency and generate reports*

Consistency the model weight results obtained by the third step is checked, and the consistency rate is examined, while the value is less than 1. These results are valid.

$X(A, C_n) = X(A, B_m)*X(B_m, C_n)$ are calculated. The $X(A, C_n)$ of the risk factors is shown in Table 4.

Table 3. Model weight calculation results.

Model name	Matrix	Eigenvalues (round-off)	Random consistency ratio	Feature vector (round-off)
Functional hierarchy model	A	1.008	0.986	[0.02 0.90 0.02 0.06]
	B1	1.013	0.713	[0.09 0.08 0.02]
	B2	7.719	0.284	[0.17 0.37 0.13 0.09 0.13 0.11]
	B3	2.000	0.741	[0.04 0.07 0.20 0.03 0.03]
	B4	2.010	0.660	[0.04 0.03 0.60 0.22 0.05]
Stakeholder hierarchy model	A	1.008	0.986	[0.02 0.03 0.03 0.91]
	B1	2.012	0.659	[0.02 0.01 0.01 0.45 0.04 0.45]
	B2	1.009	0.891	[0.02 0.01 0.96 0.01]
	B3	1.006	0.719	[0.96 0.03 0.01]
	B4	2.014	0.629	[0.06 0.16 0.01 0.05 0.03 0.16 0.01]

Table 4. Risk factor weight.

Risk factor	Function hierarchy model	Stakeholder hierarchy model	Average value
R1	0.018	0.0288	0.0234
R2	0.153	0.546	0.3495
R3	0.333	0.0006	0.1668
R4	0.0008	0.0003	0.00055
R5	0.117	0.1456	0.1313
R6	0.0016	0.0288	0.0152
R7	0.081	0.0004	0.0407
R8	0.117	0.0003	0.05865
R9	0.0024	0.0002	0.0013
R10	0.099	0.0091	0.05405
R11	0.014	0.0009	0.00745
R12	0.004	0.0455	0.02475
R13	0.0006	0.0002	0.0004
R14	0.0004	0.0003	0.00035
R15	0.0018	0.009	0.0054
R16	0.036	0.0008	0.0184
R17	0.0006	0.0273	0.01395
R18	0.0036	0.1456	0.0746
R19	0.0132	0.009	0.0111
R20	0.003	0.0091	0.00605

5 SUMMARY

As a new technology, cloud computing has brought a subversive revolution in software and hardware. Meanwhile, the structure of information system has become more complex and challenging in security. Based on analysis of risk factors of cloud computing information system, this paper conducts risk assessment of cloud computing information system by using the multi-dimensional analytic hierarchy process based on SoS. By using the idea of multidimensional engineering modeling, the paper conducts optimal expansion of the hierarchical analysis method in order to produce more comprehensive and detailed data, improve defects of analytic hierarchy process in application, and provide decision makers with better quantitative basis. The paper also discusses the maturation of risk analysis of cloud computing information system.

REFERENCES

Chi, C. & Jing, Y. 2014. *Cloud computing security system*, Beijing: Science press.

Heng, L, 2010. Cloud computing security risk assessment and analysis. *Third session of information security vulnerability analysis and risk assessment of general assembly*. Hefei.

Houmani, H. & Medromi, H. 2013. Survey: Risk assessment for cloud computing international. *Journal of advanced computer science and applications* 4(12):143–148..

Jingjing, L. & Jing, L. 2014. SoS system structure modeling method and application for the object. *Chinese Ship Research* (4):11–17.

Peiyu, L. & Donga, L. 2011. The new risk assessment model for information system in cloud computing environment. *Engineering procedia* 15:3200–3204.

Qiang, G.2013. Overview of development of foreign cloud computing. *Information technology* (6): 1–3.

Rong, J. & Zifei, M.2015. Cloud computing security risk factors and coping strategies. *Modern information* 35(1):85–90.

Yacov, Y. H. 2014. Assessing systemic risk to cloud computing technology as complex interconnected systems of systems. *Systems engineering* 18:284–299.

Yuzhen, C. 2013. Cloud Service Information Security Risk Assessment Indicators and Methods of Research. Beijing jiaotong university.

Signal and Information Processing, Networking and Computers – Chen & Huang (Eds)
© 2016 Taylor & Francis Group, London, ISBN 978-1-138-02881-4

Line outage identification based on partial measurement of Phasor Measurement Unit (PMU)

Jingyuan Guo, Tao Yang, Hui Feng & Bo Hu
Department of Electronics Engineering, Fudan University, Shanghai, China

ABSTRACT: In a large scale smart grid, it is critical to identify the line outages fast and accurately. In this paper, we propose a method to identify the line outages in a greedy fashion that has a satisfying accuracy, without introducing additional complexity, requiring only the PMU measurement of internal buses and topology information. Using sparse representation, the method can achieve real time identification. In addition, we also approach the problem from the probability inference perspective, using multiple random samples generated from an initial distribution to infer the real distribution of the line outage. Simulation tests on IEEE 118-bus systems show that the proposed method has advantages, such as high accuracy and good robustness.

Keywords: DC power flow; line outage; PMU; sparse representation

1 INTRODUCTION

In a large smart grid, a small turbulence can lead to cascading failure. One of the keys to prevent such accidents is to quickly and accurately locate the initial failure. Line outage identification is important in detecting grid status and real time line analysis. The performance of identification is closely related to the accuracy of measurement and network topologies. In order to study the line outage identification, we can model the grid as a graph $G = (V, E)$, topological relations will be determined by the state of circuit breakers and switches.

Phasor Measurement Unit (PMU) has been widely used in state estimation (Throp et al. 1985), dynamic reliability evaluation (Sun et al. 2007) and visualization (Chun-Lien & Bo-Yuan 2006). Recently, PMU has also been introduced to the grid for real time fault monitoring. Once PMU is globally deployed in the grid, state estimation and line outage detection can be easily performed. However, due to the cost consideration, currently PMUs are only deployed in certain key buses.

Directly making use of PMU to identify line outages can be modelled by an optimization problem. R. Emami (Emami & Abur 2013) assumes each line outage can be transformed into change of power injection to the two buses connected by the line. Tate & Overbye (2008) introduced algorithms to detect single and double line outages by modeling it as an optimization problem, which minimizes the PMU angle difference error. However, these algorithms have high computation complexity, and different approaches are adopted for different number of line outages. In order to improve computation efficiency, Zhu & Giannakis (2012) model the identification problem through sparse representation, identifying the line based on DC power flow equations and the topology change before and after the outage, but it requires the inverse of matrix B, which is not a full rank. He & Zhang (2011) exploits Gaussian Markov random field but it assumes that each phase angle is independent with another, and requires the global observation of voltage phase angle, which is not appropriate for the real scenario.

This paper introduces a modification to the method in (Zhu & Giannakis 2012), which they only use partial PMU observations. By dividing the grid into two parts: internal system that is PMU observable and external system that is not PMU observable only the internal bus measurements to identify line outages in the whole system is used. With DC power flow equation model, by blocking the weighted Laplace matrix, an optimization problem that uses only partially observation is derived. We apply the nature of sparsity of the problem and reconstruct the signal using compressed sensing techniques. In addition, we also solve the problem in a probability inference approach, by estimating the ML solution of the parameter of distribution. Simulations show the proposed methods have advantage in both accuracy and computation efficiency over other methods based on PMU measurements.

The rest of the paper is organized as follows, the background of DC power flow and grid topology is given in Section 2. In Section 3, the system model and line outage identification using greedy algorithm is introduced, and in Section 4, a maximum likelihood estimation algorithm is presented. The simulation and analysis are shown in Section 5, and finally the paper is concluded in Section 6.

2 BACKGROUND

2.1 DC power flow

To simplify the model analysis, here we consider DC power system only. The grid comprises of generation bus, load bus, and the transmission line. The power flow equation is defined as follows (Stott et al.2009):

$$P_k = \sum_{j=1}^{N} |V_k||V_j| \left(G_{kj} \cos\left(\theta_k - \theta_j\right) + B_{kj} \sin\left(\theta_k - \theta_j\right) \right) \tag{1}$$

where P_k is the injection power to bus k, V_k is the voltage phasor on bus k, G_{kj} is the conductance between bus k and j, B_{kj} is the susceptance between bus k and j, and k is the phase angle for bus k. Usually, the resistance of transmission circuits are significantly less than the reactance (the ratio of x/r is between 2 and 10), neglecting resistance, (1) becomes:

$$P_k = \sum_{j=1}^{N} |V_k||V_j| \left(B_{kj} \sin\left(\theta_k - \theta_j\right) \right) \tag{2}$$

On the other side, the difference in angles $\theta_k - \theta_j$ of the voltage phasors between two buses k and j is less than 10–15 degrees. It can be considered as insignificant in analysis, so we approximate (2) as the following equation:

$$P_k = \sum_{j=1}^{N} |V_k||V_j| \left(B_{kj} \left(\theta_k - \theta_j\right) \right) \tag{3}$$

In the per-unit system, the value of voltage magnitude is close to 1.0, typically between 0.9–21.05. Thus, (3) becomes

$$P_k = \sum_{j=1,j\neq k}^{N} \left(B_{kj} \left(\theta_k - \theta_j\right) \right) \tag{4}$$

Rewrite (4) in matrix form:

$$\mathbf{p} = \mathbf{B}\theta \tag{5}$$

where the $N \times N$ matrix **B** is Laplace matrix of the grid. Equation (5) shows the real power injection is dependent on the difference of phasor angle of buses. Compared to AC power flow, DC power flow can simplify computation without losing much accuracy, and its form

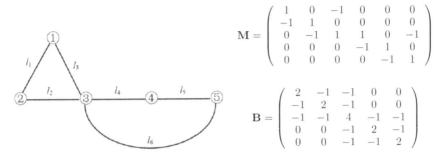

Figure 1. Illustration of a simple grid graph with its incidence matrix and Laplace matrix.

suits system state estimation and accident analysis, thus, gained much application in grid fault analysis.

2.2 Grid topology

Consider a power grid with N buses and L transmission lines, which constitutes a graph $G = (V, E)$, where V is the set of all nodes and E is the set of all edges. According to graph theories, Laplace matrix **B** is related to the bus-line incidence matrix **M**. **M** is formed with columns $\{m_l\}_{l=1}^{L}$, each of which represents an edge that connects two nodes, with values 1 or -1 at the place of corresponding buses, and 0 elsewhere. Incorporating the reactance, the relationship between **B** and **M** can be written as:

$$\mathbf{B} = \mathbf{MD}_x\mathbf{M}^T = \sum_{l=1}^{L}\frac{1}{x_l}\mathbf{m}_l\mathbf{m}_l^T \qquad (6)$$

where the diagonal matrix D_x represents reactance, with the lth element corresponding to the susceptance of the lth line. According to (6), Laplace matrix **B** is the sum of L matrices of $m_l m_l^T$. Each edge $<m, n>$ is mapped to the four elements in **B**: B_{mm}, B_{mn}, B_{nm}, and B_{nn}. Notably, the $N \times N$ matrix **B** is singular, it has rank $N - 1$. Figure 1 gives a simple illustration, with the **M** and **B** shown above.

3 GREEDY ALGORITHM

3.1 System model

This section models the system with voltage phase angle information measured from PMU and grid topology, and tries to identify the correct outage line. It can be solved by establishing the model according the following equation (Tate & Overbye.2008):

$$E^* = \arg\min_{E \in \varepsilon}\|\Delta\theta_{observed} - f(E)\| \qquad (7)$$

where ε represents the set of all events to detect. If there is only one outage line, ε is all possible outage line. $f(E)$ is the change of phase angle due to certain event in ε, $\Delta\theta_{observed}$ is the measured difference of voltage phase angle. The problem is to search for the event E^* in ε that matches with the measured voltage phase angle change.

According to previous section, in DC power flow model, power injections to each node and voltage phase angle satisfy (5), and the linear power flow equation satisfies (Zhu & Giannakis.2012):

$$\mathbf{p}' = \mathbf{B}'\theta' = \mathbf{p} + \eta = \mathbf{B}\theta + \eta \qquad (8)$$

where B' is the Laplace matrix after line outage. Equation (8) shows the line outage causes minor turbulence in power injection, which equals to the original power injection plus a white noise vector η with zero mean and covariance matrix $\sigma_\eta^2 I$ (Schellenberg et al. 2005). An outage line in the grid corresponds to a certain zero column in the incidence matrix. Thus, the Laplace matrix after line outage becomes:

$$\mathbf{B}' = \mathbf{M}'\mathbf{D}'_x(\mathbf{M}')^T = \sum_{l \notin \Delta E} \frac{1}{x_l} \mathbf{m}_l \mathbf{m}_l^T + \sum_{l \in \Delta E} 0 \cdot \mathbf{m}_l \mathbf{m}_l^T \tag{9}$$

where ΔE is the set of outage lines. Define $\Delta B = \sum_{l \in \Delta E} \frac{1}{x_l} \mathbf{m}_l \mathbf{m}_l^T$, so $\mathbf{B}' = \mathbf{B} - \Delta\mathbf{B}$, substituting into (8) we get $(\mathbf{B} - \Delta\mathbf{B})\theta' = \mathbf{B}\theta + \eta$, which is

$$\mathbf{B}(\theta' - \theta) = \Delta\mathbf{B}\theta' + \eta \tag{10}$$

The right side of the equation can be written as:

$$\Delta\mathbf{B}\theta' = \sum_{l \in \Delta E} \frac{1}{x_l} \mathbf{m}_l \mathbf{m}_l^T \theta' = \sum_{l \in \Delta E} s_l \mathbf{m}_l = \sum_{l \in \Delta E} s_l \mathbf{m}_l + \sum_{l \notin \Delta E} 0 \cdot \mathbf{m}_l = \mathbf{M}s \tag{11}$$

where $s_l = \frac{\mathbf{m}_l^T \theta'}{x_l}$ and s is a L dimensional sparse vector with sparsity level card (ΔE), only having nonzero value s_l at the changed line l. Thus, (10) becomes

$$\mathbf{B}\Delta\theta = \mathbf{M}s + \eta \tag{12}$$

The number of line outage is not great in the grid, usually only one or two line outage takes place (card$(\Delta E) \ll L$), which suits the sparse scenario. Thus, an appropriate equation for the model can be derived, turning the problem into:

$$E^* = \underset{E \in \varepsilon}{\arg\min} \|\Delta\theta - (\mathbf{B}^{-1}\mathbf{M}s)\|^2 \tag{13}$$

Under this condition, line outage identification problem turns into a sparse signal identification problem. Sparse representation is a subject with extensive research recently, and there are already plenty of high efficiency and stable algorithms for sparse signal recovery.

3.2 *Global line outage identification using internal system measurement*

PMUs are only located in certain nodes. The internal system has PMUs installed, therefore, can be directly observed; whereas the external system cannot. On the other hand, the grid Laplace matrix B is singular, solving the inverse introduces error. So we try to solve the problem and improve the accuracy of global line outage identification by using internal system observation. This section blocks the weighted Laplace matrix, using only partial observation. It satisfies the real scenario of not having PMUs located in each node, and avoids having to solve the inverse of matrix B.

According to the sparse observation Equation (12), since only the internal measurement of can be achieved, it makes sense to divide into two parts: the measurable internal part I and the unmeasurable external part E. Similarly, partitioning the Laplace matrix $\mathbf{B}$ and incidence matrix $\mathbf{M}$ in (12), we get:

$$\mathbf{B} = \begin{bmatrix} \mathbf{B}_I & \mathbf{B}_{IE} \\ \mathbf{B}_{EI} & \mathbf{B}_E \end{bmatrix} \quad \mathbf{M} = \begin{bmatrix} \mathbf{M}_I \\ \mathbf{M}_E \end{bmatrix} \tag{14}$$

$\mathbf{M}_I$ corresponds to the part of internal nodes in $\mathbf{M}$, $\mathbf{M}_E$ corresponds to the part of external nodes in $\mathbf{M}$, the four blocks of $\mathbf{B}$ are $\mathbf{B}_I = \mathbf{M}_I \mathbf{D}_x \mathbf{M}_I^T$, $\mathbf{B}_{IE} = \mathbf{M}_I \mathbf{D}_x \mathbf{M}_E^T$, $\mathbf{B}_{EI} = \mathbf{M}_E \mathbf{D}_x \mathbf{M}_I^T$ and $\mathbf{B}_E = \mathbf{M}_E \mathbf{D}_x \mathbf{M}_E^T$.

So (12) can be rewritten as:

$$\begin{bmatrix} \mathbf{B}_I & \mathbf{B}_{IE} \\ \mathbf{B}_{EI} & \mathbf{B}_E \end{bmatrix} \begin{bmatrix} \Delta\theta_I \\ \Delta\theta_E \end{bmatrix} = \begin{bmatrix} \mathbf{M}_I \\ \mathbf{M}_E \end{bmatrix} s + \begin{bmatrix} \eta_I \\ \eta_E \end{bmatrix} \tag{15}$$

Eliminating $\Delta\theta_E$, we have:

$$(\mathbf{B}_I - \mathbf{B}_{IE}\mathbf{B}_E^{-1}\mathbf{B}_{EI})\Delta\theta_I = (\mathbf{M}_I - \mathbf{B}_{IE}\mathbf{B}_E^{-1}\mathbf{M}_E)s + \eta_I - \mathbf{B}_{IE}\mathbf{B}_E^{-1}\eta_E \tag{16}$$

Unlike (12), no inverse for B is needed, rather for $\mathbf{B}_E^{-1}$. The rank of $\mathbf{B}$ is $N-1$, so $\mathbf{B}_E$ is full rank and reversible. This would avoid introducing error by doing pseudo-inverse. Let $\mathbf{y} = (\mathbf{B}_I - \mathbf{B}_{IE}\mathbf{B}_E^{-1}\mathbf{B}_{EI})\Delta\theta_I$, $\mathbf{A} = (\mathbf{M}_I - \mathbf{B}_{IE}\mathbf{B}_E^{-1}\mathbf{M}_E)$, $\eta' = \eta_I - \mathbf{B}_{IE}\mathbf{B}_E^{-1}\eta_E$, (16) becomes:

$$\mathbf{y} = \mathbf{A}s + \eta' \tag{17}$$

By solving sparse vector s, line outage is found. It also suits for completely observable systems, in which one can only pick one node as external, leaving others as internal nodes, so $\mathbf{B}_E$ degrades into a single value b_E, no inverse operation is needed.

3.3 Line outage identification based on OMP

With sparse equation derived, we need dictionary **A**, measurement **y** and corresponding way to reconstruct the sparse signal. In (17), the dimension of **y** is far smaller than that of s, there could be numerous solutions and is hard to reconstruct the original signal. However, since s is K-sparse, theoretically, we can solve the following norm optimization, reconstruct **s** from **y**:

$$\hat{\mathbf{s}} = \min\|\mathbf{s}\|_0 \quad s.t. \quad \mathbf{y} = \mathbf{A}s \tag{18}$$

where $\|\mathbf{s}\|_0$ is the l_0-norm of s. Donoho et al. has shown that is a NP-hard problem (Donoho 2006). Given this, basis pursuit and greedy methods are developed (Mallat & Zhang 1993), in which Orthogonal Matching Pursuit (OMP) can result in fast greedy solution fast in a given sparsity level. Here, we adopt this method to solve the sparse problem of line outage identification.

4 PROBABILITY INFERENCE ALGORITHM

4.1 Line outage identification based on maximum likelihood

To solve (17) in a greedy fashion as proposed in Algorithm 1 could lead to possible error aggregation. To mitigate the effect, we approach the problem again by using the samples generated from a prior distribution, and hope to find an approximated distribution of the outage line.

In order to incorporate probability inference, first, we modify the sparse solution **s** in (17) into a binary vector **b**, where $\{b_l\}_{l=1}^L$ is a binary vector, $b_l = 1$ only when the lth line belongs to $\Delta\varepsilon$, $b_l = 0$ else. So it can be modeled as a Bernoulli distribution.

$$\Delta\mathbf{B} = \sum_{l=\Delta\varepsilon}^L \frac{1}{x_l}\mathbf{m}_l\mathbf{m}_l^T = \mathbf{MD}_x diag(b)\mathbf{M}^T \tag{19}$$

So the right side of (10) changes to

$$\Delta\mathbf{B}\theta' + \eta = \mathbf{MD}_x diag(b)\mathbf{M}^T\theta' + \eta = \mathbf{MD}_x diag(\mathbf{M}^T\theta')b + \eta \tag{20}$$

Writing $\mathbf{y} = \mathbf{B}(\theta' - \theta)$ and $\mathbf{A} = \mathbf{MD}_x diag(\mathbf{M}^T\theta')\mathbf{b}$, so (10) becomes

$$\mathbf{y}' = \mathbf{A}'b + \eta' \tag{21}$$

179

Suppose **b** is Bernoulli distributed with probability **v**, each line $b_l = 1$ with probability v_l, the joint distribution is:

$$P(\mathbf{b}|\mathbf{v}) = \prod_{l=1}^{L} v_l^{b_l}(1-v_l)^{1-b_l}$$

For N samples of $b\{\boldsymbol{b}^1, \boldsymbol{b}^2, \boldsymbol{b}^3, \dots, \boldsymbol{b}^N\}$,

$$P(\mathbf{b}^1, \mathbf{b}^2, \mathbf{b}^3, \dots, \mathbf{b}^N|\mathbf{v}) = \prod_{i=1}^{N}\prod_{l=1}^{L} v_l^{b_l^i}(1-v_l)^{1-b_l^i} \tag{22}$$

Thus, we get the likelihood function

$$L(v) = \ln P(\mathbf{b}^1, \mathbf{b}^2, \mathbf{b}^3, \dots, \mathbf{b}^N|\mathbf{v}) = \sum_{i=1}^{N}\sum_{l=1}^{L} b_l^i \ln v_l + (1-b_l^i)\ln(1-v_l) \tag{23}$$

Differentiate (23) by v,

$$\frac{\partial L(v)}{\partial v_l} = \sum_{i=1}^{N} \frac{b_l^i}{v_l} - \frac{(1-b_l^i)}{1-v_l} \tag{24}$$

So the ML estimation of v_l is

$$(\hat{v}_l)_{ML} = \frac{1}{N}\sum_{i=1}^{N} b_l^i \tag{25}$$

However, (25) doesn't consider the effectiveness of the samples. In fact, only the samples that satisfy (21) have high validity, so we choose only those samples that satisfy (21) to update, and the ML estimation of v_l becomes:

Algorithm 1 OMP Algorithm for line outage identification

Input:
$N \times L$ observation matrix **A**, $N \times 1$ observation signal **y**, sparsity level K;
Output:

kth-approximation $\hat{s}_k$ to **s**, index set Λ_m containing K elements in $\{1,\dots,L\}$;
Begin
INITIALIZE: Residual $\mathbf{r_0} = \mathbf{y}$, index set $\Lambda_0 = \emptyset$, $k = 1$;
Repeat:
1: Find the index l that has the maximum inner product between column $\boldsymbol{a}_l$ of **A** and residual **r**, $l_k = argmax|\boldsymbol{a}_l^T\boldsymbol{r}_k|$;
2: Update the index set $L_k = L_{k-1} \cup l_k$, record the set of reconstruction atoms found in the dictionary $\boldsymbol{L}_k = \boldsymbol{L}_{k-1} \cup (\boldsymbol{a}_l)_k$;
3: Find the approximation of s using LS $\hat{s}_k = argmin\|\mathbf{y} - \mathbf{L_k}\hat{s}\|$;
4: Update the residual, $r_k = \mathbf{y} - \mathbf{L_k}\hat{s}_k$, $k = k + 1$;
   Until Determine if $k > K$. If so, stop repeating; else go to step 1
5: **return** $\hat{s}_k$;

$$(\hat{v}_l)_{ML} = \frac{\sum_{i=1}^{N}\mathbf{1}(\| \mathbf{y}' = \mathbf{A}'\mathbf{b}^i \| < \varepsilon)\, b_l^i}{\sum_{i=1}^{N}\mathbf{1}(\| \mathbf{y}' = \mathbf{A}'\mathbf{b}^i \| < \varepsilon)} \tag{26}$$

Threshold selection: To ensure the efficiency of the Monte Carlo simulation, a multi-level algorithm is applied to generate a sequence of non-increasing thresholds $\gamma^{(1)} \geq \gamma^{(2)} \geq \cdots \gamma^{(t)} \geq \cdots$ until convergence is reached. A common method to determine the intermediate threshold at the ith iteration is to assign

$$\gamma^{(t)} = \tilde{F}_{(\rho N)} \tag{27}$$

where $\tilde{F}_{(k)}$ represents the kth order statistic of the performance sequence $F(\boldsymbol{b}_1^t), F(\boldsymbol{b}_2^t), \dots, F(\boldsymbol{b}_N^t)$, and $F(\boldsymbol{b}_k^i) = \boldsymbol{y} - \boldsymbol{A}\boldsymbol{b}^i$. Thus, the updating equation becomes:

$$\hat{v}_l^{(t)} = \frac{\sum_{i=1}^{N} \mathbf{1}(\|\boldsymbol{y} = \boldsymbol{A}\boldsymbol{b}^i\| < \gamma^{(t)}) \, b_{l\,(t)}^i}{\sum_{i=1}^{N} \mathbf{1}(\|\boldsymbol{y} = \boldsymbol{A}\boldsymbol{b}^i\| < \gamma^{(t)})} \tag{28}$$

and the algorithm is developed in Algorithm 2.

Algorithm 2 ML Algorithm for line outage identification

Input:
 Number of samples N, the percentage of samples used for updating each time ρ, smoothing parameter α;
Output:
 The best sample $\boldsymbol{b}_n^{(t)}$;
 Begin
 INITIALIZE: Initial probability distribution $\hat{\boldsymbol{v}}^{(0)} = \{\hat{v}_l^{(0)}\}_{i=1}^{L}$, where $\hat{v}_l^{(0)} = 0.5$, iteration time $t = 1$;
 Repeat:
 1: Generate N samples $\{\boldsymbol{b}_n^{(t)}\}_{n=1}^{N}$ from distribution $P(\boldsymbol{b}|\hat{\boldsymbol{v}}^{(t)})$;
 2: Ensure that each sample $\boldsymbol{b}_n^{(t)}$ is feasible by randomly adding or removing the necessary 1s;
 3: Calculate the fitness values $\|\boldsymbol{y} - \boldsymbol{A}\boldsymbol{b}_n^{(t)}\|$;
 4: Compute $\gamma^{(t)}$ by using (27);
 5: Update $\hat{\boldsymbol{v}}^{(t)}$ by using (28);
  Until Determine if $k > K$. If so, stop repeating; else go to step 1
 6: **return** $\hat{s}_k$;

---

5 SIMULATION

To test the accuracy of the proposed algorithm, we adopt the IEEE 118-bus system. First, we test the system with global observation, and then we divide the IEEE 118-bus system into internal and external system, using only the partial observation of internal system. In the partial observation test, internal nodes are selected from {1–45, 113, 114, 115, 117}, while the rest are external {46–112, 116, 118}. Figure 2 shows the interconnection between the internal system and external system. In all experiments, we use MATPOWER tool kit (Zimmerman et al. 2011) to generate voltage phase angle and power measurements. Other methods include optimal exhaustive search (ES) (Tate & Overbye 2009) and OMP by directly solving inverse of **B** (Zhu & Giannakis 2012). We also use more realistic AC power flow measurements for comparison.

5.1 *Single line outage identification with complete information*

We tested 186 single line outages one by one. In addition, we tested algorithm 1 under different noise levels, corresponding to 0%, 1%, 2%, and 5% of the average injection power. Table 1 shows the single line outage detection of IEEE 118-bus system with all bus PMU measurements. The experiment shows all three methods have satisfactory results with all bus

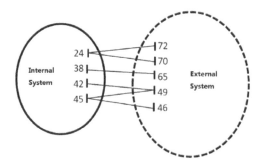

Figure 2.   Illustration of internal and external system of IEEE 118-bus system.

Table 1.   Single line outage detection of IEEE 118-BUS system with all bus PMU measurements.

Model (Noise intensity) Methods	ES	Inverse	Proposed
DC(0%)	100%	100%	99.5%
DC(1%)	89.8%	89.8%	90.3%
DC(2%)	86.0%	86.0%	88.2%
DC(5%)	73.7%	73.7%	75.3%
FAC(0%)	95.2%	95.2%	95.2%
AC(1%)	87.1%	87.1%	88.2%
AC(2%)	83.3%	83.3%	84.9%
AC(5%)	72.6%	73.1%	73.7%

PMU measurements. Identification accuracy of DC model is slightly higher than that of AC model. Accuracy of all three methods goes down with the increasing noise power.

5.2   Single line outage identification with partial information

For better and approximate real PMU distribution, we tested all three methods to identify single line outage using only internal system observation, and under different levels of noise power as well. Table 2 shows accuracy of all three methods decrease because of the loss of observation, but the proposed method still keeps the best performance. With respect to the noise robustness, ES method clearly degrades with the increasing noise level, whereas the Inverse method keeps the fairly good performance.

5.3   Multiple line outages identification with partial information

In this experiment, 200 combinations of double line outages are randomly chosen. All three methods use AC power flow data, with noise level at 0%, 1%, and 2%. Table 3 shows the results, it can be seen that certain drop in accuracy exist in all methods, but the proposed method maintains best performance.

Moreover, we tested triple and quadruple line outages. The triple outage lines are chosen as {18,66,110}, in which both line 66 and line 110 are connected to external nodes. When sparsity level is set as 3, the three line outages found are {110,66,18}, which is correctly identified; when sparsity level is set as the number of all lines L = 186, we get the distribution shown in Figure 3(a), it can be seen peak values only exist at the outage lines, with all other places near zero values, which is also correct identification. If we add another outage line to 4 as {18,66,99,110}, when sparsity level is set as 4, the four line outages found are {110,98,66,18}, with minor differences from the correct identification. Similarly, when sparsity level is set as L, we get the distribution shown in Figure 3(b).

Table 2. Single line outage detection of IEEE 118-BUS system with partial bus PMU measurements.

Model (Noise intensity) Methods	ES	Inverse	Proposed
DC(0%)	65.6%	67.7%	**99.5%**
DC(1%)	34.4%	59.1%	**90.3%**
DC(2%)	25.3%	54.8%	**88.2%**
DC(5%)	12.9%	50.0%	**75.3%**
AC(0%)	64.0%	66.1%	**66.7%**
AC(1%)	33.3%	57.5%	**59.1%**
AC(2%)	24.2%	52.7%	**54.3%**
AC(5%)	11.8%	48.9%	**50.5%**

Table 3. Double line outage detection of IEEE 118-BUS system with partial bus PMU measurements.

Model (Noise intensity) Methods	ES	Inverse	Proposed
AC(0%)	54.5%	58.5%	**59.5%**
AC(1%)	41.5%	53.0%	**54.5%**
AC(2%)	32.0%	46.5%	**47.0%**

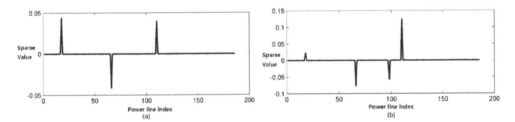

Figure 3. Detection result of three and four line outages (Distribution of sparse value).

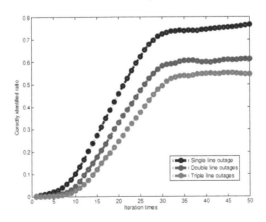

Figure 4. Detection result of line outages using ML algorithm.

We also tested algorithm 2, where 1,000 random line outages are generated on the IEEE 118-bus system. The number of samples N = 100, three cases (single line outage, double line outage and triple line outage) are tested with 50 times of iteration. See Figure 4 for the result. The correctly identified ratio of single and double line outages are roughly the same as in Table 2 and Table 3, but the triple line outage has higher accuracy.

6 CONCLUSION

Modification of the line outage identification using partial observation is introduced, considering PMUs are only partially located in the grid. The nature of the sparsity of the problem is utilized to build the model, as well as the entrenched relationship between power change and voltage phase angle. Various simulations are implemented to test complete observation and partial observation, AC model and DC model. The results show the proposed method is both more accurate and robust to noise, and can be extended to multiple line outages.

REFERENCES

Abdelaziz, A.Y. Mekhamer, S.F. & Ezzat, M. July 2012. Line outage detection using support Vector Machine (SVM) based on the Phasor Measurement Units (PMUs) technology. *Power and Energy Society General Meeting*.

Chun-Lien, S. & Bo-Yuan, J. Oct. 2006. Visualization of large-scale power system operations using phasor measurements. *Power Systems Technology*.

Donoho, D.L. 2006. For most large underdetermined systems of linear equations the minimal l1-norm solution is also the sparsest solution. *Communications on pure and applied mathematics*. 59(6): 797–829.

Emami, R. & Abur, A. 2013. External system line outage identification using phasor measurement units. *IEEE Transactions on Power Systems*. 28(2): 1035–1040.

He, M. & Zhang, J. June 2011. A dependency graph approach for fault detection and localization towards secure smart grid. *IEEE Transactions on Smart Grid*. 2(2): 342–351.

Mallat, S.G. & Zhang, Z. 1993. Matching pursuits with time-frequency dictionaries. *IEEE Transactions on Signal Processing*. 41(12): 3397–3415.

Schellenberg, A. Rosehart, W. & Augado, J. 2005. Cumulant-based probabilistic optimal power flow (P-OPF) with Gaussian and gamma distributions. *IEEE Transactions on Power Systems*. 20: 773–781.

Stott, B. Jardim, J. & Alsac, O. 2009. DC power flow revisited. *IEEE Transactions on Power Systems*. 24(3): 1290–1300.

Sun, K. Likhate, S. Vittal, V. Kolluri V.S. & Mandal, S. Nov. 2007. An online dynamic security assessment scheme using phasor measurements and decision trees. *IEEE Transactions on Power Systems*. 22(4): 1935–1943.

Tate, J.E. & Overbye, T.J. November 2008. Line outage detection using phasor angle measurements. *IEEE Transactions on Power Systems*. 23(4): 1644–1652.

Tate, J.E. & Overbye, T.J. Jul. 2009. Double line outage detection using phasor angle measurements. *Power & Energy Society General Meeting*.

Throp, J.S. Phadke, A.G. & Karimi, K.J. Nov. 1985. Real time voltage-phasor measurement for static state estimation. *IEEE Transactions on Power Apparatus and Systems*. 104(11): 3098–3106.

Tropp, J.A. 2004. Greed is good: Algorithmic results for sparse approximation. *IEEE Transactions on Information Theory*. 50(10) 2231–2242.

Zhu, H. & Giannakis, G.B. Nov. 2012. Sparse Overcomplete Representations for Efficient Identification of Power Line Outages. *IEEE Transactions on Power Systems*. 27(4): 2215–2214.

Zimmerman, R.D. Murillo-Snchez, C.E. & Thomas, R.J. 2011. MATPOWER: Steady-state operations, planning, and analysis tools for power systems research and education. *IEEE Transactions on Power Systems*. 26(1): 12–19.

Signal and Information Processing, Networking and Computers – Chen & Huang (Eds)
© *2016 Taylor & Francis Group, London, ISBN 978-1-138-02881-4*

Mitigating Primary User Emulation attacks in Cognitive Radio networks using advanced encryption standard

Huichao Jiang, Xiao Jun Jing, Songlin Sun & Hai Huang
School of Information and Communication Engineering, Beijing University of Posts and Telecommunications, Beijing, China

Yan Li
International School, Beijing University of Post and Telecommunications, Beijing, China

Xiaohan Wang
School of Software and Microelectronics, Beijing University, Beijing, China

Dongmei Cheng
305 Hospital of PLA, Beijing, China

ABSTRACT: Cognitive Radio (CR) can ease the problem of spectrum scarcity by allowing secondary users to coexist with incumbent users in licensed spectrum bands, while it causes no interference to incumbent communications. In this paper, we identify a threat to CR that is, the Primary User Emulation (PUE) attack. To deal with the threat, we put forward an advanced encryption standard scheme. When a user wants to occupy the specific spectrum, it must transmit an AES-encrypted reference signal to the secondary user. By allowing the transmitter and the receiver to share a key, the reference signal can be regenerated at the receiver and can be used for accurate identification of authorized primary users. Our simulation results suggest the effectiveness of the proposed scheme in mitigating PUE attacks in certain situations.

Keywords: spectrum sensing; cognitive radio; primary user emulation attack; advanced encryption standard

1   INTRODUCTION

In S. Haykin (2005), emerging wireless applications are increasing the demand for spectrum, and in order for better use of spectrum, the Federal Communications Commission (FCC) re-examined the problem of spectrum usage. Realizing the seriousness of the spectrum scarcity, the FCC considers the unauthorized operation of licensed frequency bands, under the condition of no interference to licensed users.

This paper focuses on the dynamic spectrum access, where secondary users are opportunistically accessing spectrum bands when the primary user is inactive. In R. Chen et al. (2008), secondary users must be able to identify primary users, and vacate the spectrum when a primary user becomes active. It is extremely difficult to distinguish the two signals when the CRs are surrounded by hostile. In D. Cabric et al. (2004), an attacker may imitate a primary user signal character which leads to secondary users to mistakenly believe that the attacker is a primary user. This refers to the PUE attack. To prevent such attacks, a scheme is needed to accurately distinguish between legal primary uses and secondary uses who imitate primary users, so as to enhance the reliability of the spectrum sensing.

There exist techniques to distinguish the two signals, such as matched filter, cyclostationary feature detection and energy detection (S. Capkun et al. 2006, K. Challapali et al. 2004). When using energy detection, a secondary user is able to identify the signals of other secondary users but cannot identify the signals of the primary users. When a secondary user detects a signal which is recognized by the secondary user, it considers that the signal is a secondary user's signal; or it judges that the signal a primary user's signal. In this simple scheme, a PUE attacker can imitate the primary user by sending unrecognized signals in an authorized band, thereby blocking other secondary users to access the spectrum. When cyclostationary feature detection is used, devices can identify the characteristics of primary user. However, to beat cyclostationary detector, an attacker may have its signals indifferent from primary user signals through making signals have the same cyclic spectrum characters as primary user signals.

This paper presents an AES-encryption scheme. If a primary user needs to use its licensed spectrum bands that are occupied by secondary users currently, it must firstly transmit an AES-encrypted reference signal to the secondary user who can regenerate the reference signal, then identify the authorized primary users. The technique makes sense in both civilian and military aspects.

The rest of the organizations are as follows. In Section 2, we describe the PUE attack. In Section 3, we present the AES-encryption scheme. In Section 4, analytical system evaluation is provided. Simulation results are shown in Section 5. And in Section 6, we conclude the paper discussing its future work.

2 THE PUE ATTACK

In PUE attacks, the enemy only operates in free bands. Therefore, the purpose of the enemy is not to interfere with the primary users, but to get resources which are used by legal users. In S. Shankar et al. (2005), according to the motive of the attack, a PUE attack could be divided into a selfish PUE attack or a malicious PUE attack.

2.1 Selfish PUE attacks

The goal of this attack is to increase its available spectrum. When selfish attackers discover an idle frequency band, they block other secondary users to access the band through sending signals that mimic the features of primary user signals. These attackers are likely to be two selfish users who want to set up the specific connection.

2.2 Malicious PUE attacks

The purpose of the attack is to prevent the dynamic spectrum access of legal secondary users—i.e. deter secondary users to detect and use idle bands, leading to denial of service. An attacker is likely to hinder the dynamic access process at the same time in multiple bands through using two DSA mechanisms. The first one needs a user to wait for a period of time before operating in the discovered idle band to ensure that it is idle really. In G. Barnett et al. (2005), studies that already exist indicates that the time delay cannot be ignored. The second one needs a user to continuously perceive whether the operating band exists in primary user signals and to instantly go to other band when such signals are tested.

Now we consider an attack scenario where malicious SU represented as SUm makes the network unavailable for others SUs. In the Figure 1 below it can be that SU1 appears and performs sensing. Based on the sensing results it transmits and then leaves the network. While SU1 was transmitting, SU2 appeared and performed sensing and entered into wait state. Once SU1 leaves the network, a malicious SU emulates the PU and pretends to be PU. SU2 senses again which is at point in time when SU1 is no more using the network but a malicious SU is pretending to be PU thus making SU2 to enter in other waiting interval. Ultimately SU2 leaves the network without getting a chance to transmit. Similarly SU3 appears

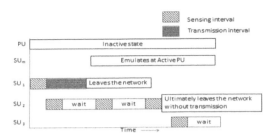

Figure 1. Representation of attack scenario by malicious SU making network resources unavailable.

and senses while emulation by malicious SU is under way. SU3 senses for network but finds it to be unavailable.

3 THE AES-ENCRYPTION SCHEME

In this section, we present our proposed AES-encryption scheme for robust. In the proposed system, the authorized user produces a pseudo-random AES-encrypted reference signal. At the receiver, the reference signal is generated to be used for the detection of the primary user. In the next subsections, we discuss the AES-encrypted transmitter and receiver in more details.

3.1 AES-encrypted transmitter

The transmitter gets the reference signal through two steps: first, producing a pseudo-random (PN) sequence and then use the AES algorithm to encrypt the sequence. More specifically, generated Linear Feedback Shift Register (LFSR) can generate a pseudo-random (PN) sequence. Maximum-length LFSR sequences can be achieved by tapping the LFSR according to primitive polynomials. The maximum sequence length that can be achieved with the primitive polynomials is $2^m - 1$, where m stands for the number of registers. Without loss of generality, a maximum-length sequence is assumed throughout this paper.

Once the maximum-length sequence is generated, it will enter the AES encryption algorithm, as illustrated in Figure 2.

We use x to represent the PN sequence, and the output (i.e. the reference signal) of the AES algorithm is shown as follows:

$$s = E(k, x) \tag{1}$$

where k is the encryption key, and $E(.,.)$ expresses the AES encryption.

3.2 AES-encrypted receiver

The secondary user is the AES-encrypted receiver. It can regenerate the ciphertext, we assume that the secret key is available at the receiver and the PN sequence can be regenerated. A cross-correlation detector is employed which is used to evaluate the correlation between the received signal $r = s + n$ and the reference signal s, where n represents noise. The cross-correlator output can be represented as:

$$R_{rs} = \sum_{i=1}^{N} \frac{r_i \cdot s_i}{N} \tag{2}$$

where N is the signal length, s_i and r_i are the i th symbols of the reference and received signals, respectively.

187

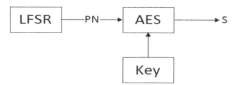

Figure 2.   AES-encryption process.

To distinguish between the actual TV transmitter and an attacker emulating the primary signal, the receiver compares the cross-correlator output to a predefined threshold λ. We have two cases:

• If the cross-correlator output is greater than or equal to λ, i.e.

$$R_{rs} \geq \lambda \qquad (3)$$

then the receiver concludes that the primary user is present.
• If the cross-correlator output is less than λ, i.e.

$$R_{rs} < \lambda \qquad (4)$$

then the receiver concludes that the primary user is absent.

This detection problem can be described by the following two hypotheses,

H_0: the primary user is absent
H_1: the primary user is present

The performance of the detection process is evaluated through the false alarm rate and miss detection probability, will be discussed in Section 4. The optimal threshold that minimizes the miss detection probability subject to a false alarm rate constraint is also obtained.

4   ANALYTICAL EVALUATION

4.1   *False alarm rate and miss detection probability*

In this subsection, we analyze the system performance through the miss detection probability and the false alarm rate. When the primary user is suppose to be absent but is present, the miss detection probability (P_m) can denote this situation, i.e.

$$P_m = P\left(H_0|H_1\right) \qquad (5)$$

When the primary user is absent but is assumed present, the false alarm rate (P_f) can denote this situation, i.e.

$$P_f = P_r\left(H_1|H_0\right) \qquad (6)$$

As can be seen from (2), the cross-correlator output is the summation of N random variables. Since N is very large, then from the central limit theorem, we can model the cross-correlator output as a Gaussian random variable. That is, under H_1, $R_{rs} \sim N(\mu_1, \sigma_1^2)$, where μ_1 and σ_1^2 are the mean and the variance of the cross-correlator output, respectively, under H_1.
The miss detection probability P_m can be obtained as:

$$P_m = P_r\{R_{rs} < \lambda | H_1\}$$

$$= \int_{-\infty}^{\lambda} \frac{1}{\sigma_1\sqrt{2\pi}} e^{-\frac{(X-\mu_1)^2}{2\sigma_1^2}} dX$$

$$= 1 - \int_{\lambda}^{\infty} \frac{1}{\sigma_1\sqrt{2\pi}} e^{-\frac{(X-\mu_1)^2}{2\sigma_1^2}} dX \tag{7}$$

$$= 1 - Q\left(\frac{\lambda - \mu_1}{\sigma_1}\right)$$

Note that under H_1, the received signal is represented as $r_i = s_i + n_i$, where s_i is the ith primary symbol, and $n_i \sim N(0, \sigma_n^2)$. We suppose that the primary signal has zero mean and power P_s, i.e. $E\{s_i^2\} = P_s \; \forall i$. We further assume that both $\{s_i\}$ and $\{n_i\}$ are i.i.d. sequences, and $\{s_i\}$ and $\{n_i\}$ are independent.

Under these assumptions, the mean μ_1 can be obtained as follows:

$$\mu_1 = E\left\{\frac{\sum_{i=1}^{N}(s_i + n_i)s_i}{N}\right\}$$

$$= E\left\{\frac{\sum_{i=1}^{N} s_i^2}{N}\right\} \tag{8}$$

$$= P_s$$

The variance σ_1^2 is obtained as:

$$\sigma_1^2 = E\{R_{rs}^2\} - \mu_1^2$$

$$= E\left\{\frac{\sum_{i=1}^{N}(s_i + n_i)s_i \sum_{j=1}^{N}(s_j + n_j)s_j}{N^2}\right\} - P_s^2$$

$$= \frac{1}{N^2} E\left\{\sum_{i=1}^{N}\sum_{j=1}^{N}\left[s_i^2 s_j^2 + n_i n_j s_i s_j\right]\right\} - P_s^2$$

$$= \frac{1}{N^2}\left[\sum_{i=1}^{N}\left(E\{s_i^4\} + E\{n_i^2\}E\{s_i^2\}\right)\right. \tag{9}$$

$$\left. + \sum_{\substack{i=1 \\ j\neq i}}^{N}\sum_{j=1}^{N}\left(E\{s_i^2\}E\{s_j^2\} + E\{n_i n_j\}E\{s_i s_j\}\right)\right] - P_s^2$$

$$= \frac{1}{N}\left[E\{s_i^4\} + P_s(\sigma_n^2 - P_s)\right]$$

Similarly, under H_0, $R_{rs} \sim N(\mu_2, \sigma_2^2)$, where μ_2 and σ_2^2 are the mean and the variance of the cross-correlator output under H_0, respectively. The false alarm rate P_f can be defined as:

$$P_f = P_r\{R_{rs} > \lambda | H_0\}$$

$$= \int_{\lambda}^{\infty} \frac{1}{\sigma_2\sqrt{2\pi}} e^{-\frac{(X-\mu_2)^2}{2\sigma_2^2}} dX \tag{10}$$

$$= Q\left(\frac{\lambda - \mu_2}{\sigma_2}\right)$$

The received signal under H_0 is represented as $r_i = n_i$; since with AES encryption even if the malicious user is present, it is regarded as noise. In this case, let $n_i \sim N(0, \sigma_{n,0}^2)$, where $\sigma_{n,0}^2$ is the noise variance under H_0.

Therefore, the mean μ_2 is obtained as:

$$\mu_2 = E\left\{\sum_{i=1}^{N} n_i s_i\right\}$$
$$= 0$$

(11)

and σ_2^2 is obtained as:

$$\sigma_2^2 = E\{R_{rs}^{2}\} - \mu_2^2$$
$$= \frac{1}{N^2} E\left\{\sum_{i=1}^{N}\sum_{j=1}^{N}[n_i s_i n_j s_j]\right\}$$
$$= \frac{1}{N^2}\sum_{i=1}^{N} E\{n_i^2 s_i^2\}$$
$$= \frac{\sigma_{n,0}^2 P_s}{N}$$

(12)

In the next subsection, we obtain the optimal threshold that minimizes the miss detection probability under a constraint on the false alarm rate.

4.2 Evaluation of the optimal threshold

The objective is to make the miss detection probability P_m minimum under a false alarm constraint $P_f \le \beta$. That is, our problem can be formulated as follows:

Minimize P_m

Subject to $P_f \le \beta$

(13)

In order to have $P_f \le \beta$, using equation we get:

$$P_f = Q\left(\frac{\lambda - \mu_2}{\sigma_2}\right) \le \beta$$

(14)

It follows that

$$\lambda \ge \sigma_2 Q^{-1}(\beta) + \mu_2$$

(15)

where $Q^{-1}(\cdot)$ is the inverse Q-function.

It can be seen from (7) that in order to minimize the miss detection probability, λ should be as low as possible. Therefore, we choose the lowest λ that satisfies the false alarm constraint. Thus, the optimal λ that minimizes P_m and satisfies the false alarm constraint is

$$\lambda_{opt} = \sigma_2 Q^{-1}(\beta) + \mu_2$$

(16)

Substituting by λ_{opt} in (7), we get the miss detection probability of the proposed approach. More specifically, we get

$$P_m = 1 - Q\left(\frac{\sigma_2 Q^{-1}(\beta) + \mu_2 - \mu_1}{\sigma_1}\right)$$
$$= 1 - Q\left(\alpha Q^{-1}(\beta) + \gamma\right)$$

(17)

where, $\alpha = \frac{\sigma_2}{\sigma_1}$ and $\gamma = \frac{\mu_2 - \mu_1}{\sigma_1}$. Therefore, we have the following result: to make the miss detection probability the minimum subject to the constraint $P_f \leq \beta$, we need to set the threshold of the cross-correlator of the AES-encrypted receiver to $\lambda_{opt} = \sigma_2 Q^{-1}(\beta) + \mu_2$, where μ_2 and σ_2 are mean and standard deviation of the cross-correlator output when the primary user is absent.

5 SIMULATION

5.1 The impacts of PUE attacks

We implement tests to demonstrate the bad impacts of PUE attacks. In the simulation, 200 secondary users including legal users and attackers are distributed in a 2000 square meter area. Three TV towers work as primary signal transmitters. Every TV tower has five 7 MHz bands. The quantity of malicious PUE attackers changed from 1 to 30, and the number of selfish PUE attackers changed from 1 to 30 pairs. Figure 3(a) and 3(b) respectively are the test results of these two attacks. The y-axis is the number of link bandwidth every secondary user can access. The results indicate that a selfish PUE attack can occupy the bandwidth of legal secondary users and a malicious PUE attack can greatly reduce the available bandwidth of legal secondary users.

5.2 The probability of false alarm and miss detection

Using λ_{opt}, we obtain the false alarm rate and miss detection probability numerically and compare them with the theoretical results. We use the false alarm constraint $\beta = 0.1$ and $\beta = 0.01$. Figure 4 shows the false alarm rate is calculated by both theoretically and by simulation. It can be seen that the theoretical calculations and the numerical simulations are almost identical, and the predefined false alarm constraint is satisfied. This shows that with the proposed AES-encrypted DTV approach, primary user emulation attacks can be effectively mitigated.

The probability of miss detection is shown in Figure 5. It can be seen that the proposed AES-encrypted DTV approach achieves zero miss detection probability even under very low SNR values. As shown in the same figure, the theoretical analysis and simulation results are identical.

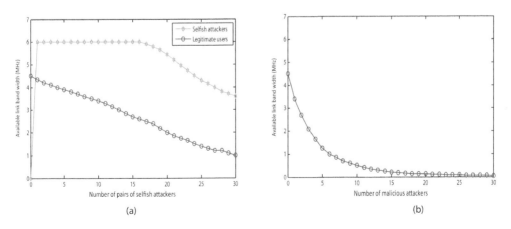

Figure 3. Simulation showing the impact of PUE attacks. (a) Impact of selfish PUE attacks. (b) Impact of malicious PUE attacks.

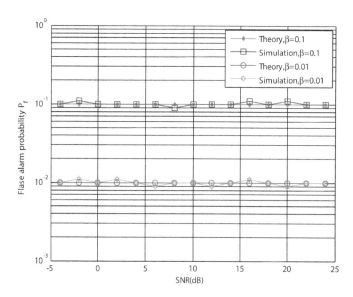

Figure 4. The probability of false alarm P_f.

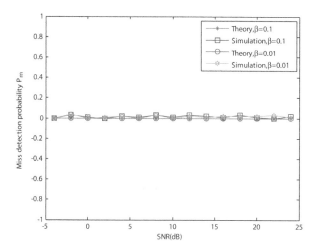

Figure 5. The probability of miss detection P_m.

6 CONCLUSION

In this paper, we propose an AES-encryption scheme to mitigate PUE attacks in CR networks. In our proposed model, if the primary users need to use its licensed spectrum bands that are occupied by secondary users currently, it must first transmit an AES-encrypted reference signal to the secondary user. By allowing the transmitter and the receiver to share a key, the secondary user can regenerate the reference signal which is used to identify authorized primary users. Through both theoretical derivation and simulation examples, it is shown that with the AES-encryption scheme, the primary user can be accurately detected under primary user emulation attacks.

ACKNOWLEDGMENTS

This work is supported by project NSFC 61471066 and the open funding project of State Key Lab of Virtual Reality Technology and Systems at Beihang University under Grant No. BUAA-VR-15 KF-19.

REFERENCES

Cabric, D., Mishra, S.M. & Brodersen, R.W. 2004. Implementation issues in spectrum sensing for cognitive radios. *Proc. Thirty-Eighth Asilomar Conf. Signals, Systems and Computers*: 772–776.

Capkun, S. & Hubaux, J.-P. 2006. Secure positioning in wireless networks. *IEEE J. Select. Areas Commun.*, 24(2): 221–232.

Challapali, K., Mangold, S. & Zhong, Z. 2004. Spectrum agile radio: Detecting spectrum opportunities. *Proc. 6th Annual Int'l Symp. Advanced Radio Technologies.*

Chen, R., Park, J. & Reed, J. 2008. Defense against primary user emulation attacks in cognitive radio networks. *IEEE J. Sel. Areas Commun.*, 26: 25–37.

Haykin, S. 2005. Cognitive radio: brain-empowered wireless communications. *IEEE J. Select. Areas Commun.*, 23(2): 201–220.

Olivieri, M.P., Barnett, G., Lackpour, A., Davis, A. & Ngo, P. 2005. A scalable dynamic spectrum allocation system with interference mitigation for teams of spectrally agile software defined radios. *In Proceedings of the IEEE International Symposium on New Frontiers in Dynamic Spectrum Access Networks, Baltimore, MD, USA.*

Shankar, S., Cordeiro, C. & Challapali, K. 2005. Spectrum agile radios: utilization and sensing architectures. *Proc. IEEE DySPAN*: 160–169.

Signal and Information Processing, Networking and Computers – Chen & Huang (Eds)
© *2016 Taylor & Francis Group, London, ISBN 978-1-138-02881-4*

Research and realization of music recommender algorithm based on hybrid collaborative filtering

Haiying Che
Beijing Institute of Technology, Beijing, China

Zishi Wang
China Three Gorges Corporation, Beijing, China

ABSTRACT: Digital music has penetrated every aspect of people's lives and massive resources of online digital music have maintained a rapid growth rate because of the leap of Internet. The traditional music information retrieval systems, therefore, could not meet the growing needs of users. This paper combines SVD with memory-based collaborative filtering, compensates the negative effects of sparse matrices, and then forms a hybrid recommendation algorithm. Experimental results in "R3 Yahoo!Music" shows the advantage of our proposal in accuracy.

Keywords: music recommender system; KNN-based collaborative filtering; SVD; hybrid collaborative filtering

1 INTRODUCTION

Digital music has penetrated every aspect of people's lives with the rapid development and improvement of multimedia technology in recent years (Cunningham et al. 2003). Meanwhile, massive resources of online digital music have maintained a rapid growth rate because of the leap of the Internet. The traditional music information retrieval systems, therefore, becomes less able to meet the growing needs of users, and the problem rises on how to choose music that meets the user needs, that is suitable for them from a huge scale of data, and then make a playlist for them (Park et al. 2006). To this end, the music recommendation system came into being; this allows the user to explore music that suits their tastes, but also made music service websites or applications that can accurately push their inventory to the target audience. Unlike random play, a music recommendation system will consider the smooth switch between songs, the user's personal preferences and real-time response. Today, this academic problem has also been maturely designed and implemented in the industrial field; web music recommendation applications such as Last.fm, Pandora and Spotify are all relatively successful (Lops et al. 2010). The music recommendation systems nowadays mainly focus on these two questions:

The first one is the prediction of scores, which predicts the possible rating score of the target user to a piece of music. The second question is to recommend a music playlist to the user according to the predicted scores from high to low.

The prediction of scores is the core step among all of these. Music recommendation systems, similar to other kinds of recommendation systems, solve this problem with a recommendation algorithm, based on content or collaborative filtering.

In recent years, real-time music recommendation systems for user groups appeared and became popular due to the social usage of music in public, such as music in bars, discos, squares and other public places (Alter et al. 2000). This kind of system, different from the music recommendation system for personal users, is based mainly on context-awareness. With the continuous development of mobile Internet, group recommendation may be one of the important directions for the future.

The outcome of traditional content-based recommendation in the field of music recommendation was not ideal, due to the fact that users always have very distinct emotions to music, and whether they like a piece of music or not is determined by their appreciation level (Park et al. 2015). The results of collaborative filtering are better, since it focuses more on the behavioral data of users. The collaborative filtering algorithm, however, has the disadvantage of a cold-start and bad performance when dealing with sparse data (Canny 2002). In response to these problems, this paper combines SVD with memory-based collaborative filtering, compensates the negative effects of sparse matrices, and then forms a hybrid recommendation algorithm. It proves that the hybrid recommendation model has a higher accuracy than any other model through the experiments on public data sets.

This paper first makes a comprehensive analysis of music recommendation, and combines two kinds of KNN-based algorithms, and then proposes a SVD & KNN hybrid collaborative filtering algorithm. In experiments, SKHCF provides better recommendations than the original ones.

2 RELATED WORKS

2.1 Music recommendation

Generally, there are two approaches for music recommendation (G. Adomavicius & A. Tuzhilin, 2005): Content-based algorithm analyzes the acoustic characteristics of music, and calculates its similarity with songs which users like, and then includes the most similar songs in the recommendation list. Pandora uses this kind of algorithm. Pandora proposed the concept of music genome in order to calculate the degree of similarity between music where genes contain melody, rhythm, instruments, arranger, lyrics and other unique characteristics of each song (Yang et al. 2013). Collaborative filtering music recommendation algorithm, compared to content-based one, it does not analyze the attributes of music objects, but recommends music based on the tendency of user preferences (Reischach et al. 2009). It identifies users with similar tastes, which are always referred to as the nearest neighbors, and recommend music by analyzing the similarities between their historical behaviors. On the basis of these two kinds of algorithm, music recommendation systems usually use methods of context-aware to get the reviews and feedback of users to enhance the accuracy of system.

2.2 Collaborative filtering

Collaborative filtering is a popular technology in information filtering and information systems. Its core idea is to find a group of users whose preferences are similar with the target user by analyzing his/her interests tendency (Kim et al. 2010), and then predict his/her rating score according to the scores rating by his neighbors, and at last generate a recommendation list to the target user from the final ordering of the predicted scores. Unlike traditional content-based filtering, collaborative filtering does not analyze the unique attributes of music, but the tendency of user's interests, therefore, collaborative filtering is often used to recognize whether a particular user is interested in an item.

There are two kinds of collaborative filtering algorithm (Wang 2012): K Nearest Neighbors (KNN) based collaborative filtering and model-based collaborative filtering. KNN-based CF includes item-based CF and user-based CF, and model-based CF includes Bayesian network, Boltzmann machine, singular value decomposition (SVD) and so on.

3 SVD AND KNN HYBRID COLLABORATIVE FILTERING

In the area of music recommendation, collaborative filtering is the best solution because the content of music items is relatively complex. In industry world, the most commonly used collaborative filtering approach now is KNN-based collaborative filtering. This method, however, may lead to a descending in accuracy for the sparseness of the user-item matrix. In fact, the user-item matrix is always relatively sparse, if we can fill the matrix before we calculate the similarity while ensuring the filling values are relatively accurate, and then the accuracy of the final prediction will be significantly improved.

3.1 Hybrid KNN algorithm

KNN-based collaborative filtering usually consists of three steps.

The first one is to collect data. The historical behavior of users is very important for collaborative filtering. All the data, including reading, listening, browsing, and the purchasing, collection or other actions, are available for collaborative filtering algorithm. The result of this step is an m × n user-item rating matrix R, which can typically be used as the inputs of collaborative filtering algorithm and shown as below.

$$R = \begin{bmatrix} r_{11} & r_{12} & \cdots & r_{1n} \\ r_{21} & r_{22} & \cdots & r_{2n} \\ \cdots & \cdots & \cdots & \cdots \\ r_{m1} & r_{m2} & \cdots & r_{mn} \end{bmatrix} \tag{1}$$

The second one is to calculate the similarities between users and items.

The last step is to generate a recommendation list according to the results of the second step.

However, there are some problems in traditional KNN-based collaborative filtering. User-based collaborative filtering focuses more on users, so it is always influenced by the tendency which may not meet the need of the target user. While item-based collaborative filtering does not use the social connections, therefore the result may not be very accurate.

In this paper, we combine these two kinds of KNN-based collaborative together and propose a hybrid KNN-based collaborative filtering algorithm to remedy the disadvantages of each method.

First, we predict the target user's rating of items in User-based collaborative filtering and Item-based collaborative filtering, and then we introduce a weighting factor t and calculate the weighted sum of two predicted scores above. The formula is written as follows.

$$r_{u,i} = t \times r_u + (1-t) \times r_i \tag{2}$$

Here, $r_{u,i}$ is the predicted score of user u to item I, r_u is the result of user-based collaborative filtering and r_i is the result of item-based collaborative filtering. While t is the control parameter introduced with the range of [0, 1].

3.2 Optimized SVD

The most serious problem of KNN-based collaborative filtering is the result of similarity calculation may be not accurate if the rating matrix R is relatively sparse, which may influence the final result of recommendation. In this paper, we use singular value decomposition to solve this problem.

We use SVD to decompose an m × n matrix R into 3 matrices U, S, M, the formula is written as follows:

$$r_{u,i} = t \times r_u + (1-t) \times r_i \tag{3}$$

U is an m × m matrix, and S is an m × n matrix of which diagonal values are non-negative real numbers while the other values are all zero. M^T is the transposed matrix of M, and it is an n × n matrix. Figure 1 shows this process.

The formula of SVD model is shown below:

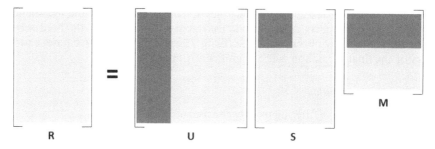

Figure 1. SVD matrix remodelling.

$$\hat{r}_{ij} = u_i \cdot m_j = \sum_{f=1}^{F} u_{if} i_{jf} \qquad (4)$$

$\hat{r}_{ij}$ is the predicted score, and u_{if} is the degree that user i preferences feature f, and i_{jf} is the degree that item i contains feature f. Therefore, the dot product of these two vectors is the predicted score. We can get u_{if} and i_{jf} through the optimization of the target parameter E, which is shown below.

$$E = \frac{1}{2} \sum_{(i,j) \in R} \left(r_{ij} - u_i m_j^T \right)^2 + \lambda \left(\|u_i\|^2 + \|m_j\|^2 \right) \qquad (5)$$

However, users' rating scales can be significantly different in the actual rating process, such as someone is customary to give high scores, while others have very high standard and will not give a high score no matter how perfect the item is. Therefore, we introduce two parameters to eliminate this bias and the formula 4 will be like:

$$\hat{r}_{ij} = d_i + d_j + \mu_i \cdot m_j = d_i + d_j + \sum_{f=1}^{F} \mu_{if} m_{jf} \qquad (6)$$

Here, d_i means the user rating deviation and d_j means item rating deviation. So the formula 5 will be like:

$$E = \frac{1}{2} \sum_{(i,j) \in R} \left(r_{ij} - d_i - d_j - u_i m_j^T \right)^2 + \lambda \left(\|u_i\|^2 + \|m_j\|^2 + d_a^2 + d_b^2 \right) \qquad (7)$$

In this paper, we use gradient descent to accomplish the optimization of target parameter. After the analysis of multiple data sets, we find that users' rating set, i.e. rows in user-item matrix, is relatively sparse compared to the item's one. Therefore, the users' influence may be weakened in this case. In this paper we propose an optimization method to fill the matrix, we add a correction value for each user recessive factor U_i and the formula is written as:

$$U_i = U_i + \frac{\sum_{k=1}^{m} I_{ik} M_k}{\sum_{k=1}^{m} I_{ik}} \qquad (8)$$

U_i represents a line of user recessive factor matrix and on behalf of how user i likes the various factors. I is a transfer matrix, the value of I_{ik} will be 1 if user i had scored item k, otherwise the value will be 0. Therefore the correction value represents the mean value of all the scores

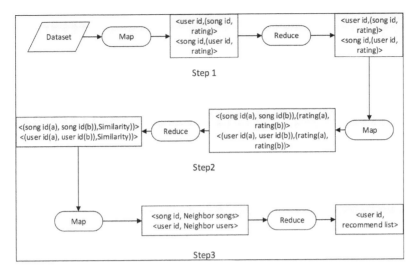

Figure 2.　Realization of H-KNN algorithm with distributed system.

that use i had rated. All the items that user i had not rated would be filtered because of the existence of I_{ik}.

After all the process above, we could get a non-singular user-item matrix.

3.3　SVD and KNN hybrid collaborative filtering

In this paper, we propose a SVD & KNN Hybrid Collaborative Filtering (SKHCF) algorithm to solve the problem that the recommendation result may be not so accurate if the user-item matrix is relatively sparse. The procedure of this algorithm is as follows:

1. Construct a m × n user-item matrix **R**, if user u had rated item i, then $r_{u,i}$ will be the rating value, otherwise $r_{u,i}$ will be 0.
2. Construct user recessive factor matrix **U** and item recessive factor matrix **M** using SVD.
3. Matrix **U** multiplies matrix **M**, and we can get a non-singular user-item matrix R_1.
4. Use H-KNN algorithm to generate a recommendation list.

3.4　Optimization based on distributed system

The overhead of KNN algorithm is always huge, what's more, and in this paper, we combined two kinds of KNN algorithm together which means we have to calculate the similarities twice. Therefore, we use distributed system to optimize the performance of SKHCF. The procedures are listed in Figure 2. Take the data set "R3.Yahoo!Music" we used in this paper for example, the data format is like "user id <TAB> song id <TAB> rating".

As a result in Figure 3, the running time of algorithm keeps shortening as the number of nodes increase. Formula 9 is introduced for comparison.

$$p = \frac{T_1}{T_n} \tag{9}$$

T_1 represents the running time when we just use one node, and T_n represents the running time when we use n node. Therefore, running time becomes shorter when the value of p becomes higher.

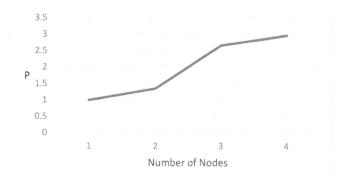

Figure 3. Relationship between number of nodes and running time of algorithm.

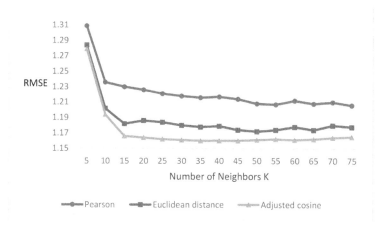

Figure 4. KNN-based collaborative filtering algorithm.

4 EXPERIMENTAL RESULTS

We have analyzed the recommendation process and conducted the subjective test so as to show the usefulness of SKFCF.

In this experiment, we use the data set named "R3. Yahoo!Music" from Yahoo's Webscope team. This data set contains two sets of users' ratings. The first set of data was generated by 10000 Yahoo users through their daily usage of "Yahoo!Music" service, and each user had at least 10 ratings record. The second set of data was collected from an online test formed by randomly generated songs. This test was hold by "Yahoo!Research" and it contained 54000 ratings from 5400 users while each user generated correctly 10 ratings. At last, the whole data set contained 3117704 ratings from 15400 users for 1000 songs.

The assessment method of this experiment is RMSE (root mean square error), the formula is written as:

$$RMSE = \frac{\sqrt{\sum_{u,i \in T}(r_{ui} - \hat{r}_{ui})^2}}{|T|} \tag{10}$$

T is the test set, r_{ui} is the rating data in T, $\hat{r}_{ui}$ is the predicted rating of the target user. We tested three kinds of similarity calculation: adjusted cosine, Euclidean distance and Person, the results are shown in Figure 4.

Table 1. The best value of parameters in gradient descent.

Parameter name	Value
Factors	15
Learning rate	0.05
Penalty factor	0.1
Iteration times	20

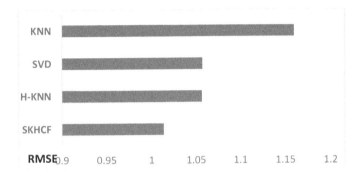

Figure 5. RMSE of each algorithm.

From Figure 4, we can recognize that the method of adjusted cosine performed best. After a few of experiments, we finally confirm the best parameters for gradient descent, which is shown in Table 1.

Figure 5 shows the result of four algorithms include KNN, SVD, H-KNN and SKHCF. From Figure 5, we can find that SKHCF performs best.

5 CONCLUSION

This paper proposes a SVD & KNN hybrid collaborative filtering algorithm. SKHCF provides a method to solve the problem that the recommendation result may not be accurate when data set is relatively sparse. In experiments, we have confirmed the usefulness of SKHCF by analyzing the recommendation process and performing subjective tests.

REFERENCES

Adomavicius, G. & Tuzhilin, A. 2005. Toward the next generation of recommender systems: A survey of the state-of-the-art and possible extensions. IEEE T Knowl Data En, vol. 17, no. 6, pp. 734–749.

Alter, O., Brown, P.O. & Botstein, D. 2000. Singular value decomposition for genome-wide expression data processing and modeling [J]. Proceedings of the National Academy of Sciences. 97(18):10101–10106.

Canny, J. 2002. Collaborative Filtering with Privacy via Factor Analysis [J]. Proceedings of Annual International Acm Sigir Conference on Research & Development in Information Retrieval. 18(1):238–245.

Cunningham, S.J., Reeves, N. & Britland, M. 2003. Ethnographic study of music information seeking [C]. Proceedings of the 2nd ACM/IEEE-CS Joint Conference on Digital Libraries.

Kim, J.K., Kim, H.K. & Oh, H.Y., et al. 2010. A group recommendation system for online communities [J]. International Journal of Information Management. 30(3):212–219.

Lops, P., Gemmis, M.D. & Semeraro, G. 2010. Content-based Recommender Systems: State of the Art and Trends [J]. Recommender Systems Handbook, 2010:73–105.

Park, H.S., Yoo, J.O. & Cho, S.B. 2006. A context-aware music recommendation system using fuzzy bayesian networks with utility theory [M]. Fuzzy systems and knowledge discovery. Springer Berlin Heidelberg, 2006: 970–979.

Park, Y., Park, S. & Jung, W., et al. 2015. Reversed CF: A fast collaborative filtering algorithm using a k-nearest neighbor graph [J]. Expert Systems with Applications. 42:4022–4028.

Reischach, F.V., Guinard, D. & Michahelles, F., et al. 2009. A Mobile Product Recommendation System Interacting With Tagged Products [C]. Pervasive Computing and Communications, PerCom 2009. IEEE International Conference on. 2009:1–6.

Wang, G. 2012. Survey of personalized recommendation system [J]. Computer Engineering & Applications.

Yang, D., Zhang, D. & Yu, Z., et al. 2013. A sentiment-enhanced personalized location recommendation system [C]. Proceedings of the 24th ACM Conference on Hypertext and Social Media. ACM.

Signal and Information Processing, Networking and Computers – Chen & Huang (Eds)
© *2016 Taylor & Francis Group, London, ISBN 978-1-138-02881-4*

Research of AdaBoost robustness based on Learning Automata

Siyu An, Ying Guo, Xinyi Guo, Yan Yan & Shenghong Li
Department of Electronic Engineering, Shanghai Jiao Tong University, Shanghai, China

ABSTRACT: AdaBoost is one of the widely used techniques in the area of pattern recognition. It searches the samples that are classified incorrectly and subsequently pays more attention to them at the next iteration. While the dataset is noisy, according to the iterative mechanism of AdaBoost, the noisy samples may be paid more attention at, which might lead to bad behavior. As a method of reinforcement learning, learning automata could search the optimal state adaptively in a random environment. In this way, we believe the algorithm based on learning automata shows good behavior towards noisy dataset. In this paper, we improve the performance of AdaBoost based on learning automata by adjusting the weight through a stochastic optimization method. The experiments indicate that the algorithm is able to restrain the negative effect caused by noise effectively.

Keywords: AdaBoost; Learning Automaton; noise; continuous action-set

1 INTRODUCTION

1.1 *AdaBoost*

In recent years, the techniques of machine learning have made great progress. The AdaBoost algorithm is the most representative method of boosting techniques (Freund & Schapire, 1997). This method is well known with high accuracy and simple rules, and it has been widely used in machine learning. Up to now, researchers have applied this method to classification (Lin et al., 2008), text detection (Lee et al., 2011), target detection (Hu et al., 2014). The core idea of AdaBoost is "adaptive enhancement", which is characterized in that it could find the points which are easier to be misclassified, and consider them as the samples with more information. For the misclassified samples, the probabilities of them in the next iteration are supposed to increase and the probabilities of the others are supposed to decrease. In this way, we could search the sample points with more information and build weak classifiers.

However, the high sensitivity of noise is always a disadvantage of AdaBoost. In AdaBoost, the exponential loss function is excessively focused on samples that are difficult to be classified correctly, while in noisy dataset, these samples are sometimes the noisy and singular points, which might cause over-fitting and other problems.

Since AdaBoost has been proposed, researchers have done lots of work to solve the problem above. Some researchers add the slack variables representing the credibility, so as to use the soft margin classification (Ratsch et al., 1999). Alleviating the over-fitting by validation sets is another effective method (Bylander et al., 2006). In this paper, we propose a learning automaton based algorithm to solve this problem.

1.2 *LA (Learning Automata)*

LA (Learning Automata) is a reinforcement learning algorithm, which is an adaptive system to adjust the behavior according to the random environment. Through the interaction with environment, a LA could finally attach an optimal state (Thathachar & Sastry, 2002).

With the characteristic of searching the optimal solution in the random environment, LA shows good behavior on the noisy dataset.

The reinforcement learning algorithm based on LA has been widely used in pattern recognition and is proved to have strong resistance to noise. A three-layer network learning model based on LA has been proposed and it has good behavior on Iris dataset (Thathachar et al., 1995). As the actions of automata could be continuous, CALA (continuous action learning automaton) is widely used in practical application. For example, CALA has been combined with the stochastic optimization of functions (Beigy et al., 2006), and it can be used to optimize the classification hypothesis (Sastry et al., 2010). Experiments show the CALA has good convergence property in practical application.

2 RELATED WORK

2.1 Classical AdaBoost

AdaBoost is a widely used technology in pattern recognition. The algorithm has made a great success in complexity and accuracy compared with previous methods. Classical AdaBoost algorithm is briefly described as follows:

First we set $T = \{(x_1, y_1), (x_2, y_2), ..., (x_N, y_N)\}$ as the training set, where $y_n \in \{-1, 1\}$. We assign a weight for each sample $\omega_n^1 = \frac{1}{N}, n = 1, ..., N$. The normalized result of ω_n^k indicates the probability with which a point is selected at each iteration. In the process, we build weak classifier C_k based on the selected points. And the points are selected from the training set according to the probability distribution which we can get on the basis of ω_n^k. In this step, the accuracy of C_k is supposed to be higher than 50%, and then we could obtain the hypothesis h_n^k. After that we calculate the error rate e_k, and subsequently we update the weight of each point ω_n^k as $\frac{\omega_n^{k-1}}{Z_n^k} \begin{cases} e^{-\alpha_n}, when\, h_n^k = y_n \\ e^{\alpha_n}, when\, h_n^k \neq y_n \end{cases}$, where y_n is the label of classification and Z_n^k is the parameter of normalization, $\alpha_n = \frac{1}{2}\ln\frac{1-e_k}{e_k}$. We can see when the points are classified correctly; their weights would decrease in the next iteration. On the contrary, the weights would increase in the next iteration, and making the points more likely to be selected in the next iteration. This process continues until it reaches the end condition (usually the number of iterations to reach a max value). Finally we build a reliable classifier through the combination classification hypothesis $H_n = sign\left[\sum_{k=1}^{k=k_{max}} \alpha_k h_n^k\right]$.

From the description we find only one parameter is needed in the process above. When there's no noise exists in the dataset, people always set a large number of iterations to obtain an accuracy classification model. However when we consider datasets with noise, the large number of iterations is likely to cause over-fitting.

2.2 LA and CALA

2.2.1 LA (Learning Automaton)

LA is a kind of reinforcement learning mechanism. It accepts the feedback of environment and tries to reach its optimal state by adaptive learning. A LA could be defined as (A, P, R, T), where $A = \{\alpha_1, \alpha_2, ..., \alpha_r\}$, $r < \infty$ is the set of actions, α_k is the action at instant k, R is the set of environment feedback. P (k) is the state of LA at time k and it is indicated by probability, T is the algorithm to update the probability. The environment is defined as (A, R, D), where $D = \{d_1, d_2, ..., d_r\}$ is the feedback probability, and $d_i(k) = E\left[\beta(k) \mid \alpha(k) = \alpha_i\right]$. As the LA changes the state according to the feedback of environment, we have the equation $P(k+1) = T\left(P(k), \alpha(k), \beta(k)\right)$.

Consider a LA and the stochastic environment connected with it. In the instant k, the LA generates an action α_k on the basis of the current probability distribution P (k). Then according to α_k, the environment generates a feedback $\beta(k)$. After that, the LA updates the probability distribution P (k+1) for the next iteration. In this way, when the environment is stable, the output of LA would converge to one action which could get the most rewards from the environment.

2.2.2 CALA (Continuous Action Learning Automaton)

The LA could be divided into FALA and CALA on the basis of finite and continuous actions. CALA is an algorithm which processes the action as a continuous value. The parameter of its action could be selected on the real line. In this paper, we build our algorithm mainly by CALA.

For a CALA, the probability distribution function is defined as a Gaussian distribution $N(\mu(k), \sigma(k))$, and the initial state is $N(\mu(0), \sigma(0))$. In the kth iteration, the LA selects x(k) according to $N(\mu(k), \sigma(k))$, and then we obtain the feedback $\beta_{x(k)}, \beta_{\mu(k)}$ from the environment, where the value of $\beta_{x(k)}, \beta_{\mu(k)}$ could be 0 or 1.

3 THE PROPOSED ALGORITHM

In this section, we propose a new AdaBoost algorithm based on CALA, which improves the classification accuracy of AdaBoost in noisy environment. We call the proposed algorithm CAdaBoost for short. In our algorithm, we use CALA to express the weights of sample points and take the normalized weight, which is defined as the expectation of distribution function, as the chosen probability of sample points in each iteration. Then we construct weak classifiers based on these sample points. We define the feedback $\beta_{\mu,n}$ as 1 to the misclassified point and 0 to the correct one. Then we build the weak classifier based on x(k), which is selected according to $N(\mu, \sigma^2)$. And in the same way we get second feedback from environment β_{xn}. After that, the probability distribution is updated by (1) and (2).

The detailed procedure of the improved algorithm is described as follows. The labeled training set is defined as $T = \{(x_1, y_1), (x_2, y_2), ..., (x_N, y_N)\}$, where $y_n \in \{-1, 1\}$. $P_n(k)$ identifies the probability that training point n is selected at time k. The probability distribution of sample n at time k is defined as $N(\mu(k), \sigma(k))$, where $n = 1, ..., N, k = 1, ..., T$.

Step 1: Initialize the probability distribution of samples as $P_n(1) = \frac{1}{N}, \mu_n(1) = \mu_0, \sigma_n(1) = 0$, $n = 1, ... N$.

Step 2: Take $\mu_n(k)$ as the probability sample selected in the kth iteration and train a weak classifier C_μ^k based on m samples. Then we calculate the hypothesis of each point as $h\mu,n(k)$, and obtain the error rate $\in_k$.

Step 3: Compare the classified result $h_{\mu,n}(k)$ and the label y_n. And get the feedback

$$\beta_{\mu,n}(k) = 1 - L(h_{\mu,n}(k), y_n) \tag{1}$$

$$L(a, b) = \begin{cases} 0 & when\ a = b \\ 1 & when\ a \neq b \end{cases}$$

Step 4: Take $x_n(k - 1)$ as the probability each sample is to select based on the probability distribution $N(\mu_n(k - 1), \sigma_n(k - 1))$, and build a weak classifier by m samples. Then we get the classified result $h_{x,n}(k)$.

Step 5: Compare the classified result $h_{x,n}(k)$ and label y_n, then we calculate the feedback in the same way as $\beta_{\mu,n}(k)$

$$\beta_{x,n}(k) = 1 - L(h_{x,n}(k), y_n) \tag{2}$$

Step 6: Update the expectation and variance of each sample according to the rules of CALA

$$\mu_n(k+1) = \mu_n(k) + \lambda \frac{(\beta_{x,n}(k) - \beta_{\mu,n}(k))}{\phi(\sigma_n(k))} \frac{(x_n(k) - \mu_n(k))}{\phi(\sigma_n(k))} \tag{3}$$

205

$$\sigma_n(k+1) = \lambda \frac{(\beta_{x,n}(k) - \beta_{\mu,n}(k))}{\phi(\sigma_n(k))} \left[\left(\frac{x_n(k) - \mu_n(k)}{\phi(\sigma(k))} \right)^2 - 1 \right] - \lambda K(\sigma_n(k) - \sigma_L) + \sigma_n(k) \qquad (4)$$

The value of λ could range from 0 to 1. As the resolution ratio, it indicates the speed to update. K is positive and large enough, and σ_L is the lower bound of variance. After a sufficient time period of iteration, the probability distribution function of action would converge and σ would be small enough and close to σ_L.

Step 7: Normalize the sample weights, and then take the expectation and variance in consideration that whether $|\bar{\sigma} - \sigma_L| < \varepsilon_1$ or $|\overline{\mu_k - \mu_{k-1}}| < \varepsilon_2$ as the stop condition. If the condition is not satisfied, we repeat step 2 to 6, otherwise we output the final hypothesis $H_n = sign\left(\sum_{k=1}^{T} \ln\left(\frac{1-\epsilon_k}{\epsilon_k} \right) h_{\mu,n}(k) \right)$.

In the process of getting feedback, the point is regarded as with rich information if the label of it is different from the hypothesis of weak classifier. In this case we set $\beta = 1$, and increase the probability the point is selected in the next iteration. So we reward the misclassified point and vice versa. The only termination condition of classical AdaBoost is usually the number of iterations reaches upper limit, but this condition has to be implemented under the premise which the algorithm is hardly over-fitting. While we could not obtain the premise when applying AdaBoost to noisy dataset, we choose the expectation and variance of probability distribution to judge whether the process is to stop in this paper.

4 EXPERIMENT

4.1 Dataset and parameter

In this part, we will present the experiment of our algorithm and compare it with the classical algorithm. In our experiment, classical AdaBoost algorithm and the proposed algorithm both use the decision-tree stumps as weak sub-classifiers. The noise added is the uniform Gaussian noise. In our experiment, the sample points are labeled falsely with certain probability to simulate the addition of Gaussian noise. The probability of the label flipped is 0.1, 0.2, 0.3 corresponding to the rate of 10%, 20%, 30% of Gaussian noise. We use 8 datasets from UCI database, and all the experiments are conducted with binary classification tests.

Figure 1 is the behavior of classical AdaBoost algorithm on a UCI dataset named statlog. We could see that on the dataset without noise, the classical algorithm performs well. Figure 2 is the error rate after we add noise, the property is significantly worse as the noise rate increases. And if the noise exists, as the number of iteration increases, the classical AdaBoost would be over-fitting and can't work well as the situation without noise.

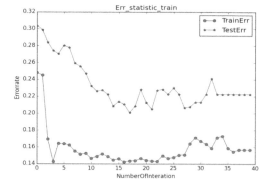

Figure 1.   The simulation result of classical AdaBoost.

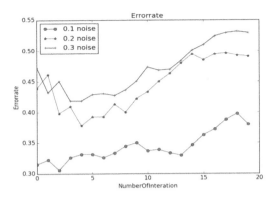

Figure 2.   The comparison among different noise rate.

Table 1.   The behavior comparison of AdaBoost and CAdaBoost on UCI datasets.

Dataset	10% noise		20% noise		30% noise	
	AdaBoost	CAdaBoost	AdaBoost	CAdaBoost	AdaBoost	CAdaBoost
Iris	0.289	0.100	0.304	0.271	0.346	0.296
Parkinson	0.359	0.115	0.371	0.263	0.384	0.360
Seeds	0.166	0.086	0.357	0.212	0.474	0.351
Banknote	0.267	0.217	0.406	0.297	0.413	0.346
BalanceScale	0.298	0.243	0.402	0.398	0.474	0.414
User	0.269	0.115	0.351	0.254	0.380	0.377
VotingRecords	0.236	0.172	0.359	0.273	0.407	0.278
Stalog	0.359	0.313	0.396	0.335	0.422	0.350

4.2   *Result and analysis*

In this part, we provide the performance of our algorithm. We can see from Table 1 that as the noise rate increases, the performance of classical AdaBoost is becoming worse, while the proposed algorithm performs better. As the number of iteration is large enough, we can see the behavior of classical AdaBoost is awful while CAdaBoost is better to some extent.

5   CONCLUSION

AdaBoost is a kind of boosting algorithm in the area of machine learning, but its bad behavior on datasets with noise is one of its major defects. Many researchers focus on finding an algorithm which could be applied to complex environments and is robust to noise. In this paper, we improve the behavior of AdaBoost on noisy datasets by introducing CALA to classical algorithm because LA is designed to search the optimal result in the random environment. The experiment shows the proposed algorithm has better performance on various datasets.

ACKNOWLEDGMENTS

This research work is funded by the National Science Foundation of China (61271316), 973 Program of China (2013CB329605), Shanghai Key Laboratory of Integrated Administration Technologies for Information Security.

REFERENCES

Beigy, H. & Meybodi, M. 2006. A new continuous action-set learning automaton for function optimization. Journal of the Franklin Institute, 343, 27–47.

Bylander, T. & Tate, L. Using Validation Sets to Avoid Overfitting in AdaBoost. FLAIRS Conference, 2006. 544–549.

Freund, Y. & Schapire, R.E. 1997. A decision-theoretic generalization of on-line learning and an application to boosting. Journal of computer and system sciences, 55, 119–139.

Hu, W., Gao, J., Wang, Y., Wu, O. & MAYBANK, S. 2014. Online Adaboost-based parameterized methods for dynamic distributed network intrusion detection. Cybernetics, IEEE Transactions on, 44, 66–82.

Lee, J.J., Lee, P.H., Lee, S.W., Yuille, A. & Koch, C. Adaboost for text detection in natural scene. 2011 International Conference on Document Analysis and Recognition, 2011. IEEE, 429–434.

Lin, K., Yan, R., Duan, H., Yao, J. & Zhou, C. Objective Classification Using Advanced Adaboost Algorithm. Fuzzy Systems and Knowledge Discovery, 2008. FSKD'08. Fifth International Conference on, 2008. IEEE, 525–529.

R Tsch, G., Onoda, T. & M Ller, K.R. 1999. Regularizing Adaboost.

Sastry, P., Nagendra, G. & Manwani, N. 2010. A team of continuous-action learning automata for noise-tolerant learning of half-spaces. Systems, Man, and Cybernetics, Part B: Cybernetics, IEEE Transactions on, 40, 19–28.

Thathachar, M. & Phansalkar, V.V. 1995. Convergence of teams and hierarchies of learning automata in connectionist systems. Systems, Man and Cybernetics, IEEE Transactions on, 25, 1459–1469.

Thathachar, M. & Sastry, P.S. 2002. Varieties of learning automata: an overview. Systems, Man, and Cybernetics, Part B: Cybernetics, IEEE Transactions on, 32, 711–722.

Signal and Information Processing, Networking and Computers – Chen & Huang (Eds)
© 2016 Taylor & Francis Group, London, ISBN 978-1-138-02881-4

On applying Confidence Interval Estimator to the pursuit learning schemes: Various algorithms and their comparison

Hao Ge, Jianhua Li, Shenghong Li & Yan Yan
Shanghai Jiao Tong University, Shanghai, China

Yuyang Huang
Shanghai Starriver Bilingual School, Shanghai, China

ABSTRACT: Rate of convergence is a metric of great importance in Learning Automata (LA) applications. The introduction of estimators in Learning Automata plays a key role in accelerating the convergence rate. However, the conventional Maximum Likelihood Estimator (MLE) suffers from several principle weaknesses. In this paper, we incorporate the pursuit concept with a newly proposed estimator—Confidence Interval Estimator (CIE). Different learning philosophies have been experimented upon, and we present a quantitative comparison of them. Numerical results confirmed that CIE based algorithms outperform their MLE counterparts.

Keywords: estimator Learning Automata; Maximum Likelihood Estimator; Confidence Interval Estimator; pursuit algorithm

1 INTRODUCTION

Learning Automata (LA) are self-adaptive learning machines that are operating in stochastic environments. The goal of LA is to maximize the long term reward through interacting with the environment. Typically, it takes hundreds of iterations for a learning automaton to get converged, depending upon the complexity of the environment. In the earlier decades, the earlier forms of LA (known as FSSA and VSSA) have a relatively slow rate of convergence. In order to address this problem, Thathachar and Sastry (Thathachar & Sastry 1984, Thathachar & Sastry 1985) proposed the first estimator based LA algorithm—pursuit algorithm, in which estimates for the reward probability of each possible action are maintained and used in the probability updating equations. Pursuit algorithm significantly improved the convergence speed of LA working in stochastic environments, making pursuit algorithm converge many times faster than other algorithms at that time. However, as we revealed in Ge et al. (2015), MLE suffers from several principal weaknesses.

MLE is inaccurate when the sample size is insufficient. The idea of introducing the estimator in learning automaton algorithms is to store past experience for better exploitation. However, the philosophy of learning in a stochastic environment is to balance exploration and exploitation. At the beginning, the LA needed more exploration to get a better understanding of the environment. When a certain amount of experiences has been acquired from interactions, the focus should gradually drift towards exploitation. As pointed out previously, in the context of estimator learning automata, the performances of estimator learning algorithms are strictly dependent on the reliability of the estimator's contents, for the reason that LA updates the probability distributions under the guidance of the estimate of each action's reward probability. The estimator gives directions of exploration and exploitation in the learning process, which is the key mechanism to balance the trade-off between exploring the stochastic environment and exploiting the learned experiences. In fact, algorithms with an MLE prefer exploitation to exploration, because little attention is paid to those actions that have been selected only a few times. Therefore, MLE may be misleading in the initial few iterations and is not appropriate to be used as an estimator method. In Papadimitriou et al. (2004), Papadimitriou made the earliest attempt to improve the traditional MLE. He imposed a controlled randomness to the MLE estimator and proposed the SE_{ri} algorithm. SE_{ri} improved the performance of pursuit algorithm by around 70% and has been the fastest algorithm since. In this paper, we make another attempt at improving the estimator of learning automata.

The paper incorporates the classical estimator based learning automata algorithm with the confidence interval estimator technique and presents an experimental result of the convergence rate in five benchmark environments. In addition, like the pioneers have done in Oommen & Agache (2001), we shall present a more comprehensive comparison between pursuit algorithms with different estimators, as well as different learning philosophies.

The paper is structured as follows. In Section 2, a brief introduction of the Pursuit Algorithm as well as Confidence Interval Estimator is given, and the CIE based algorithms with different learning philosophies are presented. Results of a large number of comparison experiments are provided in Section 3, and Section 4 concludes the paper.

2 PURSUIT ALGORITHM WITH THE CONFIDENCE INTERVAL ESTIMATOR

In this section, we will describe the pursuit algorithm and the Confidence Interval Estimator (CIE) briefly.

2.1 *Maximum likelihood estimator and confidence interval estimator*

Within the framework of learning automata systems, the environment will respond to the selected action with a feedback $\beta(t)$ in every cycle t. If the feedback takes binary values— 0 and 1, where 1 stands for a reward and 0 for a penalty, the environment is referred as to a P-model environment.

In P-model environments, the number of rewarded times k out of the number of selected times n for any action follows a binomial distribution $k \sim B(n, p)$. In order to estimate the reward probability p for each action, according to maximum-likelihood estimation, we can calculate the estimated $\hat{p}$ simply by $\frac{k}{n}$. According to the *Law of Large Numbers*, $\hat{p}$ approaches p arbitrarily close when the sample size tends to infinity.

However, one main drawback of MLE is its inaccuracy when the sample size is insufficient. The author proposed an alternative estimator to address this issue (Ge et al. 2015). The proposed estimator is called *Confidence Interval Estimator (CIE)*. The estimate of reward probability of each possible action α_i can be calculated by the following equation

$$\hat{d}_i = \left\{ 1 + \frac{Z_i - W_i}{(W_i + 1) F_{2(W_i+1),2(Z_i-W_i),0.005}} \right\}^{-1} \tag{1}$$

where Z_i and W_i is the number of action α_i have been selected and rewarded, respectively, $F_{2(W_i+1),2(Z_i-W_i),0.005}$ is the 0.005-right-hand tail probability of F distribution with $2(W_i+1)$ and $2(Z_i - W_i)$ degree of freedom, respectively.

CIE overcomes the drawbacks of MLE, and gives more chances to the under-sampled actions to be further investigated. In Ge et al. (2015), CIE is proved to be successful when applied to Generalized Pursuit Algorithm (Agache & Oommen 2002) and it also shows that CIE holds the valuable property of converging to the ture value almost surely when the sample size tends to infinity.

To explore the applicability of CIE to other estimator based algorithms, in this paper, we incorporate CIE with the conventional Pursuit learning automata.

2.2 Discretized pursuit reward-inaction algorithm with CIE

We simply replace the MLE in the original Discretized Pursuit Reward-Inaction Algorithm with CIE: **Algorithm** $DP_{ri}(CIE)$

Parameters

n	resolution parameter
$W_i(t)$	the number of events that the ith action has been rewarded up to time instant t, for $1 \leq i \leq r$
$Z_i(t)$	the number of events that the ith action has been selected up to time instant t, for $1 \leq i \leq r$

$\Delta = 1/r/n$ smallest step size

Method
Initialize $p_i(0) = 1/r$, for $1 \leq i \leq r$
Initialize $W_i(t)$ and $Z_i(t)$ by selecting each action a number of times

Repeat
Step 1: At time t, choose an action $a(t) = a_i$ according to probability distribution $P(t)$.
Step 2: Receive a feedback $\beta(t)$ from environment and update $W_i(t)$ and $Z_i(t)$.
Step 3: Update $d_i(t)$ according to the upper bound of 99% Clopper-Pearson "exact" confidence interval of action i's reward probability.
Step 4: Denote m as the index of the action with the maximal reward probability estimate.
Step 5: Update the probability distribution vector $P(t)$ according to the following equations:

If $\beta(t) = 1$ **Then**
$$p_j(t+1) = \max_{j \neq m}\{p_j(t) - \Delta, 0\}$$
$$p_m(t) = 1 - \sum_{i \neq j} p_j(t+1)$$

End Repeat
End Algorithm Algorithm $DP_{ri}(CIE)$

This algorithm can be proved to be ϵ-optimal. The proof is quite similar to the one in Ge et al. (2015) and is omitted here for brevity.

2.3 Pursuit inaction-penalty algorithm

The only difference between *Inaction-Penalty* philosophy and *Reward-Inaction* philosophy lies in the condition of the probability updating equations to be executed. Probability updating occurs every time the automaton has been rewarded in *Reward-Inaction* philosophy, on the other hand, it occurs every time the automaton has been punished in *Inaction-Penalty* philosophy.

As a result, the updating equations in step 5 of Algorithm $DP_{ip} - CIE$ should be:

If $\beta(t) = 0$ **Then**
$$p_j(t+1) = \max_{j \neq m}\{p_j(t) - \Delta, 0\}$$
$$p_m(t) = 1 - \sum_{i \neq j} p_j(t+1)$$

2.4 Pursuit reward-penalty algorithm

Reward-Penalty philosophy is distinct from both the two aforementioned philosophies. Probability updating takes place every cycle, regardless of the feedback.

As a result, the condition of updating equations in step 5 of Algorithm $DP_{rp} - CIE$ could be removed:

$$p_j(t+1) = \max_{j \neq m}\{p_j(t) - \Delta, 0\}$$
$$p_m(t) = 1 - \sum_{i \neq j} p_j(t+1)$$

3 EXPERIMENTAL RESULTS

In this section, the aforementioned CIE based pursuit algorithms with different learning philosophies are simulated in five well-known 10-action benchmark environments, which are widely used in the recent literature (Papadimitriou et al. 2004, Zhang et al. 2013, Zhang et al. 2014, Ge et al. 2015) and the numerical results of the simulation are given.

The reward probabilities for actions of the five benchmark environments are listed as follows:

E_1: {0.65,0.50,0.45,0.40,0.35,0.30,0.25,0.20,0.15,0.10}.
E_2: {0.60,0.50,0.45,0.40,0.35,0.30,0.25,0.20,0.15,0.10}.
E_3: {0.55,0.50,0.45,0.40,0.35,0.30,0.25,0.20,0.15,0.10}.
E_4: {0.70,0.50,0.30,0.20,0.40,0.50,0.40,0.30,0.50,0.20}.
E_5: {0.10,0.45,0.84,0.76,0.20,0.40,0.60,0.70,0.50,0.30}.

In all the simulations performed, the automaton is considered to have converged if the probability of choosing an action is greater than or equal to 0.999. Besides, it is considered to have converged correctly if it converges to the action with the highest reward probability.

Before a large number of simulations was performed, parameter tuning was necessary to find the "best parameter" in each environment. The resolution parameter is considered to be "best", provided that the automaton converges correctly in a sequence of 750 experiments. And the value of "best parameter" is averaged over 20 independent tunings.

And after we get the "best" parameter, 25,000 independent simulations are carried out to get averaged iterations and accuracy. Discretized Pursuit algorithm with two different estimators are used for the comparison and the numerical results of Reward-Inaction philosophy, Inaction-Penalty philosophy and Reward-Penalty philosophy are shown in Table 1, Table 2 and Table 3, respectively.

We use the coefficient of variation to evaluate the gaps among the MLE based pursuit algorithm with different philosophies and the gaps among the CIE based pursuit algorithm with different philosophies. The results are shown in Table 4.

It is noted that the metric *improvement* shown in Table 1–3 are obtained by calculating:

$$\frac{Iterations_{\{DP \text{ with } MLE\}} - Iterations_{\{DP \text{ with } CIE\}}}{Iterations_{\{DP \text{ with } MLE\}}}$$

In Oommen & Agache (2001), Oommen drew the conclusion that $DP_{ri} > DP_{rp} > CP_{ri} > CP_{rp}$ with regard to their convergence speed, which is supported by our experiments. Besides this, the following conclusions can be made from the above tables. Generally speaking, we can

Table 1. A comparison of the pursuit algorithm with Reward-Inaction philosophy in the five ten-action benchmark environments.

| Env. | DP_{ri} with MLE | | | DP_{ri} with CIE | | | |
	Parameter	Iterations	Accuracy	Parameter	Iterations	Accuracy	Improvement
E_1	n = 298	1092	0.995	n = 33	480	0.996	56.04%
E_2	n = 642	2484	0.994	n = 61	935	0.996	64.43%
E_3	n = 2312	9572	0.993	n = 232	3039	0.994	68.25%
E_4	n = 218	798	0.996	n = 31	434	0.998	45.61%
E_5	n = 855	2333	0.994	n = 107	899	0.996	61.47%

Table 2. A comparison of the pursuit algorithm with Inaction-Penalty philosophy in the five ten-action benchmark environments.

Env.	DP_{ip} with MLE			DP_{ip} with CIE			Improvement
	Parameter	Iterations	Accuracy	Parameter	Iterations	Accuracy	
E_1	n = 440	1913	0.994	n = 33	608	0.997	68.22%
E_2	n = 964	4079	0.993	n = 58	1050	0.997	74.26%
E_3	n = 3509	14309	0.992	n = 260	3495	0.995	75.57%
E_4	n = 280	1436	0.995	n = 26	555	0.998	61.35%
E_5	n = 593	5320	0.994	n = 41	1256	0.997	76.39%

Table 3. A comparison of the pursuit algorithm with Reward-Penalty philosophy in the five ten-action benchmark environments.

Env.	DP_{rp} with MLE			DP_{rp} with CIE			Improvement
	Parameter	Iterations	Accuracy	Parameter	Iterations	Accuracy	
E_1	n = 701	1362	0.994	n = 62	517	0.997	62.04%
E_2	n = 1558	3092	0.993	n = 98	897	0.996	70.99%
E_3	n = 5570	11451	0.991	n = 465	3147	0.994	72.52%
E_4	n = 483	1000	0.995	n = 48	438	0.997	56.20%
E_5	n = 1427	2841	0.993	n = 139	932	0.996	67.19%

Table 4. Evaluation of the gaps among the MLE based pursuit algorithm with different philosophies and the gaps among the CIE based pursuit algorithm with different philosophies.

	Coefficient of Variation (CV)	
Environment	DP with MLE	DP with CIE
E_1	0.2875	0.1231
E_2	0.2501	0.0829
E_3	0.2025	0.0738
E_4	0.3025	0.1445
E_5	0.4569	0.1917

rank the six algorithms as $DP_{ri}(CIE) > DP_{rp}(CIE) > DP_{rp}(CIE) > DP_{ri}(MLE) > DP_{rp}(MLE) > DP_{rp}(MLE)$ with regard to their convergence speed. Specifically, pursuit algorithms with CIE outperform their MLE counterparts considerably. The performances are improved from 45.61% up to 75.57% under different learning philosophies. An interesting phenomenon, as revealed in Ge et al. (2015), is that the more complex the environment is, the better improvement the CIE based algorithm can achieve. A complex environment usually needs more iterations for an automaton to get converged. For example, E_3 needs more iterations than any other environment, and is thus considered to be a complex environment, and obviously CIE based pursuit algorithms attain better improvements in E_3. What's more, CIE narrows the gaps among different philosophies as demonstrated in Table 4.

4 CONCLUSION

In this paper, we devoted efforts to improve the estimator of learning automata. We incorporated the Confidence Interval Estimator with pursuit algorithm, and the CIE based algorithm with different learning philosophies had been extensively simulated in five 10-action benchmark environments. The results showed that the CIE based algorithms outperformed their MLE based counterparts. Moreover, CIE narrowed the gaps among different learning philosophies.

ACKNOWLEDGMENT

This research work is funded by the National Science Foundation of China (61271316), 973 Program of China (2013CB329605), Shanghai Key Laboratory of Integrated Administration Technologies for Information Security.

REFERENCES

Agache, M. and B.J. Oommen. 2002. Generalized pursuit learning schemes: new families of continuous and discretized learning automata. *IEEE Transactions on Systems, Man, and Cybernetics, Part B: Cybernetics, 32*(6): 738–749.

Ge, H., W. Jiang, S. Li, J. Li, Y. Wang, and Y. Jing. 2015. A novel estimator based learning automata algorithm. *Applied Intelligence 42*(2): 262–275.

Oommen, B.J. and M. Agache. 2001. Continuous and discretized pursuit learning schemes: various algorithms and their comparison. *IEEE Transactions on Systems, Man, and Cybernetics, Part B: Cybernetics 31*(3): 277–287.

Papadimitriou, G.I., M. Sklira, and A.S. Pomportsis. 2004. A new class of ε-optimal learning automata. *IEEE Transactions on Systems, Man, and Cybernetics, Part B: Cybernetics 34*(1): 246–254.

Thathachar, M.A.L. and P.S. Sastry. 1984. A class of rapidly converging algorithms for learning automata. In *Proceedings of the IEEE International Conference on Cybernetics and Society*, pp. 602–606. IEEE.

Thathachar, M.A.L. and P.S. Sastry. 1985. A new approach to the design of reinforcement schemes for learning automata. *IEEE Transactions on Systems, Man and Cybernetics 15*(1): 168–175.

Zhang, J., C. Wang, and M. Zhou. 2014. Last-position elimination-based learning automata. *IEEE Transactions on Cybernetics 44*(12): 2484–2492.

Zhang, X., O.-C. Granmo, and B.J. Oommen. 2013. On incorporating the paradigms of discretization and bayesian estimation to create a new family of pursuit learning automata. *Applied Intelligence 39*(4): 782–792.

Signal and Information Processing, Networking and Computers – Chen & Huang (Eds)
© 2016 Taylor & Francis Group, London, ISBN 978-1-138-02881-4

Modified HARQ and ARQ for LTE broadcast

X.P. Zhu
Wireless Theories and Technologies laboratory, Beijing University of Posts and Telecommunications, Beijing, China
Corporate R&D, Qualcomm, Beijing, China

D.C. Yang
Wireless Theories and Technologies laboratory, Beijing University of Posts and Telecommunications, Beijing, China

J. Wang, X.X. Zhang & J.Q. Zhang
Corporate R&D, Qualcomm, San Diego, CA, USA

ABSTRACT: HARQ and ARQ retransmission are supported for unicast in LTE, but not for broadcast due to uplink feedback cost and efficiency. This paper proposes group HARQ and RaptorQ based IR-ARQ as RLC-BM (RLC broadcast mode) for efficient retransmission in LTE broadcast. With group HARQ, UEs share the same resource for uplink NACK feedback and eNB decides whether to retransmit per the energy of received NACK. Instead of directly retransmitting the lost packet by ARQ of RLC AM, we propose eNB to broadcast new packets with RaptorQ encoder generated repair packets of a packet group. Based on simulation and analysis, this paper further proposes when to use group HARQ, RaptorQ IR-ARQ or both.

Keywords: LTE broadcast; MBMS; PTM; retransmission; HARQ; RLC

1 INTRODUCTION

Wireless channel is intrinsic broadcast. Broadcast is efficient in delivering the same content to a group of UEs (user equipment) (G.W. Kent et al. 2014). In 3GPP TS 36.300 and 3GPP TS 26.346, MBMS (Multimedia Broadcast Multicast Service) was defined for LTE broadcast. MBMS has two kinds of radio bearers: MBSFN (Multimedia Broadcast multicast service Single Frequency Network, 3GPP TS 36.300) and SC-PTM (Single Cell Point to Multipoint, 3GPP TR 36.890).

In unicast transmission, HARQ is effective in improving the spectral efficiency and RLC AM (RLC Acknowledged Mode) is effective in reducing the residual packet error rate to improve the TCP and application performance. However, neither HARQ nor RLC-AM is supported for MBMS due to the cost of uplink feedback and efficiency.

Retransmission for broadcast has been discussed by academia for quite long time. Early in 1980s, Reed-Solomon based PTM-IR (point to multipoint incremental redundancy) retransmission was proposed (Metzner 1984). In recent years, NC-ARQ (Network Coded ARQ) was proposed by many papers, e.g. Sorour (2009). NC-ARQ requires every UE to feedback the number of lost packets to the network. All these proposals didn't consider the cost of UL feedback. In the practical mobile communication system, over consuming UL resource leads to DL performance downgrading. This is even more serious in some typical UL/DL configuration of LTE-TDD, e.g. configuration 2–5, in which UL is often the bottleneck of the system. This is one of the key reasons that MBMS does not support HARQ and ARQ retransmission.

To avoid the feedback cost, we propose that the UEs have failed in decoding to feedback NACK on a shared PUCCH channel. The eNB estimates the number of UEs that have not received the packet by energy detection of NACK on the shared PUCCH. HARQ is efficient in reducing the PER (packet error rate) to 0.1%. To further reduce the PER, ARQ retransmission on higher protocol layer e.g. RLC should be used. Based on the estimated number of UEs, the eNB can infer the PER level and decide whether to retransmit. Current ARQ for unicast directly retransmits the lost packet. This paper proposes RaptorQ based IR-ARQ (Incremental Redundancy ARQ) in RLC layer.

The rest of the paper is organized as follows: the second part provides background information on MBMS and unicast retransmission of LTE; group HARQ is proposed in the third part; RaptorQ based IR-ARQ is proposed in fourth part; the fifth part analyzes the performance of using group HARQ and RaptorQ based IR-ARQ together in the delay constrained scenario; the sixth part is the conclusion of the study.

2   BACKGROUND

Figure 1 is the logical architecture of MBMS. MCE is the controller of each MBSFN area. For a MBMS service, MCE decides which bearer type (MBSFN or SC-PTM) to use.

2.1   *MBSFN*

MBSFN bearer broadcasts to a MBSFN area, which consists of a group of synchronized adjacent cells. MCE (Multi-Cell/Multicast Coordination Entity) is the controller of each MBSFN area and configures each cell for synchronized broadcast. Each MBSFN area has multiple PMCHs (Physical Multicast Channel). On each PMCH, each member cell broadcasts the same content at the same time using the same MCS. The signals from the MBSFN member cells are combined by receiving UE just like multi-path signal.

MBSFN transmission does not support HARQ/ARQ feedback. But MBSFN usually uses FEC e.g. Raptor code to ensure the coverage goal, e.g. 95% UEs could receive the content with less than 1% PER. With well-tuned MCS and FEC, the coverage goal can be ensured without transmission. For the service e.g. file downloading requiring zero or very low packet

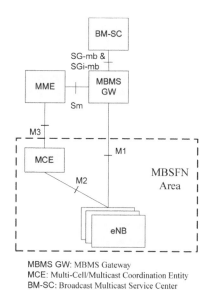

MBMS GW: MBMS Gateway
MCE: Multi-Cell/Multicast Coordination Entity
BM-SC: Broadcast Multicast Service Center

Figure 1.   MBMS logical architecture.

error rate, service layer retransmission i.e. unicast fix is supported using Associated Delivery Procedure.

2.2 *SC-PTM*

In unicast, transmission is scheduled by eNB on PDSCH (Physical Downlink Shared Channel). The scheduling information is sent on PDCCH (Physical Downlink Control Channel) addressed with UE identifier: RNTI (Radio Network Temporary Identifier). SC-PTM enables a group of UEs to receive common content in a cell sharing the same RNTI.

SC-PTM is a complementary bearer type of MBMS. Therefore the network architecture is the same as MBSFN as showed in Figure 1. The MCE decides to carry a service over SC-PTM or MBSFN.

In current LTE standard (up to 3GPP release 13), SC-PTM does not support HARQ/ARQ retransmission and link adaption.

2.3 *Performance comparison*

Figure 2 shows the per cell spectral efficiency of MBSFN, SC-PTM and unicast under different number of UEs based on simulation in Nokia Network 2015 3GPP TR36.890. It is based on 3GPP D3 model at 800 MHz and the detail parameters can be found in appendix. The simulation shows that:

- SC-PTM with feedback is more efficient than unicast when the number of receiving UEs is 2 or more
- SC-PTM without feedback is more efficient than unicast when the number of receiving UEs is more than 6
- Large area MBSFN is more efficient than unicast when the number of receiving UE is more than 2 per cell.
- Large area MBSFN is more efficient than SC-PTM when the number of receiving UE is more than 4 per cell.

3 GROUP HARQ

HARQ combines FEC and ARQ together to improve the spectral efficiency by leveraging the previous transmissions in decoding. HARQ has several variants. LTE uses IR (incremental redundancy) type of HARQ for unicast. HARQ requires UE to feedback ACK/NACK in PUCCH (Physical Uplink Control Channel). The HARQ can be extended to broadcast by

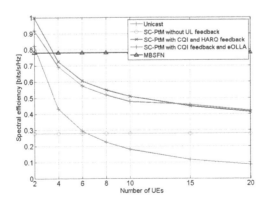

Figure 2.   Spectrum efficiency (in 3 center cells) for unicast, SC-PTM and MBSFN transmission in 21 cells.

each UE sending individual feedbacks (ACK/NACK) through its PUCCH. However, many UEs sending feedback (ACK/NACK) at the same time would lead to PUCCH channel overloaded and impact the DL throughput and latency performance. To avoid this issue, we propose the UEs that failed decoding broadcast content only send NACK through shared PUCCH, as shown in Figure 3.

In this way, eNB receives NACK from shared PUCCH with aggregated energy of multiple UEs. By measuring the energy, eNB knows whether there are any UEs failed in decoding and roughly how many UEs failed in decoding. It is difficult and inefficient to make all the UEs correctly decode the broadcasted content using HARQ. PER the experience of unicast, HARQ has highest efficiency when the residual PER is around 0.001 and initial transmission PER is around 0.1. In broadcast reception with number of M UEs, to achieve same initial transmission PER, the initial transmission PER (P0) per UE should be:

$$(1 - P_0)^M = 0.9 \tag{1}$$

$$P_0 = 1 - 0.9^{1/M} \tag{2}$$

The decoding failure probability of i-th retransmission per UE can be specified as P_i. With IR, the coding redundancy is increased in each decoding. The value of P_i depends on the SNR, channel type and MCS. In unicast, a typical assumption is: each retransmission downgrades PER for an order of magnitude, i.e. to 1/10 of its original PER. If we inherit the same rule to broadcast,

$$P_i = 10^{-i} P_0 \tag{3}$$

For one UE, the probability that the UE correctly receives the packet after n retransmission is:

$$P_{sr}(n) = (1 - P_n) \prod_{i=0}^{n-1} P_i = (1 - 10^{-i} P_0) 10^{-n(n-1)/2} P_0^n \tag{4}$$

If the worst UE requires N retransmission, the rest M-1 UEs receive the packet with up to N retransmission. The probability of M UEs receives the packet after up to N retransmissions is:

$$P_{rM}(N) = M P_{sr}(N) [1 - P_0 + \sum_{n=1}^{N} P_{sr}(n)]^{M-1} \tag{5}$$

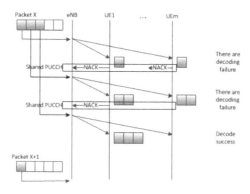

Figure 3.   Group HARQ.

218

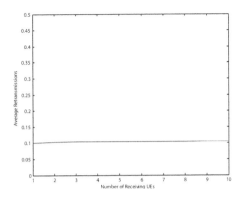

Figure 4.   Average HARQ retransmissions with M receiving UEs.

For one UE, the PER after n retransmissions are:

$$P_r(n) = \prod_{i=0}^{n} P_i = 10^{-n(n+1)/2} P_0^{n+1} \tag{6}$$

To avoid endless retransmission, HARQ usually has a maximum retransmissions restriction: $N_{R\max}$. So, the HARQ residual PER is:

$$P_{re} = P_r(N_{R\max}) \tag{7}$$

The average retransmission for a packet is:

$$N_r = \sum_{N=1}^{N_{R\max}} N P_{rM}(N) \tag{8}$$

The average delay per packet is:

$$T_{HARQ} = (N_r + 1) RTT_{HARQ} \tag{9}$$

RTT_{HARQ} is 8 ms in LTE FDD. The maximum delay is $RTT_{HARQ} N_{R\max}$. Figure 4 shows the average HARQ retransmissions with M receiving UEs based on formula (8). The result of simulation perfectly aligns with the formula. The number of HARQ retransmissions is almost independent of the number of receiving UEs because the initial transmission error probability is adapted as per the number of UEs by formula (2).

4   RAPTORQ BASED IR-ARQ

The acknowledging mode of LTE unicast, lost packet is retransmitted with ARQ by RLC. The intention is to reduce the PER to 10–5 or lower so that TCP can work efficiently. Broadcast system doesn't use TCP protocol. MBMS uses Raptor code as FEC to overcome the random packet loss in the broadcast transmission. When the number of receiving UEs is large, FEC can achieve the PER goal by setting suitable MCS and FEC redundancy. When the number of receiving UEs is not large, e.g. only several receiving UEs, due to the random channel status change and UE mobility, the PER goal cannot be efficiently met with the semi-statically configured FEC redundancy and MCS. Similar to HARQ, the RLC ARQ feedback and RaptorQ FEC can be combined to better adapt to channel change. RaptorQ is

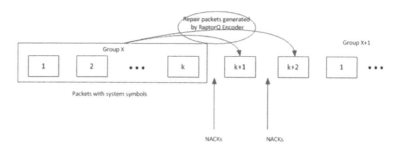

Figure 5.   RaptorQ based IR-ARQ.

a relatively new fountain code with much better performance than Raptor (Bouras C. et al. 2012). RaptorQ decoding failure probability is:

$$pf_{RQ}(n,k) = \begin{cases} 1, & \text{if } n < k \\ 0.01 \times 0.01^{n-k}, & \text{if } n \ge k \end{cases} \tag{10}$$

n is the number of symbols received by UE. k is the number of source/systematic symbols. When k is large, the overhead n-k is ignorable, i.e. approaching the overhead of ideal fountain code. In the following analysis, we approximately assume UE could decode when $n \ge k$.

Figure 5 shows the principle of RaptorQ based IR-ARQ. k packets are group as input to RaptorQ encoder. After the group with k packets is sent by eNB, UEs not received all the k packets feedback NACK to eNB. eNB retransmits the repair packets generated by RaptorQ encoder until no more NACK is received or the maximum retransmission reached. Then, eNB starts to broadcast next group of packets.

In N transmissions, the probability of UE correctly received n packets is:

$$P_u(n) = C_N^n P_r^{N-n} (1 - P_{re})^n \tag{11}$$

When the worse UE receives n = k packets, all the UEs would be able to decode all the k source packets with the N-k retransmissions. The average retransmissions are:

$$N_{rARQ} = (N - k) M P_N(k) (\sum_{n=k}^{N} P_N(n))^{M-1} \tag{12}$$

The average delay is:

$$T_{ARQ} = N_{rARQ} RTT_{ARQ} + k T_{HARQ} \tag{13}$$

If HARQ retransmission is not used, $T_{HARQ} = RTT_{HARQ}$. RTT_{ARQ} is usually much larger than HARQ RTT. The typical value is in the order of 100 ms. It is too long for delay sensitive services like voice.

5   COMPARISON AND COOPERATION OF GROUP HARQ AND RAPTORQ BASED IR-ARQ

Each MRB (radio bearer of MBMS) has a QoS profile with delay requirement: $T_{ARQ} < T_{MRB}$. With delay constrain, the transmission efficiency can be optimized by:

$$(k, N_{R\max}) = \underset{T_{ARQ} < T_{MRB}}{Arg \min} (N_{rARQ} + k)(1 + N_r) \tag{14}$$

220

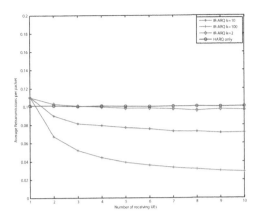

Figure 6.   Comparison of group HARQ and RaptorQ IR-ARQ.

Figure 6 is a comparison of retransmission efficiency between **HARQ** and **IR-ARQ**, when k = 2, 10, and 100. The average retransmissions per packet of RaptorQ based **IR-ARQ** is almost always less than group **HARQ** when the number of receiving UEs is larger than 2. Even when k = 2, the RaptorQ based **IR-ARQ** is better than group **HARQ** when the number of receiving UEs is greater than 4. Considering the average retransmissions per packet are usually less than 0.1%, if N_{Rmax} is unlimited, the average delay in formula (13) could be approximately expressed as:

$$
\begin{aligned}
T_{ARQ} &= 0.1\,kRTT_{ARQ} + 1.1\,kRTTH_{HARQ} \\
&= k(0.1\,RTT_{ARQ} + 1.1\,RTT_{HARQ}) \\
&\approx 18.8\,k\,(ms)
\end{aligned}
\tag{15}
$$

6   CONCLUSION

HARQ and ARQ can be used in LTE broadcast with modification. Group HARQ using shared PUCCH channel for NACK feedback resolves the issue of UL feedback cost in using HARQ for broadcast. RaptorQ based IR-ARQ enables efficient ARQ retransmission. This IR-ARQ can be supported as a new broadcast mode of RLC, i.e. RLC-BM. For delay sensitive service like voice, group HARQ should be used. For delay tolerant services, RaptorQ based IR-ARQ is more efficient. For the services with delay requirement in between, optimal packet group size and maximum retransmission can be derived to minimize the retransmissions.

REFERENCES

3GPP 2015. TR 36.890 Technical Specification Group Radio Access Network; Evolved Universal Terrestrial Radio Access (E-UTRA); Study on single-cell point-to-multipoint transmission.
3GPP 2015. TS 23.246, Technical Specification Group Services and System Aspects; Multimedia Broadcast/Multicast Service (MBMS); Architecture and functional description.
3GPP 2015. TS 36.300 Evolved Universal Terrestrial Radio Access (E-UTRA) and Evolved Universal Terrestrial Radio Access Network (E-UTRAN) Overall description; Stage 2.
Bouras, C., N. Kanakis, V. Kokkinos & A. Papazois, 2012. Evaluating RaptorQ FEC over 3GPP Multicast Services, *Wireless Communications and Mobile Computing Conference (IWCMC)*, 2012 8th International.
Lv, Z.X., K. Xu & Y.Y. Xu, 2012. A Practical HARQ Scheme with Network Coding for LTE-A Broadcasting System, *Wireless Communications & Signal Processing (WCSP)*, 2012 International Conference.

Metzner, J., 1984. An improved broadcast retransmission protocol, *IEEE Trans. Commun.*, vol. COM-32, pp. 679–683, June.

Nokia Networks, 2015. R2-151592 Performance evaluation of UL feedback schemes for SC-PTM 3GPP TSG-RAN WG2 Meeting #89bis Bratislava, Slovakia.

Sorour, S., S. Valaee, 2009. A network coded ARQ protocol for broadcast streaming over hybrid satellite systems, *Personal, Indoor and Mobile Radio Communications*, 2009 IEEE 20th International Symposium on.

Walker, G.K., J. Wang, C. Lo, X.X. Zhang & G. Bao, 2014. Relationship Between LTE Broadcast/eMBMS and Next Generation Broadcast Television, *IEEE transactions on broadcasting*, vol. 60, no. 2

Signal and Information Processing, Networking and Computers – Chen & Huang (Eds)
© *2016 Taylor & Francis Group, London, ISBN 978-1-138-02881-4*

Secure transmission with Artificial Noise in massive MIMO systems

Zhi Ma, Songlin Sun & Xinran Zhang
Beijing University of Posts and Telecommunications, Beijing, China

ABSTRACT: This is a review article that introduces a secure transmission with Artificial Noise (AN) in Low Power Nodes (LPNs) based massive Multiple-Input-Multiple-Output (MIMO) systems. With the future generations of cellular network rapidly developing, in Heterogeneous Network (HetNet), the provision of physical layer security is a promising aspect in multi-cell massive MIMO systems. We investigated secure downlink transmission in a multi-cell massive MIMO system with generation of Artificial Noise (AN), in which a mobile station can be served by multiple Base Stations (BSs) with LPNs in the presence of a passive multi-antenna eavesdropper. Thereby, we consider two different AN shaping matrices, namely, the conventional AN shaping matrix, where the AN is transmitted in the null space of the matrix formed by all user channels, and a random AN-shaping matrix, which avoids the complexity associated with finding the null space of a large matrix. Furthermore, we summarize several existing precoding algorithms in the LPNs with two different AN-shaping matrices.

Keywords: HetNet; massive MIMO; precoding; physical layer security

1 INTRODUCTION

Security has become a common and vital issue in wireless networks, with the broadcast nature of the medium developing in recent times. In the past, the traditional approach to achieve network security relied on cryptographic encryption implemented in the application layer which was not perfect. This method is based on certain assumptions regarding computational complexity, and is potentially vulnerable and inefficient, as is shown in A. Mukherjee et al. (2014) and R. Q. Hu et al. (2011). However, physical layer security, which complements cryptographic methods, has drawn significant research and industrial interest recently. The pioneering work on physical layer security in R.Q. Hu et al. (2011) uses the most classical three-terminal network, which consists of a transmitter (Alice), an intended receiver (Bob), and an eavesdropper (Eve). It was shown in R.Q. Hu et al. (2011) that a source-destination pair can exchange perfectly secure messages with a positive rate as long as the desired receiver enjoys better channel conditions than the eavesdropper(s). Meanwhile, there are more and more studies concentrating on the provision of physical layer security in multi-antenna multi-user networks in (A. Khisti & G. Wornell, 2010a), (A. Khisti & G. Wornell, 2010b), (E. Ekrem & S. Ulukus, 2011), (F. Oggier & B. Hassibi, 2011), and G. Geraci et al. (2012). Although the problem of the secrecy capacity region for multiuser networks is still left to be solved, it is interesting to investigate the achievable secrecy rates of such networks for certain practical transmission strategies. Eavesdroppers are typically passive in order to hide their existence, and thus their Channel State Information (CSI) cannot be obtained by Alice in M. Pei et al. (2012). In this case, multiple transmit antennas can be exploited to enhance secrecy by simultaneously transmitting both the information-bearing signal and Artificial Noise (AN) in (S. Goel & R. Negi, 2008). In fact, precoding is a process that helps the AN invisible to Bob while degrading the decoding performance of possibly present Eves in M. Pei et al. (2012), (S. Goel & R. Negi, 2008), (X. Zhou & M.R. McKay, 2010), and (A. Mukherjee &

A.L. Swindlehurst, 2011). But due to imperfect channel estimation, robust beamforming designs were presented in M. Pei et al. (2012), (A. Mukherjee & A.L. Swindlehurst, 2011), and (W. Xu & X. Dong, 2012) in various scenarios including secrecy channels.

Recently, a new promising design approach for cellular networks, known as massive or large-scale Multiple-Input Multiple-Output (MIMO), has been proposed in (T.L. Marzetta, 2010) and F. Rusek et al. (2013), where Base Station (BS) antenna arrays are equipped with an order of magnitude of more elements than what are used in current systems, i.e., a hundred antennas or more. Deploying large-scale antenna arrays at the existing macro Base Stations (BSs) is expected to dominate the implementation of next generation HetNets in F. Rusek et al. (2013). A typical cell of HetNet includes eNodeBs and Low Power Nodes (LPNs), e.g., access points, and these two types of nodes can cover different sizes or layers by being categorized by different transmission powers in S. Sun et al. (2015). Massive MIMO systems offer an abundance of BS antennas, while multiple transmit antennas can be exploited for secrecy enhancement. Whereas in conventional MIMO systems, the AN is transmitted in the null space of the channel matrix, in E. Larsson et al. (2014), the complexity associated with computing the null space may not be affordable in the case of a massive MIMO and simpler AN-shaping methods may be needed. Finally, most of the related work and our work are distinguished by considering a multicell setting where not only the data signals but also the AN cause intercell interference, which has to be carefully taken into account for system design. We put AN into a signal-based null-space so as to enhance the security of the entire precoding process in multicell massive MIMO systems.

The rest of this paper is organized as follows. We first introduce the channel model of a massive MIMO and two different AN-shaping matrix designs for the considered secure massive MIMO system. Then in Section 3 we review some existing MIMO equalization algorithms and some of its drawbacks. In section 4, we summarize several precoding methods in LPNs with an AN-based null-space. Finally, Section 5 is the conclusion of this paper.

2 SYSTEM MODEL

2.1 *The LPNs*

Figure 1 shows a two-tier macro cell consisting of a BS and three LPNs in J. Hoydis et al. (2013), where the BS is equipped with N_{BS} antennas serving K single-antenna users and $S(S \geq 0)$ LPNs are deployed arbitrarily in the macro cell forming an overlay layer. Each LPN is equipped with M antennas with $1 \leq M \leq 8$. The coverage area that was depicted in dashed circles of the LPNs is constrained due to power limits. This paper supposes that the BS can get high QoS targets with less power limits for the entire cell. The number of antennas, N_{BS} at BS can scale from eight to hundreds, i.e., massive MIMO. The single-antenna equipment

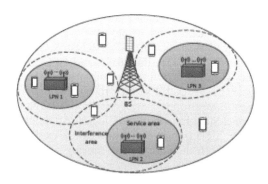

Figure 1.   The layout of heterogeneous networks.

of a user can either be served by the BS or one of the LPNs though they can be affected by the BS or neighboring LPNs.

Let $\mathbf{h}_j$ denotes the stationary channel between a certain LPN j and the users it serves, where

$$\mathbf{h}_j = \begin{bmatrix} \mathbf{h}_{j,1} \\ \mathbf{h}_{j,2} \\ \cdots \\ \mathbf{h}_{j,K_j} \\ \mathbf{h}_{j,K_j+1} \\ \cdots \\ \mathbf{h}_{j,K_j+L_j} \end{bmatrix}, \mathbf{h}_j \in C^{(K_j+L_j) \times M}, \mathbf{h}_{j,k} \in C^{1 \times M}. \tag{1}$$

$\mathbf{h}_0$ denotes the stationary channel between macro eNB and the users it serves. The k-th row vector $\mathbf{h}_{j,k}$ is the channel between the j-th LPN and the k-th user. The k-th row vector $\mathbf{h}_{0,k}$ is the channel between the macro eNB and the k-th user. Let $K_j + L_j \leq M$; K_j denotes the number of users of the j-th LPN, L_j denotes the number of users interfered by the j-th LPN and served by other nodes. The precoding vector of the i-th user is denoted by $\mathbf{T}_{j,i} \in C^{M \times 1}$. s_0 is the transmitted signal at the BS and $s_{j,i}$ is the information symbol, which is sent to the i-th user with $i \leq K_j$. Thus, the received signal of the k-th user can be written as:

$$y_{j,k} = \mathbf{h}_{j,k} \sum_{i=1}^{K_j} \mathbf{T}_{j,i} s_{j,i} + \mathbf{h}_{0,k} \mathbf{T}_{0,k} s_0 + \omega_{j,k}, \tag{2}$$

where $w_{j,k}$ is the additive white Gaussian noise of zero mean and variance σ^2.

The SINR of the k-th user in the small cell of the j-th LPN is given by:

$$SINR_{j,k} = \frac{\left| \sqrt{p} h_{j,k} T_{j,k} \right|^2}{\sum_{m=1}^{M} \sum_{i=1}^{N_t-K} \mathbb{E}\left[\left| \sqrt{q} h_{mk} v_{mi} \right|^2 \right] + \sum_{\{m,l\} \neq \{j,k\}} \mathbb{E}\left[\left| \sqrt{p} h_{mk} T_{ml} \right|^2 \right] + \sigma_{jk}^2} \tag{3}$$

To improve the capacity $\log_2(1 + SINR)$, interference between users should be effectively avoided. There are two different types, one is the interference from the remaining users in the same BS/LPN which can be ignored through orthogonality of precoding vectors, and the other one is the interference from the other BS/LPNs around. Thus, we will focus on the interference from the other BS/LPNs.

2.2   Two different designs of AN-shaping matrix

In this paper, we consider two different designs for the AN-shaping matrix V_j. Null-Space Method: For a conventional (non-massive) MIMO, V_j is usually chosen to lie in the null space of the estimated channel $\hat{h}_{jk}$ i.e., $\hat{h}_{jk} V_n = 0_{N_t-K}^T, k = 1,...,K$, which is possible as long as $N_t > K$ holds [9]. We refer to this method as N in the following. If a perfect CSI is available, i.e., $\hat{h}_{jk} = \tilde{h}_{jk}$, the N method prevents impairment of the users in the local cell by the AN generated by the local BS. However, in case of pilot contamination, AN leakage to users in the local cell is unavoidable. More importantly, for the large values of N_t and K typical for massive MIMO systems computation of the null space of $\hat{h}_{jk}, k = 1,...,K$ is computationally expensive. This motivates the introduction of a simpler method for generation of the AN-shaping matrix. Random Method: in this case, the columns of V_n are mutually independent random vectors. We refer to this method as R in the following. Here, we construct the columns of V_j as $v_{ji} = \tilde{v}_{ji} / \|\tilde{v}_{ji}\|$ where the $\tilde{v}_{ji}, i = 1,...,N_t - K$ are mutually independent Gaussian random vectors. Note that the R-method does not even attempt to avoid AN leakage to users in the local cell. However, it may still improve the ergodic secrecy rate as the precoding vector for the desired user signal, T_{jk}, is correlated with the user channel, $\tilde{h}_{jk}$, whereas the columns of the AN-shaping matrix are not correlated with the user channel. But when N approaches infinity, the two designs are similar which makes the R-method an attractive alternative for massive MIMO systems due to its simplicity.

In the next section, we will analyze the existing MIMO precoding methods, as well as the proposed novel null-space based scheme.

2.3 Downlink data transmission

In the local cell, the BS intends to transmit a confidential signal s_{jk} to the k-th MT. The signal vector for the K MTs is denoted by $s_j = [s_{j1}, ..., s_{jK}]^T \in C^{K\times 1}$ Before transmission, each signal vector s_n is multiplied by a transmit beam-forming matrix $t_j = [T_{j1}, ..., T_{jk}, ..., T_{jK}] \in^{N_t \times K}$. In this paper, we utilize matched-filter precoding, $T_{jk} = \hat{h}_{jk}^H / \|\hat{h}_{jk}\|$ for simplicity instead of ZF and MMSE precoding. Furthermore, we assume that the eavesdropper's CSI is not available at the local BS. The AN vector, $z_j = [z_{j1}, ..., z_{j(N_t-K)}]^T \sim (0_{N_t-K}, I_{N_t-K})$ is multiplied by an AN-shaping matrix $V_j = [v_{j1}, ..., v_{ji}, ..., v_{j(N_t-K)}] \in C^{N_t \times (N_t-K)}$ with $\|v_{ji}\| = 1, i = 1, ..., N_t - K$. The signal vector transmitted by the local BS is given by:

$$x_j = \sqrt{p}T_j s_j + \sqrt{q}V_j z_j = \sum_{k=1}^{K} \sqrt{p}T_{jk}s_{jk} + \sum_{i=1}^{N_t-K} \sqrt{q}v_{ji}z_{ji}, \qquad (4)$$

where p and q denote the transmit power allocated to each MT and each AN signal, which are both identical value, s, here. The M-1 cells adjacent to the local cell transmit their own signals and AN. The received signals at the k-th MT in the local cell, y_{jk}, and at the eavesdropper, y_{eve}, are given by:

$$y_{jk} = \sqrt{p}h_{jk}T_{jk}s_{jk} + \sum_{\{m,l\}\neq\{j,k\}} \sqrt{p}h_{mk}T_{ml}s_{ml} + \sum_{m=1}^{M} \sqrt{q}h_{mk}V_m z_m + j_{jk} \qquad (5)$$

$$y_{eve} = \sqrt{p}\sum_{m=1}^{M} H_m^{eve}W_m s_m + \sqrt{q}H_m^{eve}V_m z_m + n_{eve} \qquad (6)$$

3 TRADITIONAL PRECODING SCHEMES

In this section, we review and analyze three existing fundamental precoding algorithms for mitigating MIMO channel interference, including Zero-Forcing (ZF), minimum mean square error (MMSE) [?], and maximum ratio transmission (MRT) [?].

3.1 Review of the existing algorithms

3.1.1 ZF precoding
ZF precoding is to completely eliminate the interference without consideration of the noise, which implements a pseudo-inverse of the channel status matrix. The ZF precoding matrix T_{ZF} is defined as:

$$T_{ZF} = \frac{1}{\beta}h^h(hh^h)^{-1}, \qquad (7)$$

where $\beta = \sqrt{\frac{tr(BB^h)}{P_{tr}}}, B = h^h(hh^h)^{-1}$.

ZF precoding is a practical technique, especially under the circumstance of the transmitter having only a limited amount of CSI from each user. Based on the feedback information from all users, the BS performs scheduling and transmits signals to the selected users through ZF precoding technology.

3.1.2 MMSE precoding

MMSE precoding is the optimal linear precoding [?], which leads to the least symbol distortion in the sense of least square error. This precoder is achieved by using the Lagrangian optimization method due to the constraint of average power at each transmitting antenna.

The precoding matrix $\mathbf{T}_{MMSE}$ is given as:

$$\mathbf{T}_{MMSE} = \frac{1}{\beta}\mathbf{h}^h\left(\mathbf{h}\mathbf{h}^h + \frac{K}{P_{tr}}\mathbf{I}_K\right)^{-1},$$

(8)

where $\beta = \sqrt{\frac{tr(\mathbf{B}\mathbf{B}^h)}{P_{tr}}}$, $\mathbf{B} = \mathbf{h}^h\left(\mathbf{h}\mathbf{h}^h + \frac{K}{P_{tr}}\mathbf{I}_K\right)^{-1}$.

It can be observed that MMSE precoding approaches ZF precoding as $P_{tr} \to \infty$.

3.1.3 MRT precoding

The MRT precoding aims to maximize the receiving signal-to-noise ratio (SNR) and it is essential to a matched filter (MF) in spatial perspective. The MRT precoding matrix $\mathbf{T}_{MRT}$ is defined as:

$$\mathbf{T}_{MRT} = \frac{1}{\beta}\mathbf{H}^H,$$

(9)

where $\beta = \sqrt{\frac{tr(\mathbf{B}\mathbf{B}^h)}{P_{tr}}}$, $\mathbf{V} = \mathbf{h}^h$.

It is a practical solution that the BS radiates low signal power to the users; meanwhile, the performance and complexity of the system can be balanced and performed well. MRT approaches MMSE when $P_{tr} \to 0$.

3.2 Drawbacks of existing precoding algorithms

The existing precoding algorithms have a number of advantages in theory, but share drawbacks and suffer from common practical problems.

3.2.1 CSI acquirement

In reality, the existing algorithms have a significant problem due to constraints on precise CSI. But all current algorithms assume that the BS knows the perfect CSI of all users, which is impossible in practice. In fact, without accurate and effective CSI, extra distortion will be introduced and may cause reverse effect.

3.2.2 Computational complexity

The computational complexity required by the aforementioned algorithms is different. MRT precoding has the lowest complexity among the three linear precoding since only conjugate transpose is performed. The computing complexity of ZF and MMSE precoding is similar as both have an extra diagonal addition. The performance of three precoding algorithms is in proportion to the corresponding computational complexity. The result of the precoding algorithms also has some imperfections, which leads to suboptimal performance. For example, due to the quantization error in reality, "zero forcing" in ZF precoding cannot be truly achieved, and the received signals are still corrupted by the residual multiuser interference.

In the rest of paper, we considered both the conventional null space-based AN shaping matrix design and a novel random AN-shaping matrix design at the BS for secure downlink transmission in the presence of a multi-antenna passive eavesdropper.

4 PRECODING METHODS WITH AN IN LPNs

The two critical steps of the signal processing diagram of the proposed scheme are the acquisition of CSI for LPNs and the generation of precoding matrix based on the null space of the eavesdropper users. We will conclude every step in the following sections.

4.1 *CSI acquirement*

In practice, the LPNs can only collect the local CSI though the backward channel of their own users and cannot access the CSI of the external victim users. However, owing to the hub-spoke structure of the network, the CSI of all MIMO channels can be collected and disseminated through the backhaul link. One way is that the BS transmits part of the channel matrix regarding the victim users of each LPN. That is to send $[\mathbf{h}^h_{j,K_j+1},...,\mathbf{h}^h_{j,K_j+L_j}]^h$ to the j-th LPN, which contains $L_j \times M$ elements. Each LPN locally computes the precoding matrix $\mathbf{T}_{j,i}, i=1,2...K_j$ by precoding the algorithm proposed in the next subsection.

The other CSI acquisition method given in Figure 2 is that the BS performs the block-diagonal (BD) algorithm and then transmits the corresponding part of the null-space vector $\tilde{\mathbf{V}}_{j,i}^{(0)} = [\tilde{\mathbf{V}}_{j,i,L_j+1}...\tilde{\mathbf{V}}_{j,i,M_j}]$ to the associated LPNs. Thus, there are $L_j \times M$ elements to be transmitted, as long as the L_j dimension of $\tilde{\mathbf{V}}_{j,i}^{(0)}$ is enough for avoiding interference. This way, all the matrix decomposition computation will be performed by the BS and it is suitable in the situation that LPNs are simple and devices are of low performance.

Practically, the volume of data transmission is similar in the two methods. The null-space of victim users is obtained by singular value decomposition (SVD) computation between these two methods.

4.2 *Precoding based on null-space with AN*

This section illustrates various precoding methods with an AN-based multi-cell massive MIMO system. Massive MIMO systems offer an abundance of BS antennas, while multiple transmit antennas can be exploited for secrecy enhancement. In this section, we consider a multi-cell setting where not only the data signals cause intercell interference but also the AN, which has to be carefully taken into account for system design. In this paper, we study secure downlink transmission in multi-cell massive MIMO systems in the presence of a multi-antenna eavesdropper, which attempts to intercept the signal intended for one of the users. To arrive at an achievable secrecy rate for this user, we assume that the eavesdropper can acquire perfect knowledge of the channel state information (CSI) of all user data channels and is able to cancel all interfering user signals. Furthermore, employing random AN-shaping matrices is an attractive low-complexity option for massive MIMO systems.

For the j-th LPN, the complementary space concerning the victim users can be obtained by

$$\tilde{\mathbf{h}}_{j,v} = \left[\mathbf{h}_{j,K_j+1}^h ... \mathbf{h}_{j,K_j+L_j}^h\right]^h, \tag{10}$$

where $\tilde{\mathbf{h}}_{j,v}$ is combined by the channel matrix of all the victim users interfering with the j-th LPN. To avoid interference from the victim users, $\mathbf{T}_{j,i}$ should satisfy the following condition

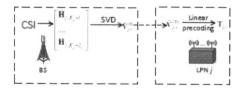

Figure 2. BS transmits null-space vector to LPN j.

228

$$\tilde{\mathbf{h}}_{j,v}\mathbf{T}_{j,i} = 0^{L_j \times 1}. \tag{11}$$

By SVD, $\tilde{\mathbf{H}}_{j,v}$ can be further written as:

$$\tilde{\mathbf{h}}_{j,v} = \tilde{\mathbf{U}}_{j,v}\tilde{\mathbf{\Lambda}}_{j,v}\tilde{\mathbf{V}}_{j,v}{}^h = \tilde{\mathbf{U}}_{j,v}\left[\sum_{j,v} 0\right]_{L_j \times M}\left[\tilde{\mathbf{v}}_{j,v,1}\tilde{\mathbf{v}}_{j,v,2}\cdots\tilde{\mathbf{v}}_{j,v,M_j}\right]^h. \tag{12}$$

Define $\tilde{\mathbf{V}}_{j,i}{}^{(0)} = [\tilde{\mathbf{V}}_{j,i,L_{i+1}}\cdots\tilde{\mathbf{V}}_{j,i,M_j}]$. Since the column vectors belonging to $\tilde{\mathbf{V}}_{j,i}{}^{(0)}$ are located in the null-space of all victim users, we will have $\tilde{\mathbf{h}}_{j,v}\tilde{\mathbf{V}}_{j,i}{}^{(0)} = 0$

For the k-th user of the j-th LPN, define the projection matrix $\mathbf{P}_{j,k}$ onto the null-space of victim users as:

$$\mathbf{P}_{j,k} = \mathbf{h}_{j,k}\tilde{\mathbf{V}}_{j,v}{}^{(0)}\left(\tilde{\mathbf{V}}_{j,v}{}^{(0)}\right)^h, \tag{13}$$

and define the projection matrix of all K_j users served by the j-th LPN as:

$$\mathbf{P}_j = \begin{bmatrix} \mathbf{P}_{j,1} \\ \mathbf{P}_{j,2} \\ \cdots \\ \mathbf{P}_{j,K_j} \end{bmatrix}, \mathbf{P}_j \in C^{K_j \times M}, \mathbf{P}_{j,k} \in C^{1 \times M}. \tag{14}$$

Define the power allocation matrix $P_{tr,j}$ as $P_{tr,j} = E[\|\mathbf{x}_j\|^2] = tr(\mathbf{T}_j^H\mathbf{T}_j)$. With P_j, the precoding matrices corresponding to the three fundamental linear algorithms can be obtained.

In particular, we rewrite the received signal at the kth MT in the local cell as:

$$y_{jk} = \mathbb{E}\left[\sqrt{p}h_{jk}T_{jk}\right]s_{jk} + n'_{jk} \tag{15}$$

$$n'_{jk} = \left(\sqrt{p}h_{jk}T_{jk} - \mathbb{E}\left[\sqrt{p}h_{jk}T_{jk}\right]\right)s_{jk} + \sum_{m=1}^{M} h_{mk}\sqrt{q}V_mz_m$$

$$+ \sum_{\{m,l\}\neq\{j,k\}} \sqrt{p}h_{mk}T_{ml}s_{ml} + n_{jk} \tag{16}$$

The received Signal-to-Interference-Noise Ratio (SINR) in the local cell of the j-th LPN is given by

4.2.1 ZF

ZF: This part investigates and summarizes ZF precoding method with AN. We can rewrite the ZF precoding matrix as:

$$T_{j,ZF} = \frac{1}{\beta_j}P_j^H\left(P_jP_j^H\right)^{-1} \tag{17}$$

$$\text{SINR}_{j,k} = \frac{\left|\sqrt{p}h_{j,k}T_{j,k}\right|^2}{\sum_{m=1}^{M}\sum_{i=1}^{N_t-K}\mathbb{E}\left[\left|\sqrt{q}h_{mk}v_{mi}\right|^2\right] + \sum_{\{m,l\}\neq\{j,k\}}\mathbb{E}\left[\left|\sqrt{p}h_{mk}T_{ml}\right|^2\right] + \sigma_{jk}^2}$$

$$= \frac{\left|\sqrt{p}h_{jk}\left[P_j^H\left(P_jP_j^H\right)^{-1}\right]_{j,k}\right|^2}{\sum_{m=1}^{M}\sum_{i=1}^{N_t-K}\mathbb{E}\left[\left|\sqrt{q}h_{mk}v_{mi}\right|^2\right] + \sum_{\{m,l\}\neq\{j,k\}}\mathbb{E}\left[\left|\sqrt{p}h_{mk}\left[P_j^H\left(P_jP_j^H\right)^{-1}\right]_{j,k}\right|^2\right] + \sigma_{jk}^2} \tag{18}$$

where $B_j = P_j^H \left(P_j P_j^H \right)^{-1}$ and $\beta_j = \sqrt{\frac{tr\left(B_j B_j^H\right)}{P_{tr,j}}}$.

4.2.2 MMSE

MMSE: This part investigates and summarizes the MMSE precoding method with AN. By the criterion of minimum mean square error, the precoding vector $T_{j,MMSE}$ can be expressed as:

$$T_{j,MMSE} = \frac{1}{\beta_j} P_j^H \left(P_j P_j^H + \frac{K_j}{P_{tr,j}} I_{K_j} \right)^{-1} \tag{19}$$

$$\text{SINR}_{j,k} = \frac{\left| \sqrt{p} h_{j,k} T_{j,k} \right|^2}{\sum_{m=1}^{M} \sum_{i=1}^{N_t-K} \mathbb{E}\left[\left| \sqrt{q} h_{mk} v_{mi} \right|^2 \right] + \sum_{\{m,l\} \neq \{j,k\}} \mathbb{E}\left[\left| \sqrt{p} h_{mk} T_{ml} \right|^2 \right] + \sigma_{jk}^2}$$

$$= \frac{\left| \sqrt{p} h_{jk} \left[P_j^H \left(P_j P_j^H + \frac{K_j}{P_{tr,j}} I_{K_j} \right)^{-1} \right]_{j,k} \right|^2}{\sum_{m=1}^{M} \sum_{i=1}^{N_t-K} \mathbb{E}[\left| \sqrt{q} h_{mk} v_{mi} \right|^2] + \sum_{\{m,l\} \neq \{j,k\}} \mathbb{E}\left[\left| \sqrt{p} h_{mk} \left[P_j^H \left(P_j P_j^H + \frac{K_j}{P_{tr,j}} I_{K_j} \right)^{-1} \right]_{j,k} \right|^2 \right] + \sigma_{jk}^2} \tag{20}$$

where $B_j = P_j^H \left(P_j P_j^H + \frac{K_j}{P_{tr,j}} I_{K_j} \right)^{-1}$ and $\beta_j = \sqrt{\frac{tr\left(B_j B_j^H\right)}{P_{tr,j}}}$.

4.2.3 MRT

MRT: This part investigates and summarizes MRT precoding method with AN. When $P_{tr,j} \to 0$, the precoding matrix can be obtained by:

$$T_{j,MRT} = \frac{1}{\beta_j} P_j^H \tag{21}$$

$$\text{SINR}_{j,k} = \frac{\left| \sqrt{p} h_{j,k} T_{j,k} \right|^2}{\sum_{m=1}^{M} \sum_{i=1}^{N_t-K} \mathbb{E}\left[\left| \sqrt{q} h_{mk} v_{mi} \right|^2 \right] + \sum_{\{m,l\} \neq \{j,k\}} \mathbb{E}\left[\left| \sqrt{p} h_{mk} w_{ml} \right|^2 \right] + \sigma_{jk}^2}$$

$$= \frac{\left| \sqrt{p} h_{jk} P_j^H \right|^2}{\sum_{m=1}^{M} \sum_{i=1}^{N_t-K} \mathbb{E}\left[\left| \sqrt{q} h_{mk} v_{mi} \right|^2 \right] + \sum_{\{m,l\} \neq \{j,k\}} \mathbb{E}\left[\left| \sqrt{p} h_{mk} P_j^H \right|^2 \right] + \sigma_{jk}^2} \tag{22}$$

where $B_j = P_j^H$ and $\beta_j = \sqrt{\frac{tr\left(B_j B_j^H\right)}{P_{tr,j}}}$.

5 CONCLUSION

In recent years, it has become more popular to consider a multi-cell massive MIMO system with matched-filter precoding and AN generation at the BS for secure downlink transmission in the presence of a multi-antenna passive eavesdropper. For the massive MIMO system, it can offer an abundance of BS antennas, while multiple transmit antennas can be exploited for secrecy enhancement. For AN generation, there are two classical types of AN, that is, the conventional null space based AN-shaping matrix design and a novel random AN-shaping matrix design. Various precoding algorithms have been developed previously for HetNet MIMO downlink systems. In particular, we can summarize that: for the considered multi-cell massive MIMO system with matched-filter precoding (1) AN generation is necessary to achieve a non-zero ergodic secrecy rate if the user and the eavesdropper experience the same path loss, (2) secrecy cannot be guaranteed if the eavesdropper has too many antennas, and (3) the proposed random AN-shaping matrix design is a promising less complex alternative to the conventional null space based AN-shaping matrix design.

REFERENCES

Ekrem, E. & Ulukus, S. 2011. The secrecy capacity region of the Gaussian MIMO multi-receiver wiretap channel. *IEEE Trans. Inf. Theory* 57(4):2083–2114.

Geraci, G., Egan, M., Yuan, J., Razi, A. & Collings, I. 2012. Secrecy sum-rates for multi-user MIMO regularized channel inversion precoding. *IEEE Trans. Commun* 60(11):3472–3482.

Goel, S. & Negi, R. 2008. Guaranteeing secrecy using artificial noise. *IEEE Trans. Wireless Commun* 7(6):2180–2189.

Hoydis, J., Hosseini, K., ten Brink, S. & Debbah, M. 2013. Making smart use of excess antennas: Massive MIMO, small cells, and TDD. *Bell Labs Technical Journal* 18:5–21.

Hu, R.Q., Qian, Y., Kota, S. & Giambene, G. 2011. Hetnets—a new paradigm for increasing cellular capacity and coverage. *IEEE Trans. Wireless Commun* 18(3):8–9.

Khisti, A. & Wornell, G. 2010a. Secure transmission with multiple antennas I: The MISOME wiretap channel. *IEEE Trans. Inf. Theory* 56(7):3088–3104.

Khisti, A. & Wornell, G. 2010b. Secure transmission with multiple antennas II: The MIMOME wiretap channel. *IEEE Trans. Inf. Theory* 56(11):5515–5532.

Larsson, E., Edfors, O., Tufvesson, F. & Marzetta, T. 2014. Massive MIMO for next generation wireless systems. *IEEE Commun. Mag* 52(2):186–195.

Marzetta, T.L. 2010. Noncooperative cellular wireless with unlimited numbers of BS antennas. *IEEE Trans. Wireless Commun* 9(11):3590–3600.

Mukherjee, A. & Swindlehurst, A.L. 2011. Robust beamforming for security in MIMO wiretap channels with imperfect CSI. *IEEE Trans. Signal Process* 59(1):351–361.

Mukherjee, A., Fakoorian, S.A.A., Huang, J. & Swindlehurst, A.L. 2014. Principles of physical-layer security in multiuser wireless networks: A survey. *IEEE Commun. Surveys Tuts* 16(3):1550–1573.

Oggier, F. & Hassibi, B. 2011. The secrecy capacity of the MIMO wiretap channel. *IEEE Trans. Inf. Theory* 57(8):4961–4972.

Pei, M., Wei, J., Wong, K.-K. & Wang, X. 2012. Masked beamforming for multiuser MIMO wiretap channels with imperfect CSI. *IEEE Trans. Wireless Commun* 11(2):544–549.

Rusek, F., Persson, D., Lau, B., Larsson, E., Marzetta, T., Edfors, O. & Tufvesson, F. 2013. Scaling up MIMO: Opportunities and challenges with very large arrays. *IEEE Signal Process. Mag* 30(1):40–60.

Sun, S., Kaoch, M. & Ran, T. 2015. Adaptive SON and cognitive smart LPN for 5G heterogeneous networks. *Journal of Mobile Networks and Applications Special Issue on Networking 5G Mobile CommunicationsSystems: Key Technologies and Challenges.*

Xu, W. & Dong, X. 2012. Optimized one-way relaying strategy with outdated CSI quantization for spatial multiplexing. *IEEE Trans. Signal Process* 60(8):4458–4464.

Zhou, X. & McKay, M.R. 2010. Secure transmission with artificial noise over fading channels: Achievable rate and optimal power allocation. *IEEE Trans. Veh. Technol* 59(8):3831–3842.

Signal and Information Processing, Networking and Computers – Chen & Huang (Eds)
© 2016 Taylor & Francis Group, London, ISBN 978-1-138-02881-4

Parameter estimation of 60-GHz millimeter wave communications using the nonlinear Power Amplifier based on PSO algorithm

Meng-yang Zhang, Xue-bin Sun, Zheng Zhou, Ying-ze Xing & Zi-ping Zhang

School of Information and Communication Engineering, Beijing University of Posts and Telecommunications, Beijing, China

ABSTRACT: At present, a nonlinear amplifier is generally found in 60-GHz millimeter wave wireless communication systems. But there are few channel estimation algorithms for the estimation of a nonlinear amplifier parameter. To solve the problem of nonlinear effects caused by nonlinear amplifiers, the Particle Swarm Optimization (PSO) algorithm is used, which is based on the field of swarm intelligence. Simulation results show that the proposed algorithm can efficiently reduce the number of times a convergence is reached, and it can also simplify the complex implementation of nonlinear parameter estimations in a 60-GHz millimeter wave wireless communication system.

Keywords: particle swarm optimization; nonlinear; parameter estimation; 60GHz

1   INTRODUCTION

Considering the high propagation attenuation characteristic of 60-GHz signals, Power Amplifiers (PA) need to amplify the modulated Radio Frequency (RF) signal to meet the level value so that the signal-to-noise ratio is enough for receivers to obtain signals in 60-GHz millimeter wave wireless communication systems. However, the operating point of the power amplifier in this situation is usually in the nonlinear region near saturation point, which may arouse serious nonlinear distortions that result in deterioration of system performance. In order to compensate the nonlinear distortions, estimating the nonlinear parameters of the power amplifiers in 60-GHz millimeter-wave wireless communication systems becomes very important. Although power amplifiers are widely applied to 60-GHz systems, to the best of our knowledge, there are few works in the literature reported on the estimation of the nonlinear parameter in 60-GHz systems. So, a Particle Swarm Optimization (PSO) algorithm based on an intelligent population field is proposed to estimate the nonlinear parameters. We get the estimated values of all the parameters via simulations and the experimental simulations validate the performance of the proposed algorithm, which can fast converge to the global optimal results effectively.

The rest of this article is organized as follows. In Section 2, the nonlinear PA model, channel model, and the nonlinear system model are introduced. In section 3, we introduced the PSO algorithm and a process to estimate the nonlinear parameters. Experimental simulations are provided in section 4. Finally, the whole investigation is concluded in section 5.

2   SYSTEM MODEL

2.1   *Nonlinear power amplifier*

In Lim Wei, Meng (2013), non-ideal factors affecting the 60-GHz system RF front-end include IQ amplitude and phase unbalance, phase noise, and nonlinear Power Amplifier (PA) characteristics. Generally, the nonlinear amplifier distortions are characterized by the amplitude Modulation-Amplitude Modulation (AM-AM) model and the Amplitude Modulation-

Phase Modulation (AM-PM) model, which are a direct reflection of the nonlinear effects of power amplifier on the amplitude and phase modulation signal (Yong, S.K. 2011). In this work, the nonlinear PA model approved by the IEEE 802.11ad Task Group (TG) is adopted. The AM-AM model $G(A)$ is expressed as:

$$G(A) = g\frac{A}{(1+(gA/A_{sat})^{2s})^{1/2s}} \qquad (1)$$

where A and $G(A)$ represent the input and output voltage amplitude, respectively. g denotes the linear gain and is typically configured to 4.65; s is the smoothness turning point and $s = 0.81$; A_{sat} is the saturation level and we practically have $A_{sat} = 0.58$ V.

The AM-PM model is given by:

$$\Psi(A) = \frac{\alpha A^{q1}}{(1+(A/\beta)^{q2})} \qquad (2)$$

Here, $\Psi(A)$ is the output voltage phase distortion in degrees when the input voltage amplitude is A. The typical values of the parameters α, β, $q1$, and $q2$ are respectively configured to 2560, 0.114, 2.4, and 2.3.

The AM-AM and AM-PM model show that the larger amplifier input signal amplitude value is the greater amplifier distortion and phase distortion is.

2.2 Channel model

Figure 1 shows the channel model we take for the system. Let us consider that the binary data blocks d_i are encoded by 16QAM. By passing the PA model, it results in a transmission signal x_k, $k = 0,...,M-1$. M is the size of the transmitted symbol block. In this single-path channel model, the received signal y_k for the k-th time moment can be written as:

$$y_k = x_{k-i} + n_k, k = 0,...,M-1 \qquad (3)$$

where x_k is the transmitted symbol for the k-th time moment. n_k is the additive complex white Gaussian noise for the k-th time moment with zero mean and variance N_0. The transmitted symbol x_k, the received signal y_k, and the noise sample n_k, respectively.

2.3 Nonlinear system model

From the above model, the nonlinear system model can be abstracted as follows:

$$y_k = f(x_k, \theta) + e, e \sim N(0, \delta^2) \qquad (4)$$

where x_k is input variable in system s, y_k is the system output variable, $\Theta = [\theta_1, \theta_2,...,\theta_k]^T$ are nonlinear parameter vectors to be estimated in this article, and e is the complex-valued Additive White Gaussian Noise (AWGN) with zero mean and variance θ^2. The model structure of

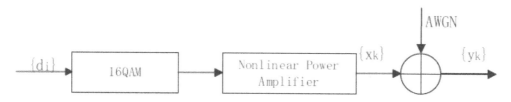

Figure 1. The channel model diagram.

f is AM-AM/AM-PM model introduced as above in part A. Let us consider we get a set of observation data (x_k, y_k), $k = 0,\ldots,M-1$, the nonlinear estimation problem is estimating the value of vector $\Theta = [\theta_1, \theta_2, \ldots, \theta_k]^T$ by summing the squared deviations, which is written as:

$$g(\theta) = \sum_{i=1}^{n} (y_k - f(x_k, \Theta))^2 \qquad (5)$$

the minimum. Thus, it is actually a nonlinear function optimization problem.

3   NONLINEAR PARAMETER ESTIMATION WITH PSO

Particle Swarm Optimization (PSO) algorithm is a very important branch of swarm intelligence. In Kennedy, J. & Eberhart, R.C. (1995), PSO is an optimization algorithm of swarm intelligence which was inspired by the foraging behavior of birds. The search for the optimal solution of the problem is realized through the individual cooperation and competition among the populations. Swarm intelligence is a new evolutionary computation technology which is produced by the behavior of the swarm intelligence in nature, society, and complex systems. It provides a new method and idea for solving the global optimization problems such as complex, constraint, nonlinear and multiple minima, etc. PSO algorithm is a kind of effective random global optimization technique. It is easy to understand, and fewer parameters are required. The simulation is easy to operate and the convergence speed is fast. It has been widely used in pattern recognition, image processing, operational research, and many other fields.

Because of the high complexity of the problem (5), it is difficult to solve by using the traditional optimization method. Therefore, a random search algorithm based on population iteration, in other words, the particle swarm optimization algorithm is proposed to solve the nonlinear parameter estimation problem of the model. When using PSO algorithm to solve the optimization problem, the proposed solution to this problem is to find a bird's position in the search space. The birds are called "particles." All of the particles include two parameters, which are determined by the optimization function and the speed of which is determined by the direction and distance. In the process of optimization, the memory and following the current optimal particle of each particle is searched in the solution space. PSO algorithm is initialized to a group of random particles, and then the optimal solution is found by iteration. During each iteration, the particles update their positions by chasing two optimal values. One is the current optimal solution found by the particle itself, which is called the individual extremum: *ibest*; the other one is the optimal solution found by the whole group, which is called global extremum: *gbest*.

The mathematical description of the search process of PSO algorithm is that a group of n particles is used to search the Q dimension space (the dimension number of each particle). Each particle can be expressed as: $x_i = (x_{i1}, x_{i2}, x_{i3}, \ldots, x_{iQ})$, the speed of each particle can be expressed as $v_i = (v_{i1}, v_{i2}, v_{i3}, \ldots, v_{iQ})$. Each particle in the search must be considered into two factors: 1. the history of optimal values of their own search p_i, $p_i = (p_{i1}, p_{i2}, \ldots, p_{iQ})$, $i = 1,2,3,\ldots,n$. 2. the global optimal value of all particles' search p_g, $p_g = (p_{g1}, p_{g2}, \ldots, p_{gQ})$ (there is only one p_g in the whole search). The particle i velocity updating formula for the $d-th$ dimension is:

$$v_{id}^{k+1} = w v_{id}^k + c_1 \eta (p_{id}^k - x_{id}^k) + c_2 \zeta (p_{gd}^k - x_{id}^k) \qquad (6)$$

The particle i location updating formula for the $d-th$ dimension:

$$x_{id}^{k+1} = x_{id}^k + r v_{id}^{k+1} \qquad (7)$$

where the v_{id}^{k+1} is the $d-th$ dimensional component of $(k+1)-th$ iteration of flight particle i velocity vector, and the x_{id}^{k+1} is the $d-th$ dimensional component of $(k+1)-th$ iteration of flight particle i location vector. The parameter w is a coefficient, which can maintain the original speed, called inertia weight. It can balance the local search ability and global search ability. If the inertia weight is larger, global search ability is stronger and the local search ability is weaker (Parsopoulos, K.E. & Vrahatis, M.N. 2002). If the inertia weight value is smaller, the local search ability is stronger, and the global search ability becomes weaker (Eberhart, R.C. & Shi, Y. 2001). Experiments show that the dynamic value w can get better results than the fixed value. A dynamic value w can be considered in the PSO search process, and it can also be dynamically changed according to the performance of PSO. At present, the Linear Decreasing Weight (LDW) strategy is proposed to be used in most fields. In function (5), c_1 is the weight coefficient of the optimal value of the particle's own tracking, which represents the particle's own understanding of so called "cognition." Cognition is an acceleration constant, which is usually set to 2 and regulates the learning of the maximum step size; c_2 is the weight coefficient of the optimal value of the particle tracking group, which represents the knowledge of the whole group of particles, so called "social knowledge" or "society," which is usually set to 2. It is also an acceleration constant, adjusting the maximum step length of learning; η, ζ are random numbers uniformly distributed between 0 and 1. Both of them are used to increase the randomness of the search; r is a factor that is added to front of the speed, which we call the "constraint factor" usually set to 1. The parameters estimation flow chart based on PSO algorithm for the specific implementation is shown in Figure 2.

4 SIMULATION RESULTS

To investigate the performance of nonlinear parameter estimation based on the proposed PSO algorithm, numerical simulations have been carried out. The 60-GHz millimeter-wave wireless communication system with the PA nonlinearity and a single-path channel is considered. Thirty pairs of input and output sample data (x_k, y_k), $k = 0,\ldots,29$ were ran-

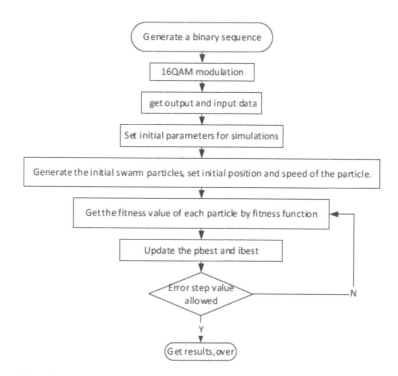

Figure 2.   Estimation flow chart based on PSO.

domly generated. The swarm size $pso_M = 6$. The inertia factor $w = 0.4$. The learning factor $c1 = c2 = 2$. A one-dimensional particle dimension is used. The maximum number of iterations is 500. The fitness function g is in accordance with the sum of squared deviations form as equation (5).

Figure 3 shows estimation results of g, which is one of the seven nonlinear parameters in the 60-GHz system PA model. In this figure, we denote six different particles with six different symbols. The initial positions of these particles are distributed randomly. With the increase of the number of iterations, their positions are changing according to the speed and position conversion formula; gradually they concentrated near the optimum value and finally all the particles have the tendency to be the optimal solution. Thus, the estimation results can be given. It can be seen that there is not only cooperation but also competitive relationship between each particle. The cooperation relationship means that each particle contributes their best position for the overall level of knowledge about optimal value. At the same time, the competition relationship means that each particle is trying to replace the best overall value to become the "winner". Moreover, there is competition in each particle itself, trying to get rid of the whole cognitive control.

Figure 4 shows the other six nonlinear parameter estimation results, which are s, As, $q1$, $q2$, $alfa$, and $beta$. In order to be shown more clearly, just one random particle trajectory in each subplot is chosen to be visible. So, as it follows from Figure 3, the position of each particle is changing according to the speed and position conversion formula gradually to the optimal position near the real value. Probably after about 20 iterations, the estimated value reaches the vicinity of the real value with the estimation error not changing anymore.

In Figure 5, considering the complexity that both the MCMC and PSO algorithm are in the same order of magnitude, we take the parameter g as an example to compare the PSO and MCMC estimation errors changing with increasing iterations. As we can see from the figure, the MCMC estimation error curve is changing in an oscillating state, which is because

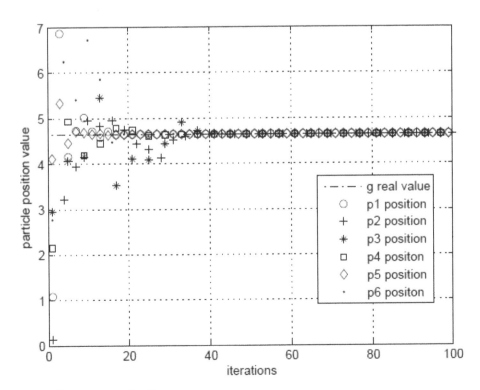

Figure 3. Estimation results of g.

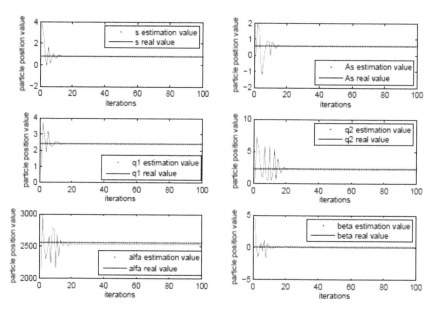

Figure 4.   Estimation results of *s*, *As*, *q*1, *q*2, *alfa*, and *beta*.

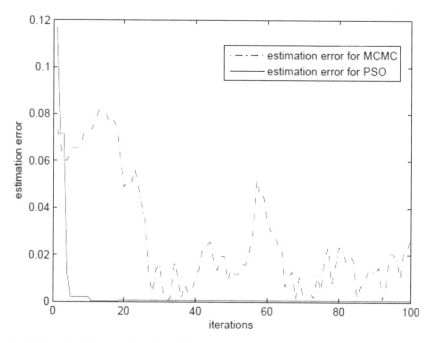

Figure 5.   Estimation flow chart based on PSO.

MCMC builds a stable Markov chain distribution iteratively. However, the PSO estimation error curve reaches convergence faster because of the "social" and "cognitive" capability of the swarm intelligence.

238

5 CONCLUSION

In this paper, we analyzed the single path channel model for nonlinear power amplifier. In the case of the known nonlinear model, we proposed a parameter estimation scheme for nonlinear amplifiers in the 60-GHz millimeter wave wireless communication system based on Particle Swarm Optimization (PSO) algorithm. Seven nonlinear parameters are estimated by experimental simulations. Estimation error changing with PSO algorithm is compared with the existing MCMC method in Sun, Shan-shan et al. (2013). The result shows that the nonlinear parameter estimation method based on PSO algorithm, which is easy to program, has little computation, and the convergence speed is faster, and it shows that the proposed algorithm is effective and feasible. Therefore, it can provide a new method for nonlinear parameter estimation in 60-GHz millimeter-wave wireless communication systems.

REFERENCES

Eberhart, R.C. & Kennedy, J. 1995. A new optimizer using particles swarm theory. *Proceedings of Sixth International Symposium on Micro Machine and Human Science*: 39–43.
Eberhart, R.C. & Shi, Y. 2001. Comparing inertia weights and constriction factors in particle swarm optimization. *Proceedings of the IEEE Conference on Evolutionary Computation*: 84–88.
Kennedy, J. & Eberhart, R.C. 1995. Particle swarm optimization. *Proceedings of IEEE International Conference on Neural Networks*, vol. 4, 1942–1948.
Lim Wei, Meng. 2013. A 60GHz power amplifier with 12.1 dBm & P1dBCP in 0.18 um SiGe BiCMOS process. *SoC Design Conference*: 17–19 November.
Parsopoulos, K.E. & Vrahatis, M.N. 2002. Recent approaches to global optimization problems through particle swarm optimization. *Natural Computing* 1(2/3): 236–306.
Shi, Y. & Eberhart, R.C. 1998. A modified particle swarm optimizer. *IEEE World Congress on Computational Intelligence*: 69–73.
Sun, Shan-shan, Sun, Xue-bin & Zhou Zheng. 2013. Joint channel estimation for nonlinear system channels. *Radio Engineering*: (12) 7–9+47.
Yong, S.K. *60GHz technology for Gbps WLAN and WPAN: from theory to practice 2011:336-348*. New York: John Wiley & Sons. Ltd.

The ICSINC2015 workshop on telecom big data based research and application

Signal and Information Processing, Networking and Computers – Chen & Huang (Eds)
© *2016 Taylor & Francis Group, London, ISBN 978-1-138-02881-4*

Analytical method of gridding user perception and market development based on big data

Mingxin Li, Jing Yuan & Jian Xu
Chong Qing Branch of China Unicom, China

ABSTRACT: A gridding analysis and evaluation method was proposed in this paper. It combined network and market development operation data, extracted key characteristic parameters, standardized data, and unified the evaluation rules. An analysis of user perception, value, behavior, and business habits was conducted. We output the evaluation scores of grids. We deeply analyzed the grid of the worst performance, and put forward effective advices and implemented them. Operators can make decisions more finely and comprehensively, according to the mining results.

Keywords: big data; gridding; related analysis; precision operation

1 INTRODUCTION

The network quality and market development are closely related to each other. Operators are facing great challenges, i.e. accurately evaluating the network quality, user—and business value, clearly positioning the marketing focus, promoting the accuracy of funds delivery, and optimizing and improving the marketing capability.

Big data analysis technologies have been developing rapidly. Telecom operators also need to analyze and explore their own and third parties' data. It helps them to enhance their operations, support, and service capabilities. Big data analysis results have been used in network planning, construction, and internet finance and so on. The traditional data analysis method is mainly aimed at the analysis of the overall situation, but lacks fineness. This paper proposed a method of analyzing and evaluating the big data restricted to a grid. The actual situation of an area in Chongqing was analyzed, and the accuracy of the analysis was improved compared to the traditional methods.

2 METHOD OF DATA ANALYSIS

2.1 *Key data source*

Compared to the traditional internet data, operators' data not only contains the business type, business content, user behavior information, but also includes the precise location information, terminal information and content rich network bearer layer signaling data. These data make it possible to do analysis more comprehensively and much further.

The main data sources are OSS(Operation Support System), BSS(Business Support System), CQT(Call Quality Test), DT(Drive Test), and complaints from customers. Network coverage, capacity, and quality related data can be extracted from OSS, DT, CQT, and complaints. BSS data include user attributes, terminal and business usage data.

2.2 Correlation analysis

First, filter and associate multi-platform data, output the related information table of users and networks. Extract key feature parameters needed from the associated data, then classify and make statistics of the parameters from the network, terminal, business and charge and other dimensions. Network parameters include good coverage, network capacity, network quality, and number of complain heats, etc. Terminal parameters include terminal brand, price and operation system. Business parameters include an index of traffic and income, etc. Charge parameters include tariff penetration, package saturation, proportion of high value packages and users, etc.

Second, standardize each key parameter, standardization method as in formula (1).

$$X'_i = \frac{X_i - \overline{X}}{\sqrt{\sum_{j=1}^{n} (X_j - \overline{X})^2 / (n-1)}} \qquad i \in [1,n], \qquad (1)$$

where X_i is the original value; $\overline{X}$ is the typical average value of X, such as the average of a country, or a province, or the areas of the same type; X'_i is the standardized result of X_i; and n is the number of X_i.

Third, standardized data are in normal distribution. Scoring rule is determined according to the probability distribution of the standard normal distribution data, in formula [2].

$$Score_{X_i} = \begin{cases} 100 & X'_i \geq 3 \\ 95 + (X'_i - 2) \times 5 & 2 \leq X'_i < 3 \\ 85 + (X'_i - 1) \times 10 & 1 \leq X'_i < 2 \\ 60 + X'_i \times 25 & 0 \leq X'_i < 1 \\ 25 + (X'_i + 1) \times 35 & -1 \leq X'_i < 0 \\ 15 + (X'_i + 2) \times 10 & -2 \leq X'_i < -1 \\ (X'_i + 3) \times 15 & -3 \leq X'_i < -2 \\ 0 & X'_i < -3 \end{cases} \qquad [2]$$

The standard score of each parameter can be calculated according to the formula [2] and the standardized result X'_i.

Fourth, calculate the weights of different dimensions and key parameters' weights of each dimension. Based on the experts' sorting results, construct the judgment matrix using an analytical hierarchy process and check whether the matrix satisfies the need of coincidence test. The weight of each parameter is calculated by using the eigenvector of the largest eigenvalue of the matrix.

Fifth, the score of each dimension and the total score can be calculated in formula [3].

$$Score_i = \sum_j w_{ij} \cdot Score_{ij}, \qquad \sum_j w_{ij} = 1, \qquad [3]$$

where w_{ij} is the weight of parameter j in the i dimension. $Score_{ij}$ is the score of parameter j in the i dimension.

2.3 Gridding

Aiming to refine the results of the assessment, divide the area to be analyzed into smaller size grids. Grids are regional, continuous, and convenient to be managed. The division is based on the characteristics of topography, economy, culture, and population, etc. The results of the grid division and the corresponding characteristics are shown in Table 1.

After the grid division is determined, resources of the network and market are matched to the corresponding grids. Network resources include BS(Base Station) resources and human

Table 1. The results of the grid division.

Grid name	Grid rank	Characteristics
Business Grid	A	High-grade commercial district, densely populated, large flows of people
North Business Grid	A	North to the Business Grid, residential area, high population density
Hi-tech Grid	A	Universities concentrated area, emerging areas of information and manufacturing industries
Industry Grid	B	Old Industrial Area
South Business Grid	B	South to the Business Grid, residential area, 2 first-class Hospitals at Grade 3
Mountain-1 Grid	C	Mountainous areas, complex topography, population dispersed, grid shape is the same as mountain range
Mountain-2 Grid	C	The formation of two mountains pass area, slightly complex terrain, population concentrated

resources. Market resources include funds, users, and human resources. The key parameters of each dimension are correlated with the grid geographic information. Operation results of the network and market are also gridded based on their geographic information. Evaluation results of the user perception and market development of each grid are evaluated according to the analysis method in 2.2.

3 GRIDDING ANALYSIS

3.1 Grid comprehensive evaluation results

According to the analysis method, the area in 2.3 was evaluated, and the results are shown in Table 2.

There are big gaps between the evaluation results of the different grids, evaluation results are basically in line with the positioning of the grid. But there are exceptions, such as Hi-tech Grid is A, its performance is worse than South Business Grid of grade B in every dimension. Empirical value deviates from the actual value, this fact proves the value and significance of the big data analysis.

As Table 2 shows, the user perception of each grid is closely related to its economic level. The more it earns the more it invests in networks, user perception is also better. Though networks are in a good condition, some problems still need to be optimized. Such as the 3G users fall back to 2G. Users' perception changes at poor coverage areas, the capacity of large, volume-based stations is limited. New technologies and new ways can be applied to enhance users' perception in network construction and optimization.

The analysis focuses on the market characteristics and problems as below.

3.2 Characteristic analysis of high value users

The age distribution of high value users is proportional to the consumption capacity, users between 30 and 40 years have the highest spending power; they have also accounted for the highest value among high value users. The next is 20~30 year old users, as shown in Table 3.

Most packages chosen by high value users are 186 Yuan or above, and 49.29% of them are iphone contract packages, nearly 1/2, as shown as in Table 4. The Top 5 terminals held by high value users are iphone5S, iphone5, iphone4S, SM-N9002(Galaxy Note 3), and iphone 6, respectively.

From the analysis results, star contract terminals appeal to high value users. One problem is that the terminal eco-system is thin, more diverse contract terminals need to be tapped to attract more users.

Table 2.   Grid comprehensive evaluation results.

Grid name	Grid rank	Network score	Terminal score	Business score	Charge score	Comprehensive score
Business Grid	A	78.29	87.35	93.28	82.06	84.20
North Business Grid	A	87.17	81.03	65.14	64.27	80.00
Hi-tech Grid	A	65.28	77.72	62.80	38.77	67.84
Industry Grid	B	87.64	70.66	35.74	44.61	70.58
South Business Grid	B	83.67	78.25	68.12	81.85	79.26
Mountain-1 Grid	C	79.42	58.74	23.68	38.38	60.52
Mountain-2 Grid	C	67.72	50.37	26.63	36.52	52.89

Table 3.   The age distribution and consumption capacity of the 3G users.

User age range	High value users	Regional users	Monthly consumption (Yuan)
20~30	31.00%	33.80%	94.31
30~40	45.00%	35.20%	96.26
40~50	19.20%	22.30%	83.26
50~60	4.70%	6.60%	74.74
60+	0.10%	2.10%	57.68

Table 4.   Main packages chosen by high value users.

Package name	Proportion
3G-iphone-286	33.61%
3G-286A	8.81%
3G-iphone-386	8.06%
3G-226A	4.96%
3G-186B	4.55%
3G-iphone-186	4.34%
3G-96A	3.79%
3G-96B	3.57%
3G-156B	2.81%
3G-126A	2.49%
3G-iphone-226	2.30%
3G-156A	2.19%

3.3   Evaluation results of market development

The main evaluation indicators of the gridding market development are shown in table 5.

As shown in table 5, the total income and revenue per BS of the Hi-tech grid are both low. On the contrary, total income and ARPU of the South Business Grid are both the highest. Market characteristics of the analyzed area are as below.

- The ARPU of 3G users is 3~4 times to 2G users, users between 30 and 40 years have the highest consumption capacity.
- The 2G terminal proportion held by 3G and 2G users is respectively 20–60%; terminal is an important factor to restrict users to use the 3/4G network.
- In some hot spots, the capacities of 3G base stations are not enough, which affects user perception.
- The number of campus users is large, but their ARPU is lower than average.
- The number of high value users are bigger than that of high package users, namely, some high value users' packages are lower than needed.

Table 5. Development of the grid market.

Grid name_rank	Proportion of income	Proportion of BS	Monthly income per BS (ten thousand Yuan)	ARPU of 3G users (Yuan)	ARPU of 2G users (Yuan)
Business Grid_A	7%	4%	1.75	91.35	29.46
North Business Grid_A	20%	13%	1.46	87.76	28.36
Hi-tech Grid_A	22%	30%	0.69	78.86	25.31
South Business Grid_B	25%	19%	1.25	93.07	28.13
Industry Grid_B	16%	16%	1.01	79.63	25.83
Mountain-1 Grid_C	5%	11%	0.49	76.06	26.00
Mountain-2 Grid_C	4%	7%	0.55	74.23	24.50
Average	/	/	0.96	84.80	26.74

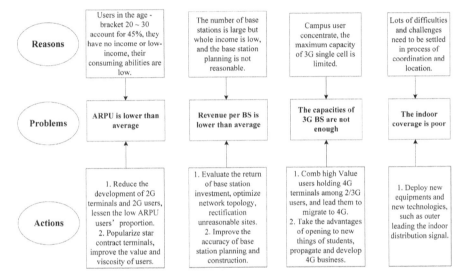

Figure 1. Suggestions for High-tech Grid.

4 PRECISION OPERATION

4.1 Adjustment suggestions

It is necessary to apply the analysis results to the operation management process. According to the evaluation results, the problems of the Hi-tech Grid were the most prominent. Therefore, the strategy of network and market development for the Hi-tech Grid was adjusted, and the specific adjustment proposals are shown in Figure 1.

Implementation of the recommendations as mentioned is under way according to the actual situation. And till now, enhancement of the indoor coverage and migration to 4G has been implemented.

4.2 Implementation effects

4.2.1 Enhancement of indoor coverage

Although, the ARPU of campus users is low, they are potential high value users. They will become loyal customers once the spending habits developed, so the construction and optimization of the campus network can't be ignored. Based on fine assessment and multi-coordination, 15 outer leading base stations have run for months, and proven to be effective. The indoor good coverage rate, namely proportion of samples of RSCP>−80 dbm, boosts from 65.71%

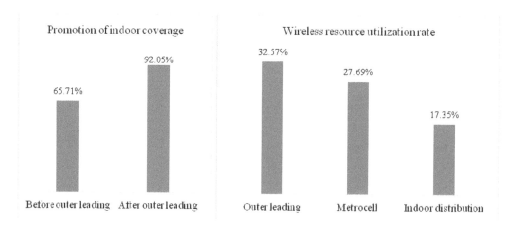

Figure 2.   Effects of indoor coverage optimization.

up to 92.05%. The wireless resource utilization rate of outer leading BS is 32.57%, far higher than that of indoor distribution system, as shown in Figure 2. Furthermore, the construction and property coordination of outer leading is more convenient than indoor distribution system, and the cost is about one tenth of the indoor distribution system.

4.2.2   *Migration to 4G*

Through the analysis, high value users holding 4G terminals in the Hi-tech Grid is about 1500, and more than 800 users have migrated to 4G in 2 months by means of telephone and SMS propagation. The transfer rate is more than 50%.

5   CONCLUSION

In this paper, an analytical method combining network quality and market development was proposed. This has been associated with the multi-dimensional data that has been matched, analyzed, and evaluated in the granularity of the grid. The final comprehensive score was calculated by 3 steps, which means standardization of gridding data, determining scoring rule and calculating weights of different dimensions and parameters. We analyzed and summarized the characteristics and problems of the network and market development of each grid, and then corresponding suggestions and solutions were put forward. The effects of adjustment have a good performance.

REFERENCES

Cerra, A., Easterwood, K. & Power, J. 2012. Transforming business-big data mobility and globalization. Wiley, UK.

Chenxi, Q.I. 2013. Big data analysis and management in telecommunications. *Telecommunications Science* 29(3): 12–16.

Han Jing, Zhang Zhijiang & Wang Jianquan. 2014. The Unified-Operation-Oriented big data strategy for telecom operators. *Telecommunications Science* 30(11): 154–158.

Li Chunhao, Sun Yonghe & Jia Yanhui. 2010. Analytic hierarchy process based on variable weights. *System Engineering-Theory & Practice* 30(4): 723–731.

Li Li, Zhang Deng & Xie longjun. 2013. A condition assessment method of power transformers based on association rules and variable weight coefficients. *Proceedings of the CSEE*: 152–159.

Shang Ye & Cheng Xiaojun. 2014. Research on network optimization framework based on big data analysis in mobile Internet Times. *Designing Techniques of Posts and Telecommunications*: 58–62.

Wang Shan, Wang Huiju & Qin Xiongpai. 2011. Architecting big data: challenges, studies and forecasts. *Chinese Journal of Computers* 34(10): 1741–1751.

Signal and Information Processing, Networking and Computers – Chen & Huang (Eds)
© *2016 Taylor & Francis Group, London, ISBN 978-1-138-02881-4*

WCDMA data based LTE site selection scheme in LTE deployment

Lexi Xu
China Unicom Network Technology Research Institute, Beijing, P.R. China

Yuting Luan
Shenyang Railway Survey Design Consulting Company, Shenyang, P.R. China

Xinzhou Cheng, Xiaodong Cao, Kun Chao, Jie Gao & Yuwei Jia
China Unicom Network Technology Research Institute, Beijing, P.R. China

Shuche Wang
Beijing University of Posts and Telecommunications, Beijing, P.R. China

ABSTRACT: LTE site selection is a vital stage for the LTE deployment. Most site selection schemes optimise the site location to improve the network performance, e.g. system coverage, successful access probability etc. However, in the practical LTE deployment, telecom operators preferentially reuse existing 2G/3G sites in order to reduce the Capital Expenditure (CAPEX) and the Operational Expenditure (OPEX). This paper proposes a novel WCDMA data based LTE site selection (WDLSS) scheme. The WDLSS scheme analyses five factors of existing WCDMA sites, including the height, the density, the signal power, the coverage control and the interference. Then the WDLSS scheme ranks each site according to the optimisation difficulty of above five factors. Finally, we apply the WDLSS scheme into the LTE deployment in a China's city. Results show it can assist telecom operators to be aware of each site's detailed problems.

Keywords: LTE; WCDMA; site selection; LTE deployment; interference

1   INTRODUCTION

Recently, telecom operators start to deploy LTE networks worldwide, in order to meet users' high data rate and low latency requirements (Ramiro et al. 2012, Xu et al. 2015). Site selection is vital for the LTE cell deployment, since effective site selection can improve the network performance and users' Quality of Service (QoS) as well as reduce the Capital Expenditure (CAPEX).

In Schultz et al. (2003), the base station planning under the Manhattan scenario is introduced. However, the planned deployment may become sub-optimal under the time-varying traffic distribution. In Jiang et al. (2008), an optimised relay station deployment scheme is designed, in order to achieve the balanced load distribution among cells in cellular networks. Wu (2011) investigates the self-optimised site selection, in order to deal with the coverage hole problem and reach the seamless cooperative coverage. In Andrews et al. (2011), the author employs the stochastic geometric model to optimise the site selection in order to reach the minimum outage probability. Most conventional site selection schemes aim to find suitable site location in order to optimise the network performance. However, it is difficult to deploy the practical LTE cell in the ideal site location, since the site construction envisages a series of practical challenges, including the building property coordination, government management, public obstacle etc. In addition, compared with the LTE cell deployment on a totally new location site, the method of utilising existing 2G/3G sites can significantly reduce the CAPEX and Operational Expenditure (OPEX).

As the leading 3G mobile system, 3GPP WCDMA is widely used across Europe and Asia (Ramiro et al. 2012). However, the characteristic of WCDMA is different from that of LTE. For example, WCDMA networks employ the soft handover, in which the inter-cell signal can become the useful signal. Whilst LTE networks employ the hard handover, in which the inter-cell signal becomes the interference and deteriorates the LTE performance (Xu et al. 2011, 3GPP 2010, Xu et al. 2009). Therefore, only a small proportion of existing WCDMA sites can be utilised to deploy LTE cells directly. This paper aims to identify the feasibility of the LTE deployment on existing WCDMA sites. In order to realise this objective, this paper proposes a novel WCDMA data based LTE site selection (WDLSS) scheme for the LTE deployment. Specifically, the WDLSS scheme analyses five factors of existing WCDMA sites, including the height, the density, the signal power, the coverage control as well as the interference. According to the characteristic and the optimisation difficulty of above five factors, the WDLSS scheme ranks each WCDMA site and outputs the detailed problems of each site.

2 KEY FACTORS ANALYSIS OF WDLSS SCHEME

Figure 1 shows the structure of the proposed WCDMA data based LTE site selection (WDLSS) scheme for the LTE deployment. In this paper, a WCDMA site refers to the corresponding WCDMA cell in this site. Section 2 introduces its five factors, including seven sub-factors.

2.1 Cell height factor

Inappropriate cell height impacts the cell coverage. For example, for a 60-meter height cell, it is difficult to control its coverage, which may result in the *cross-boundary coverage* and even the *island effect* (Xu et al. 2015, China Unicom 2013). Hence, the WDLSS scheme analyses the cell height factor, including the cell absolute height sub-factor and the cell relative height sub-factor.

2.1.1 Cell absolute height

In the city downtown, a cell's antenna may be equipped on the top of a skyscraper. Under this scenario, the antenna with large tilt angle is widely used to control the cell coverage. However, large tilt angle will deteriorate the antenna pattern and the cell coverage control

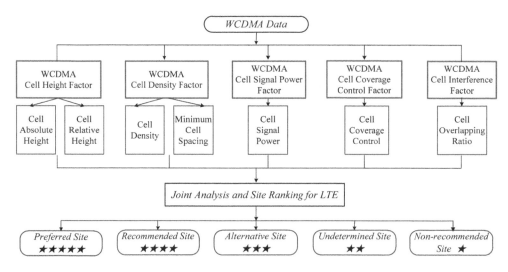

Figure 1.   Structure of the proposed WCDMA data based LTE site selection scheme.

effect. Hence, an extremely high cell may result in the *cross-boundary coverage* and the *island effect*, which may lead to handover failure and call dropping. On the contrary, for a low height cell, its adjacent buildings may shelter its electromagnetic wave propagation, which decreases its coverage capability.

In this paper, both the *Cell$_i$ absolute height attribute* and the *Cell$_i$ absolute height level* are categorized into four types, as shown in (1).

$$
\begin{aligned}
&Cell_i\ absolute\ height\ attribute \\
&\qquad\qquad\qquad / \\
&Cell_i\ absolute\ height\ level
\end{aligned}
=
\begin{cases}
Low\ cell\ /\ \textbf{\textit{Moderate}} & \text{if } H_i < H_{low} \\
Medium\ height\ cell\ /Excellent & \text{if } H_{low} \le H_i < H_{med} \\
High\ cell\ /\ \textbf{\textit{Good}} & \text{if } H_{med} \le H_i < H_{high} \\
Extremely\ high\ cell\ /\ \textbf{\textit{Poor}} & \text{if } H_{high} \le H_i
\end{cases}
\tag{1}
$$

where H_i is the height of *Cell$_i$* (unit: meter), H_{low} is the threshold of the low cell, H_{med} is the threshold of the medium height cell, H_{high} is the threshold of the high cell. For example, $H_{low} = 15\ m$, $H_{med} = 45\ m$, $H_{high} = 55\ m$ in the typical network planning (China Unicom 2013).

2.1.2 *Cell relative height*

In addition to the cell absolute height, the proposed WDLSS scheme also considers the cell relative height sub-factor. This is to avoid selecting the site, which is medium height for the cell absolute height sub-factor, whilst it is much higher than neighbouring sites.

Assuming *Cell$_i$* has J neighbouring cells indexed with j ($j \in \{1, 2 \dots J\}$) in surrounding *one square kilometer*, the average height difference between *Cell$_i$* and neighbouring cells is defined as ΔH_i^{neigh}, using (2a). Meanwhile, the height ratio of *Cell$_i$* to neighbouring cells is defined as RH_i^{neigh}, using (2b).

$$
\Delta H_i^{neigh} = H_i - \frac{\sum_{j=1}^{J} H_j}{J}
\tag{2a}
$$

$$
RH_i^{neigh} = \frac{H_i}{\sum_{j=1}^{J} H_j / J}
\tag{2b}
$$

where H_j is height of neighbouring *Cell$_j$* ($j \in \{1, 2 \dots J\}$) (unit: meter). In this paper, both the *Cell$_i$ relative height attribute* and the *Cell$_i$ relative height level* are categorized into two types, as illustrated in Table 1. *Cell$_i$* is labelled as the *relatively high cell* under two conditions, including the average height difference ΔH_i^{neigh} is larger than the *relatively high cell threshold* $H_{relhigh}$, or the height ratio RH_i^{neigh} is larger than the *height ratio threshold of relatively high cell* $RH_{relhigh}$. (e.g. $H_{relhigh} = 10\ m$ and $RH_{relhigh} = 150\%$)

2.2 *Cell density factor*

Our previous works (Xu et al. 2014, Xu Chen et al. 2015) introduce each LTE cell utilizes the co-frequency bandwidth for data transmission. Meanwhile, LTE employ the hard handover, in which a user's receiving signal from neighbouring cells becomes the inter-cell interference. This paper employs the cell density factor to analyse the poor signal power induced by the

Table 1. Evaluation criterion of cell relative height.

Cell$_i$ relative height	Cell$_i$ relative height attribute	Cell$_i$ relative height level
$\Delta H_i^{neigh} \ge H_{relhigh}$ or $H_{relhigh} \ge RH_i^{neigh}$	Relatively high cell	Poor
Other conditions	Non-relatively high cell	Excellent

sparse cell distribution, as well as the severe interference induced by the dense cell distribution. The cell density factor consists of the cell density sub-factor and the minimum cell spacing sub-factor.

2.2.1 Cell density

From the analysis above, in the LTE deployment, the cell density should be in a suitable range, in order to reduce the inter-cell interference and keep the strong signal power, thus achieving high SINR and QoS. Figure 2 shows the analysis method of the cell density. Under ±47.5 degrees of the $Cell_i$'s main lobe direction, $Cell_i$ searches neighbouring cells within R_{dens} meters. The number of neighbouring cells in this area is defined as $Cell_i$ density $Dens_i$. R_{dens} can be set according to the average inter-site distance in the downtown area (e.g. R_{dens}=900m). This paper categorizes both the $Cell_i$ density attribute and the $Cell_i$ density level into four types, as illustrated in (3).

$$
\begin{array}{l} Cell_i \ density \ attribute \\ \qquad / \\ Cell_i \ density \ level \end{array} =
\begin{cases}
Low \ density \ / \ Good & if \ Dens_i < Dens_{low} \\
Medium \ density \ / \ Excellent & if \ Dens_{low} \le Dens_i < Dens_{med} \\
High \ density \ / \ Moderate & if \ Dens_{med} \le Dens_i < Dens_{high} \\
Extremely \ high \ density \ / \ Poor & if \ Dens_{high} \le Dens_i
\end{cases}
\tag{3}
$$

where $Dens_i$ is the density of $Cell_i$, $Dens_{low}$ is the threshold of the low density, $Dens_{med}$ is the threshold of the medium density, $Dens_{high}$ is the threshold of the high density. For example, $Dens_{low} = 5$, $Dens_{med} = 18$, $Dens_{high} = 36$ (China Unicom 2013).

2.2.2 Minimum cell spacing

$Cell_i$ density sub-factor reflects $Cell_i$'s average density level. However, a medium density $Cell_i$ may have a close neighbouring cell. This neighbouring cell will result in heavy inter-cell interference towards $Cell_i$ and deteriorate $Cell_i$ performance. Hence, this paper employs the *minimum cell spacing* sub-factor to analyse the neighbouring cell, which leads to the heaviest interference.

Assuming $Cell_i$ has J neighbouring cells indexed with j ($j \in \{1, 2 \dots J\}$). $Spa_{i,j}$ (unit: meter) is the distance between $Cell_i$ and neighbouring $Cell_j$ ($j \in \{1, 2 \dots J\}$). The *minimum cell spacing* Spa_i^{min} can be calculated as $Spa_i^{min} = minimum_{j \in \{1,2\dots J\}} Spa_{i,j}$, as exemplified in Figure 3.

In this paper, both the *minimum $Cell_i$ spacing attribute* and the *minimum $Cell_i$ spacing level* are categorized into four types, as illustrated in Table 2. $Spa_{extreshort}$ is the threshold of extremely short spacing, Spa_{short} is the threshold of short spacing, Spa_{med} is the threshold of medium spacing. (e.g. $Spa_{extreshort} = 150$ m, $Spa_{short} = 200$ m, $Spa_{med} = 650$ m.)

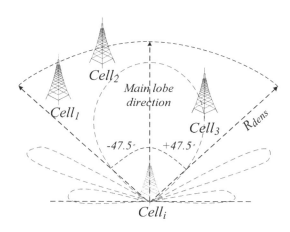

Figure 2. Diagram of LTE cell density analysis.

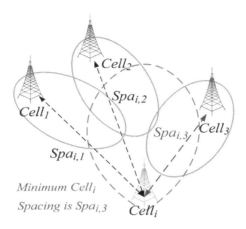

Figure 3. Diagram of minimum cell spacing.

Table 2. Evaluation criterion of minimum cell spacing.

Minimum $Cell_i$ spacing	Minimum $Cell_i$ spacing attribute	Minimum $Cell_i$ spacing level
$Spa_i^{min} < Spa_{extreshort}$	Extremely short	Poor
$Spa_{extreshort} \leq Spa_i^{min} < Spa_{short}$	Short	Moderate
$Spa_{short} \leq Spa_i^{min} < Spa_{med}$	Medium	Excellent
$Spa_{med} \leq Spa_i^{min}$	Long	Good

2.3 *Cell signal power*

Cell signal power is vital for LTE site selection, since inappropriate site may result in the weak coverage and even the coverage hole (3GPP 2010, Xu et al. 2012). In this paper, we estimate user's Reference Signal Received Power (RSRP) from the LTE cell, on the basis of user's Received Signal Code Power (RSCP) from the co-site WCDMA cell. The RSRP estimation considers two factors, including the path-loss difference induced by different frequencies, and the pilot transmission power difference between the LTE cell and the WCDMA cell.

1. Path-loss difference induced by different frequencies

According to 3GPP 25.942 (3GPP 2012), for $User_u$ served by $Cell_i$, the urban macro-cell propagation pass-loss model is shown in (4a):

$$PL_{i,u} = 40 \times (1 - 4 \times 10 - 3 \times H_i) \times Log_{10}(D_{i,u}) - 18 \times Log_{10}(H_i) + 21 \times Log_{10}(frq) + 80 \quad (4a)$$

$$\Rightarrow PL_{i,u}^{WCDMA} - PL_{i,u}^{LTE} = 1.4(dB) \quad (4b)$$

where H_i is $Cell_i$ height, $D_{i,u}$ is the distance between $Cell_i$ and $User_u$, feq is the frequency. For China Unicom, the typical WCDMA frequency frq_{WCDMA} is 2100 MHz and the typical LTE frequency frq_{LTE} is 1800 MHz (China Unicom 2013, Han et al. 2012). For a specific $Cell_i$ and $User_u$, both H_i and $D_{i,u}$ are constant. From (4a), the path-loss $PL_{i,u}^{WCDMA}$ between WCDMA $Cell_i$ and $User_u$ is *1.4 dB* larger than the $PL_{i,u}^{LTE}$ between LTE $Cell_i$ and $User_u$, as shown in (4b).

2. Pilot transmission power difference between the LTE cell and the WCDMA cell

The WCDMA pilot transmission power TP_{pilot}^{WCDMA} and the LTE reference signal transmission power $TP_{ref\ sig}^{LTE}$ are different. The typical TP_{pilot}^{WCDMA} value is 33 dBm, whilst the typical $TP^{LTE}_{ref\ sig}$ value is 15.2 dBm. Therefore, TP^{WCDMA}_{pilot} is 17.8 dB larger than $TP^{LTE}_{ref\ sig}$, namely $TP^{WCDMA}_{pilot} - TP^{LTE}_{ref\ sig} = 17.8\ dB$. According to the transmission power difference $TP^{WCDMA}_{pilot} - TP^{LTE}_{ref\ sig} = 17.8(dB)$ and the propagation path-loss difference of (4b), $User_u$'s receiving $RSCP^{WCDMA}_{i,u}$ from WCDMA $Cell_i$ is 16.4 dB larger than $User_u$'s $RSRP^{LTE}_{i,u}$ from LTE $Cell_i$, as shown in (5)

$$RSCP^{WCDMA}_{i,u} - PSRP^{LTE}_{i,u} = 16.4(dB) \tag{5}$$

According to (5), we can estimate user's receiving RSRP from LTE cell, based on this user's receiving RSCP from existing co-site WCDMA cell. In this paper, both the $Cell_i$ signal power attribute and the $Cell_i$ signal power level are categorized into three types, as illustrated in Table 3.

$RSRP_{high}$ and $RSRP_{med}$ are the thresholds of LTE high signal power and medium signal power, restively (e.g. $RSRP_{high} = -100\ dBm$, $RSRP_{med} = -115\ dBm$ China Unicom 2013)). We rank $Cell_i$ signal power attribute as *excellent signal*, if the ratio of the number of users, which meet $RSRP_{high} \leq RSRP^{LTE}_{i,u}$, is larger than 90% in LTE $Cell_i$ (namely, the ratio of the number of users, which meet $RSRP_{high} + 16.4 \leq RSRP^{WCDMA}_{i,u}$, is larger than 90% in the co-site WCDMA $Cell_i$). Similarly, we label *good signal* cells and *poor signal* cells, as shown in Table 3.

2.4 Cell coverage control

The cell coverage control impacts the network performance. As shown in Figure 4(a), the coverage control of WCDMA $Cell_i$ is good and 85% area is mainly served by $Cell_i$ without soft handover. Figure 4(b) illustrates the scenario of $Cell_i$ suffering some problems (e.g. antenna problem, circle line break etc.), which shrinks $Cell_i$'s coverage. Then, only 60% area is

Table 3.   Evaluation criterion of cell signal power.

$Cell_i$ signal power	$Cell_i$ signal power attribute	$Cell_i$ signal power level
Ratio of the number of users meeting $RSRP_{high} \leq RSRP^{LTE}_{i,u}$ is larger than 90% (Ratio of the number of users meeting $RSRP_{high} + 16.4 \leq RSCP^{WCDMA}_{i,u}$ is larger than 90%)	Excellent signal	Excellent
Ratio of the number of users meeting $RSRP_{med} \leq RSRP^{LTE}_{i,u}$ is larger than 90% (Ratio of the number of users meeting $RSRP_{med} + 16.4 \leq RSCP^{WCDMA}_{i,u}$ is larger than 90%)	Good signal	Good
Ratio of the number of users meeting $RSRP_{med} \leq RSRP^{LTE}_{i,u}$ is less than 90% (Ratio of the number of users meeting $RSRP_{med} + 16.4 \leq RSCP^{WCDMA}_{i,u}$ is less than 90%)	Poor signal	Poor

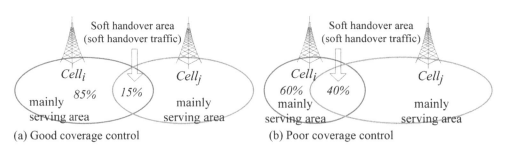

(a) Good coverage control   (b) Poor coverage control

Figure 4.   Coverage control diagram in WCDMA cellular networks.

mainly served by Cell$_i$ whilst the rest area is the *soft handover area* together with neighbouring *Cell$_j$ (j ∈ {1, 2 … J})*.

Due to the time-varying shadowing and fast fading, it is difficult to calculate the accurate areas mentioned above. However, the ratio of *Cell$_i$'s mainly serving area* to the *whole Cell$_i$'s serving area* (the *whole Cell$_i$'s serving area* equals the sum of *Cell$_i$'s mainly serving area* and *Cell$_i$'s soft handover area*) can be estimated according to the ratio of *Cell$_i$'s mainly serving traffic Traf$_i^{main}$* to *Cell$_i$'s whole traffic*. The ratio of *Cell$_i$'s mainly serving traffic RT$_i^{main}$* can be expressed as:

$$RT_i^{main} = \frac{Traf_i^{main}}{Traf_i^{main} + \sum_{j=1}^{J} Traf_{i,j}^{handover}} \tag{6}$$

where *Traf$_i^{main}$* is the traffic in *Cell$_i$'s mainly serving area*, *Traf$_{i,j}^{handover}$* is the traffic in the handover area between *Cell$_i$* and *Cell$_j$ (j ∈ {1, 2 … J})*. This paper categorizes both the *Cell$_i$ coverage control attribute* and the *Cell$_i$ coverage control level* into four types, as shown in (7).

$$
\begin{aligned}
&Cell_i \text{ coverage control attribute} \\
&\qquad\qquad / \\
&Cell_i \text{ coverage control level}
\end{aligned}
=
\begin{cases}
Poor\ coverage\ /\ Poor & if\ RT_i^{main} < RT_{low}^{main} \\
Medium\ coverage\ /\ Moderate & if\ RT_{low}^{main} \le RT_i^{main} < RT_{med}^{main} \\
Good\ coverage\ /\ Good & if\ RT_{med}^{main} \le RT_i^{main} < RT_{high}^{main} \\
Excellent\ coverage\ /\ Excellent & if\ RT_{high}^{main} \le RT_i^{main}
\end{cases}
$$

$$\tag{7}$$

where RT_{high}^{main} is the threshold of high traffic ratio, RT_{med}^{main} the threshold of medium traffic ratio, RT_{low}^{main} is the threshold of low traffic ratio. For example, $RT_{high}^{main} = 50\%$, $RT_{med}^{main} = 40\%$, $RT_{low}^{main} = 30\%$

2.5 Cell interference

As shown in Figure 5, cell overlapping is an effective method to keep the seamless coverage in LTE networks. However, severe cell overlapping results in the heavy inter-cell interference. This will reduce SINR and impact the system throughput (Xu et al. 2009, Liu et al. 2013). Figure 6 shows the throughput improvement under different SINR. In this test, the cell overlapping area is adjusted to reduce the interference, thus increasing SINR and throughput. The system throughput improvements reach 11% and 24% as well as 39%, under 1 dB (9 dB –>10 dB) and 2 dB (8 dB –>10 dB) as well as 3 dB (7 dB –>10 dB), respectively.

The proposed WDLSS scheme considers the cell overlapping for LTE site selection. According to (China Unicom 2013), the calculation of *Cell$_i$ overlapping ratio* follows two steps. Firstly, for each *User$_u$'s (u ∈ {1, 2 …, U})* Measurement Report (MR), the number of

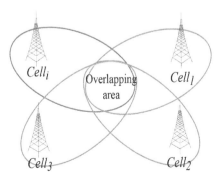

Figure 5.   Diagram of cell overlapping.

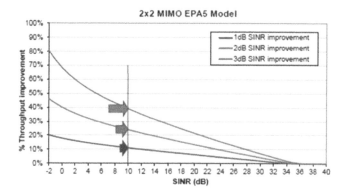

Figure 6.   Throughput improvement comparison.

neighbouring $Cell_j$ ($j \in \{1, 2 \ldots J\}$), which meets $RSCP_{j,u}^{WCDMA} - RSCP_{i,u}^{WCDMA} > -6\,dB$, is the overlapping value of this MR. $Cell_i$ overlapping ratio $R_i^{overlap}$ equals the ratio of all MRs' total overlapping values to the number of all MRs. This paper categorizes both the $Cell_i$ overlapping attribute and the $Cell_i$ overlapping level into three types, as illustrated in (8).

$$
\begin{array}{l}
Cell_i \ overlapping \ level \\
/ \\
Cell_i \ overlapping \ level
\end{array}
= \begin{cases}
Low \ overlapping \,/\, Excellent & if \ R_i^{overlap} < R_{med}^{overlap} \\
Medium \ overlapping \,/\, Good & if \ R_{med}^{overlap} \le R_i^{overlap} < R_{high}^{overlap} \\
High \ overlapping \,/\, Poor & if \ R_{high}^{overlap} \le R_i^{overlap}
\end{cases}
\tag{8}
$$

where $R_{high}^{overlap}$ is the threshold of high overlapping, $R_{med}^{overlap}$ is the threshold of medium overlapping. For example, $R_{high}^{overlap} = 10\%, R_{med}^{overlap} = 5\%$ (China Unicom 2013).

3   JOINT ANALYSIS AND SITE RANKING OF WDLSS SCHEME

Section 2 introduces five factors of the proposed WDLSS scheme. In Section 3, the WDLSS scheme jointly analyses above five factors (including seven sub-factors). Then, existing WCDMA sites are ranked into five categories for the LTE deployment, including the preferred site, the recommended site, the alternative site, the undetermined site, the non-recommended site.

3.1   Three types of key sub-factors

Since the optimisation processes of seven sub-factors are different. We divide seven sub-factors into three types, according to the optimisation difficulty when each sub-factor suffers problems.

- Type-A: Type-A sub-factors include the *cell overlapping* sub-factor (belongs to the cell interference factor), and the *cell density* sub-factor (belongs to the cell density factor). Under the severe cell overlapping scenario or the high cell density scenario, it is difficult to deal with these problems via conventional network optimisation methods, such as antenna adjustment and transmission power adjustment (Han et al. 2012). Therefore, Type-A sub-factors have the largest weight, compared with Type-B and Type-C sub-factors.
- Type-B: Type-B sub-factors include the *cell signal power* sub-factor (belongs to the cell signal power factor), and the *minimum cell spacing* sub-factor (belongs to the cell density factor), as well as the *cell coverage control* sub-factor (belongs to the cell coverage control factor).

Under the low signal power scenario, we can increase the transmission power to deal with this problem. Under the short minimum cell spacing scenario, we can adjust the direction of two cells antennas' main lobes to address this problem (China Unicom 2013). In addition, we can check/repair either antenna or circle line break to deal with the poor coverage control problem. Therefore, the weight of Type-B sub-factors is lower than the weight of Type-A sub-factors.

- Type-C: Type-C sub-factors include both the *cell absolute height* sub-factor and the *cell relative height* sub-factor (above two sub-factors belong to the cell height factor).

Under scenarios of both the absolutely high cell and the relatively high cell, we can reduce the antenna height. In addition, we can employ the electronic antenna with large tilt angle, instead of the mechanical antenna with small tilt angle (Han et al. 2012). Hence, the weight of Type-C sub-factors is lower than that of Type-B and Type-C sub-factors.

3.2 Site ranking

Figure 7 shows the site ranking flowchart for each WCDMA cell. In the 1st step, we analyse levels of seven sub-factors, as introduced in Section 2. In the 2nd and 3rd steps, if levels of all seven sub-factors are excellent, this WCDMA cell's site is labelled as the *preferred site* for LTE site selection. If this cell cannot meet requirements in the 2nd step, in the 4th and 5th steps, we check whether levels of all Type-A sub-factors are excellent (including the cell overlapping sub-factor and the cell density sub-factor). Meanwhile, we check whether levels of all Type-B sub-factors are either excellent or good. In addition, we check whether there are no poor levels of Type-C sub-factors. If this cell meets above three requirements, this WCDMA cell's site is labelled as the *recommended site*. If this cell cannot meet above three requirements simultaneously, it comes to the 6th step. This WCDMA cell's site is labelled as the *alternative site* in the 7th step, if all Type-A and Type-B sub-factors are excellent/good/moderate without any poor levels (As introduced in Section 3.1, poor levels of Type-C sub-factors are more easily to be addressed than Type-A and Type-B sub-factors. Therefore, poor levels of Type-C sub-factors are allowed for the *alternative site*).

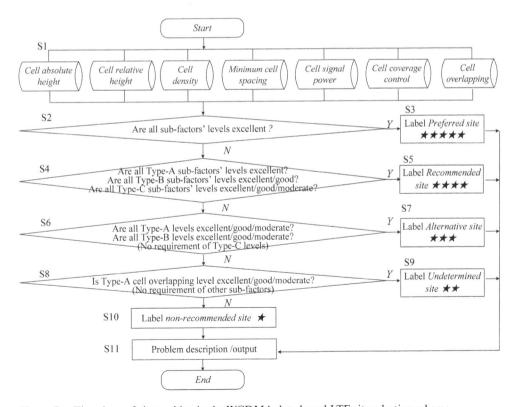

Figure 7. Flowchart of site ranking in the WCDMA data based LTE site selection scheme.

On the contrary, if the cell cannot meet requirements in the 6th step, it comes to the 8th step. In this step, if this cell's overlapping level is excellent/good/moderate, we label this WCDMA cell's site as the *undetermined site* in the 9th step. Otherwise, it is labelled as the *non-recommended site* in the 10th step. Both the *undermined site* and the *non-recommended site* suffer severe problem and are difficult to be addressed. Therefore, they have the lowest priority for LTE site selection. Finally, in the 11th step, each sub-factor's problems are recorded and output in details, including each problematic sub-factor's value and the attribute as well as the level.

4 APPLICATION OF WDLSS SCHEME

We apply the proposed WDLSS scheme into the LTE deployment of a city downtown in China (named as X-city). The implementation results are shown in Figure 8.

Employing the WDLSS scheme, 809 sites of WCDMA cells are labelled as *preferred sites* for LTE site selection, which account for 32.1% of existing sites. For these 809 sites, levels of all seven sub-factors are excellent. Therefore, they can be used to deploy LTE networks preferentially.

As shown in Figure 8, 259 sites are labelled as the *recommended sites*, accounting for 10.2% of existing WCDMA sites. In addition, 485 sites are labelled as *alternative sites*, accounting for 19.2% of existing sites. For both *recommended sites* and *alternative sites*, one or more Type-B / Type-C sub-factors (including the cell absolute height, the cell relative height, the minimum cell spacing, the cell signal power, the cell coverage control) may suffer problems. However, levels of Type-A sub-factors (including the cell overlapping and the cell density) are not poor. Hence, these sites can also deploy LTE cells under the network optimisation (e.g. antenna adjustment, transmission power adjustment). Their priorities are lower than above 809 *preferred sites*.

On the contrary, 508 WCDMA sites are labelled as *undetermined sites*, accounting for 20.1% of existing sites. These 508 sites suffer the problem of extremely high cell density, which is difficult to be addressed. In addition, 465 sites are labelled as *non-recommended sites*. These 465 sites suffer the problem of severe cell overlapping, which will result in the heavy interference and deteriorate the LTE performance significantly. Both 508 *undetermined sites* and 465 *non-recommended sites* are shown in Figure 9. These sites have the lowest priority for LTE site selection.

In order to assist telecom operators to be aware of each site's detailed problems, the WDLSS scheme also outputs the detailed information of each sub-factor, including the value and the attribute as well as the level, as exemplified in Table 4. This site detailed information output table can assist telecom operators take relevant measures to deal with these problems during the LTE deployment, thus effectively reducing the OPEX. Meanwhile, compared with the conventional manual site selection, the WDLSS scheme can save large manual costs of exploring non-recommended sites and reduce the CAPEX.

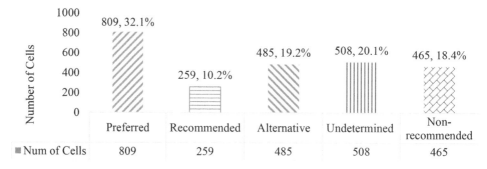

Figure 8.   Site ranking for different labels in downtown of X-city.

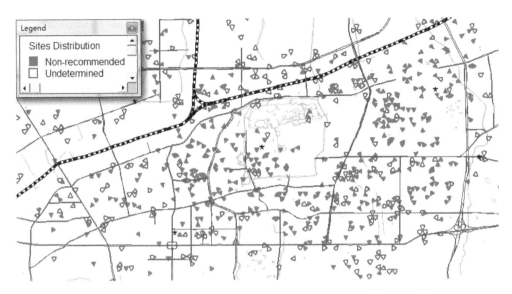

Figure 9. Distribution of *undetermined sites* (hollow) and *non-recommended sites* (solid).

Table 4. Output of detailed problems for each site (example).

Cell name	Absolute height	Relative height	Density	Mini cell spacing	Signal power	Coverage control	Overlapping	Site rank
JN38-Wad01	**Poor**	**Poor**	**Poor**	Moderate	Good	Good	**Poor**	Non-recommended
JN38-Wad01	Extremely high cell with 60 meters. Relatively high cell with 25 meters higher than neighbour cells. Extremely high cell density with 42 cells. Cell overlapping ratio is high, reaching 38%.							
JN46-SD03	Good	Good	**Poor**	Moderate	**Poor**	**Poor**	Moderate	Undermined
JN46-SD03	Extremely high cell density with 36 cells. Poor signal power and only 20% users' RSCP are larger than -99dBm. Poor coverage control, cell's mainly serving traffic ratio is only 28%.							
etc.	......	......	......	......	......	......	......	......

5 CONCLUSIONS

In the LTE deployment, the method of reusing existing WCDMA sites can reduce the CAPEX and OPEX for telecom operators. This paper proposes a novel WCDMA data based LTE site selection (WDLSS) scheme for the LTE deployment. The WDLSS scheme analyses five factors of existing WCDMA sites, including the height, the density, the signal power, the coverage control and the interference. Then the WDLSS scheme ranks each site according to the optimisation difficulty of above five factors. Finally, we apply the WDLSS scheme into the LTE deployment of a city downtown in China.

REFERENCES

3GPP TR 25.942. 2012. *Radio frequency (RF) system scenarios (Release 10)*.
3GPP TS 36.331. 2010. *Radio resource control (RRC) protocol specification (Release 9)*.
China Unicom. 2013. *China Unicom LTE network optimization guide book*. Beijing: China Unicom Press.

Daniel Christian Schultz, Bernhard Walke, Ralf Pabst, & Tim Irnich. 2003. Fixed and planned relay based radio network deployment concept. *Proc. Wireless World Research Forum, New York, USA*: 1–6.

Dantong Liu, Yue Chen, Kok Keong Chai, & Tiankui Zhang. 2013. Performance evaluation of Nash bargaining solution based user association in HetNet. *Proc. International Conference on Wireless and Mobile Computing, Networking and Communications, Lyon, France*: 571–577.

Jeffrey G. Andrews, François Baccelli, & Radha Krishna Ganti. 2011. A tractable approach to coverage and rate in cellular network. *IEEE Transactions on Communications* 59(11): 3122–3134.

Jiayi Wu. 2011. *Coverage-based cooperative radio resource allocation in mobile communication systems.* PhD thesis. Queen Mary University of London.

Juan Ramiro, & Khalid Hamied. 2012. *Self-organizing networks (SON): self-planning, self-optimization and self-healing for GSM, UMTS and LTE.* New York: John Wiley and Sons.

Lexi Xu, & Yue Chen. 2009. Priority-based resource allocation to guarantee handover and mitigate interference for OFDMA systems. *Proc. IEEE International Symposium on Personal Indoor and Mobile Radio Communications, Tokyo, Japan*: 783–787.

Lexi Xu, Xinzhou Cheng, Yue Chen, Kun Chao, Dantong Liu, & Huanlai Xing. 2015. Self-optimised coordinated traffic shifting scheme for LTE cellular systems. *Proc. EAI International Conference on Self-Organizing Networks, Beijing, China*: 1–9.

Lexi Xu, Yue Chen, John Schormans, Laurie Cuthbert, & Tiankui Zhang. 2011. User-vote assisted self-organizing load balancing for OFDMA cellular systems. *Proc. IEEE International Symposium on Personal Indoor and Mobile Radio Communications, Toronto, Canada*: 217–221.

Lexi Xu, Yue Chen, KoK Keong Chai, Tiankui Zhang, & John Schormans. 2012. Cooperative load balancing for OFDMA cellular networks. *Proc. European Wireless, Poznan, Poland*: 1–7.

Lexi Xu, Yue Chen, KoK Keong Chai, John Schormans, & Laurie Cuthbert. 2015. Self-organising cluster-based cooperative load balancing in OFDMA cellular networks. *Wiley Wireless Communications and Mobile Computing* 15(7): 1171–1187.

Lexi Xu, Yuting Luan, Kun Chao, Xinzhou Cheng, Heng Zhang, & John Schormans. 2014. Channel-aware optimised traffic shifting in LTE-Advanced relay networks. *Proc. IEEE International Symposium on Personal Indoor and Mobile Radio Communications, Washington, USA*: 1597–1602.

Peng Jiang, John Bigham, & Jiayi Wu. 2008. Self-organizing relay stations in relay based cellular networks. *Computer Communications* 31(13): 2937–2945.

Zhigang Han, Li Kong, Guoli Chen, & Fuchang Li. 2012. *LTE FDD technology principle and network planning.* Beijing: China Post and Telecommunications Press.

Signal and Information Processing, Networking and Computers – Chen & Huang (Eds)
© 2016 Taylor & Francis Group, London, ISBN 978-1-138-02881-4

A novel LTE network deployment scheme using telecom big data

Tao Zhang, Xinzhou Cheng, Lexi Xu, Xiaodong Cao,
Mingqiang Yuan & Yongfeng Wang
Department of Network Optimization and Management, China Unicom Network Technology
Research Institute, Beijing, P.R. China

ABSTRACT: In order to maintain an optimal user experience and improve the quality of mobile data service, telecom operators now are accelerating the deployment of the next generation mobile network-Long Term Evolution (LTE). The conventional LTE deployment approach seeks to cover as much ground as possible. That will pose a great challenge for operators to profit from their investments in LTE networks. In this paper, we propose a novel multi-dimensional elements LTE deployment scheme based on big data analysis to solve the problems faced by telecom operators. The method can achieve the purpose of making the investment in LTE network more cost-effective and enhancing customers' experiences. A result of case study as calculated from the real collected cellular network data also shows the performance of the proposed method.

Keywords: LTE network; telecom big data; network deployment priority; precise investment

1 INTRODUCTION

The mobile broadband industry in China has grown at a phenomenal rate in recent years. The three big mobile network operators of China are seeing an explosion of varied, high-volume and high-velocity data streams due to the increased use of smart phones and rise of social media. However, historically, operators of China have not taken advantage of monetizing the value that is inherent in the data they hold. Big Data concepts provide the means by which operators can analyze the value of information they collect and store.

When it comes to big data strategy in the field of telecommunications, the telecom industry has an advantage over others due to the sheer breadth and depth of data it collects in the course of normal business. There are many types of data and information that can be gained from cellular network such as internet data, Call Detail Records (CDRs) in H. Wang, et al. (2010), location information etc. Reliable and transparent information on mobile service usage is of great value to many stakeholders (Kivi and Antero 2007) in these industry, including marketing, business, product development and network planning. Cellular network data has been carried out to study human mobility in Becker R., et al. (2013) and has been used for urban planning in Becker R., et al. (2011) and many other services such as traffic monitor in G. Rose (2006), social structure in J. Onnela et al. (2006) and location based services in Calabrese, et al. (2010). However, rare research has been done for network planning especially in the LTE network deployment (Lexi Xu, et al. 2015).

In this paper, we explore a novel LTE network deployment scheme based on telecom big data analysis. We have found an efficient way to combine data from business support system (BSS) and Operation Support System (OSS) together to support LTE network deployment. Meanwhile, we take five dimensions into considerations to judge whether this area should deploy LTE network: the situation of customers, the 4G terminal numbers, the load of streaming service, the load of data service and the total revenue of this area. We also propose an evaluation system to score each cell site's demand for LTE network in a local network. Lastly, after the evaluation of each cell site, we present the priority cell site lists of

LTE deployment in this local network. The purpose of this scheme is to make the investment in LTE network more cost-effective and raise the utilization of network resources and the most important thing is to enhance customers' experiences.

This paper is organized as follows. In section 2 gives an overview of the data collected from mobile network and the architecture of our big data analysis system. In section 3 elaborates the reason why we choose the five impact factors for LTE network deployment and introduce the proposed scheme. The performance of the use case is revealed in section 4. The last section gives a conclusion.

2 DATASET AND BIG DATA ANALYSIS SYSTEM

This section presents an overview of the cellular network datasets and the corresponding analyzing procedures. We firstly describe the data types that are commonly collected in the course of normal business. We then give the specific description of each data type. We finalize by discussing the procedures related to the cellular network big data analysis system.

2.1 Data types

For telecom operators, most of the data is collected by various departments, spread across the organization, such as network optimization department and marketing department etc. They collect network performance data, cell-site data, device information, accounting data and other information spread across network as shown from Figure 1. Generally speaking, the data collected by operators can be divided into two main types: the data from OSS side referring to information about the network and the data from BSS side referring to the information about the customers. By combining these two types of data to let them interact closely with each other, share unified capabilities and information to enable instant insight into how the network is being used.

Each data type corresponds to a table in which each row represents a record such as a phone number or a user IDs. In this paper, the concerned information of each data type is detailed in Table 1.

2.2 Telecom big data analysis system architecture

Recent years, many big data technologies such as SQL (NoSQL) (Han J., et al. 2011) and MapReduce (Dean, et al. 2008) J programming model have been introduced for processing large-scale distributed data sets. And a wide variety of industry including telecom operators have built up their own big data system. This part we will give a brief introduction of our big data system architecture which is used to process the use case below.

Figure 2 illustrates how value is derived from our big data solution system architecture. The whole system is made up of three layers: data collection layer, platform layer and data

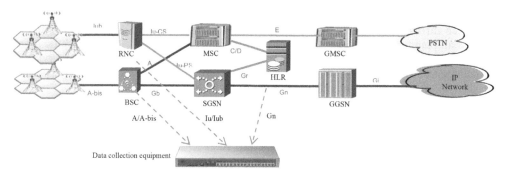

Figure 1. Cellular network data collection method.

Table 1. Data description.

Data Type	Data Description
Cell-site	Province code, city code, cell code, location area code, cell longi tude, cell latitude.
Key Performance Index (KPI) Data	Province code, city code, cell code, time, uplink RLC layer throughput, downlink RLC layer throughput, CS voice traffic.
Circuit Switched (CS) CDR	Province code, city code, date, user ID, cell ID, location area code, IMSI, IMEI, phone number, start time, end time, call duration, call type, roam type, radio access network type.
Packet Switched (PS) CDR	IMSI, Phone number, location area code, cell ID, IMEI, traffic style, start time, end time, duration, uplink traffic, downlink traffic, sum traffic, client IP, destination IP, status.
Customer Monthly Accounting Data	Month, province code, city code, user ID, vip class, IMSI, user network type, age, sex, total call minutes, total packet MBs.
Customer Monthly Voice Detail Data	Province code, city code, date, user ID, cell ID, location area code, IMSI, IMEI, phone number, start time, end time, call duration, call type, roam type, station type, base fee.
Customer Monthly Packet Detail Data	Province code, city code, date, user ID, cell ID, location area code, IMSI, IMEI, phone number, start time, end time, uplink traffic, downlink traffic, base fee.

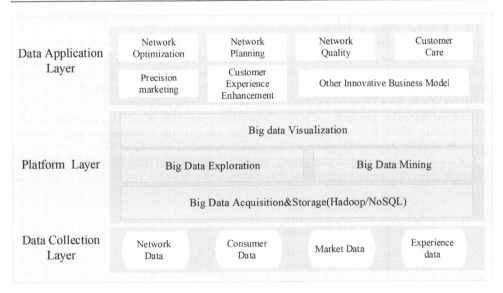

Figure 2. Big data analysis system architecture.

application layer. At the bottom of the big data system is the data collection layer aiming to collect all the data together—from structured, semi-structured and unstructured data sources, especially the large data sets from operators' existing management systems. Next, at the platform layer we developed a prototype based on Mapreduce programming model on the Hadoop platform to run analytical algorithms efficiently. Meanwhile, the interactive data visualization can also make operator officers understand information more easily and quickly. The top data application layer encompasses many business models such as network optimization, network planning and other internal and external applications. In addition, application layer provides open interfaces to support better scalability and availability for the future plan.

All in all, big data technology will help the operators transform the way they do business and enable them to create innovative new business models to maximize the value of the network resources.

3 PROPOSED LTE NETWORK DEPLOYMENT SCHEME

As the traditional way of network deployment, the network rollout is driven by capacity relief. However, this kind of method neglects the fact that even if the new generation network is deployed, very few suitable devices is available and many customers' service contracts is not permitted to camp on this network.

Therefore, in this section, we propose a novel LTE network deployment scheme by making use of operators' most valuable asset—cellular network data. Our algorithm for identifying LTE deployment candidate area has two stages. In the first stage, we calculate the 4G target user number, the 4G terminal number, the streaming service volume, the PS service total volume and the revenue contribution of each site based on the data types we mentioned above. In the second stage, firstly we score each cell site in a local network its demand for LTE network based on the value of each element in stage one. Secondly, according to the total score by descending order, we can get the site priority lists of LTE deployment in this local network. The LTE network deployment processing flow is depicted in Figure 3.

3.1 *LTE network deployment impact element analysis*

3.1.1 *Aggregated area of target customers*
In this customer-centric society, operators should transform from the network centric deployment to the customer centric deployment in Mona Matti and Tor Kvernvik (2012). Therefore, during the process of LTE network deployment, operators should know where their target customers are located and try to deploy the limited network resources precisely.

In this article, for LTE network deployment procedure, the target customers can be divided into two groups. One group is the person who already have the 4G service contract which is the operators' direct service target. And the other group is the potential 4G user who have great demand for data traffic and have the great potential to migrate to 4G.

For the 4G service contract user group, we can get the user list from the CMAD data table. The user network type column tells us this customer is a 2G or 3G or 4G user. After we get

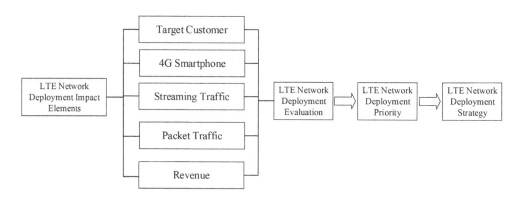

Figure 3.   LTE network deployment processing flow.

the 4G IMSI list, then we can correlate the CMAD table and the CMPD data table with the 4G IMSI list. Because the CMPD data table registers the location of the wireless cell carrying every packet service, we can then get the 4G distinct user number at every site.

For the potential 4G user group, we manage to dig into the characteristics of the 4G customers who have migrated from 2G or 3G services to 4G services. Starting from the 4G migrated subscribers' characteristics, we try to locate the potential 4G users from the existing 2,3G subscribers by data mining algorithm. As is the same to 4G service contract user group, when we digging out the potential 4G user IMSI list, we then calculate the potential 4G user deduplication number at every site.

At last, the total target user number of every cell site is the sum of 4G user deduplication number and potential 4G user deduplication number.

3.1.2 *Aggregated area of 4G smart phones*

As portable devices such as smartphones, tablets or digital cameras are getting cheaper, consumers are buying more LTE-enabled devices these days. This means that an operator might have a big potential market of LTE users waiting even before it begins to offer LTE service. So LTE-enabled Smart Phones is considered as a key elements for LTE deployment.

From the CMPD and CMVD data table, the IMEI column registers the phone information. The Type Allocation Code (TAC) is the initial eight-digit portion of the 15-digit IMEI codes used to uniquely identify wireless devices. TAC contains the information about the manufacturer and the specific model. Based on the information above, we can recognize whether it is a 2G or 3G or 4G terminal. Therefore, from the location information of the CMPD or CMVD data table, we can get the 4G terminal deduplication number at every site.

3.1.3 *Streaming traffic high load area*

Mobile video is becoming one of the most popular and widely growing services. It is recognized as the most important LTE service. Since mobile video content has higher bit rates than other mobile content types, it contributes much of the mobile traffic growth. If in some regions of the local network its video traffic is very high, high-speed LTE is expected to enhance user experience in this area.

In this article, we extract PS CDR data from Gn interface-between the Gateway GPRS Support Node (GGSN) and the Serving GPRS Support Node (SGSN) network equipment, which record all of the user transmission data. Based on the column of traffic style, sum traffic, location area code and cell ID from Table 1 PS CDR data, we can calculate the video traffic of each site in the local area.

3.1.4 *Packet traffic high load area*

As we move from a voice-centric world to a data-centric world, there is much greater demand for data, speed, and access to information. Relative to 3G, LTE/4G allows operators to provide more efficient data and better user experience. Especially in the regions of large data traffic load (Lexi Xu 2015, Lexi Xu 2012), it's probably a good time to rapidly deploy LTE to shift off a big part of data traffic and free up legacy capacity for voice services.

In operators' routine maintenance, KPI data is used to monitor the traffic load of each cell. Based on the cell ID, time, uplink Radio Link Control (RLC) layer throughput and downlink RLC layer throughput, we can calculate the daily average traffic load of each site in the local area.

3.1.5 *High revenue contribution area*

As mentioned earlier, the migration to LTE or upgrading of existing mobile networks will incur high costs. Therefore, during the process of LTE deployment, the big concern for operators is how to make the investment in LTE network more cost-effective and how to get fast Return On Investment (ROI). Then the revenue of one cell is a key elements to reflect its value for operators.

In this paper we propose a method to evaluate the revenue of each cell. Based on the service type, we divided the revenue into two parts: voice revenue and data revenue. For the voice revenue, the CMVD data table record the expense of the every phone call in each cell. As is the same to data revenue, from the CMPD data table we can get the expense of the total MBs in each cell. By adding the total voice revenue and data revenue of each cell, we can calculate the total revenue of each site in the local network.

3.2 A comprehensive LTE deployment evaluation system

As mentioned in part 3.1, we have discussed the different demand for LTE deployment from five different perspectives. In this section, we integrate the five key elements to build up a comprehensive LTE deployment evaluation system. The details of this evaluation system are elaborated as follows.

Typically, network deployment is considered in the unit of site. For each site we have a LTE deployment value $Site_{i,LTE}, i \in (1,2,..n)$ n stands for the total site number in the local network. Each $Site_{i,LTE}$ is characterized by the five key elements and expressed in the target customers V_{i1}, the 4G smart phones V_{i2}, the streaming traffic load V_{i3}, the packet traffic load V_{i4} and the revenue V_{i5}. The $Site_{i,LTE}$ can be gained from the above five elements, as shown in expression (1),

$$Site_{i,LTE} \longleftarrow [V_{i1},V_{i2},V_{i3},V_{i4},V_{i5}] \tag{1}$$

However, the site covering area is quite different between each other due to the geographical position and the density of the sites. The site covering area in downtown area is far smaller than the suburban's. In this paper, we adopt the propagation delay method to estimate the transmission radius and then we can get the actual site covering area.

The Radio Network Controller (RNC) equipment can measure the propagation delay by receiving the Random Access Channel (RACH) preamble from the terminal side. Then the RNC can estimate the cell radius using the wireless signal propagation model. If the specific time delay is within the range $(t,t + \Delta t)]$ $(t \in (0,0.12ms]$, the max distant is set to 35000 meters, $\Delta t = 0.0012 \ ms, \ total \ 100 \ interval)$, then the result will be stored in the corresponding counter. Based on all the counter result, we can get the estimated average site radius in equation (2),

$$R_{i,site} = \frac{\sum\limits_{p=0}^{99} c*(t_{i,p} + \Delta t)*N_{i,p}}{\sum\limits_{p=0}^{99} N_{i,p}} \tag{2}$$

where c is the speed of light and N is the number of sample within the range $(t,t + \Delta t)]$. Then we can calculate the area of each site in equation (3),

$$S_{i,site} = pi * R^2_{i,site} \tag{3}$$

In order to avoid the influence from the difference in site area, we use the five key elements divided by site area $S_{i,site}$ respectively, and the meaning of each $V^*_{ij} j \in (1, 2,..5)$ is shown in Table 2.

$$V^*_{i1} = \frac{V_{i1}}{S_{i,site}}, V^*_{i2} = \frac{V_{i2}}{S_{i,site}}, V^*_{i3} = \frac{V_{i3}}{S_{i,site}}, V^*_{i4} = \frac{V_{i4}}{S_{i,site}}, V^*_{i5} = \frac{V_{i5}}{S_{i,site}} \tag{4}$$

By doing so, $Site_{i,LTE}$ can be changed to $Site^*_{i,LTE}$ as equation (5):

$$Site^*_{i,LTE} \longleftarrow [V^*_{i1},V^*_{i2},V^*_{i3},V^*_{i4},V^*_{i5}] \tag{5}$$

Table 2. LTE network deployment variables.

Variable	Description and Unit
V_{i1}^*	Target customers per square kilometers (persons/Km²)
V_{i2}^*	4G smart phones per square kilometers (terminals/Km²)
V_{i3}^*	Streaming traffic load per square kilometers (MBs/Km²)
V_{i4}^*	Packet traffic load per square kilometers (MBs/Km²)
V_{i5}^*	Revenue per square kilometers (RMB/Km²)

For the different key elements of all the sites, we can get vectors like this

$$\mathbf{V}_j = (V_{1j}^*, V_{2j}^*, \ldots, V_{nj}^*)^T \tag{6}$$

where j = 1, 2, 3, 4, 5 represents the five elements.

The next step, due to the different dimension of the five key elements, we try to achieve consistency in dynamic range for a set of data through normalization processing. The normalization formula is listed as follows:

$$X_{ij}^* = \frac{V_{ij}^* - Min(\mathbf{V}_j)}{Max(\mathbf{V}_j) - Min(\mathbf{V}_j)} \tag{7}$$

After the normalization processing, the range of every elements value have been changed to [0,1] and for $Site_{i,LTE}^*$ five variables can be change to equation (6):

$$Site_{i,LTE}^* \in [X_{i1}^*, X_{i2}^*, X_{i3}^*, X_{i4}^*, X_{i5}^*] \tag{8}$$

Then, we can gain the LTE deployment value for $Site_{i,Deploy}$,

$$Site_{i,Deploy} = \sum_{j=1}^{5} X_{ij}^* \tag{9}$$

Lastly, after the evaluation of each cell site, we sort the $Site_{i,Deploy}$ on descending order to get the priority lists of LTE deployment in this local network.

4 USE CASE

To verify the performance of our proposed scheme, then we applied this approach to our large cellular network datasets by big data platform. We selected a typical provincial capital in northern east part of China as a sample city for this use case. We demonstrated the feasibility of our approach through tabulation, statistical analysis and visualization.

Firstly, we collected nearly 2 weeks of cellular network historical data between October 13, 2014 and October 28, 2014. In total, we collected 5 TB and 60 million records for this analysis. The specific data time requirements is illustrated in the Table 3. The time span of these serval data types should be alignment in time.

Next step, we used our collected data to calculate the five key elements of every site as we mentioned in part 3.1. For each key elements, we calculate its weekly average value. We evaluated each cell site in this local network its demand for LTE network based on the evaluation

Table 3. Time requirements of each data type.

Data Type	Time Requirements
Cell-site	Latest Month (e.g. September, 2014)
KPI data	At least a week (e.g. October 20, 2014 to October 26, 2014)
CS CDR	At least a week (e.g. October 20, 2014 to October 26, 2014)
PS CDR	At least a week (e.g. October 20, 2014 to October 26, 2014)
CMAD	Latest Month (e.g. September, 2014)
CMVD	At least a week (e.g. October 20, 2014 to October 26, 2014)
CMPD	At least a week (e.g. October 20, 2014 to October 26, 2014)

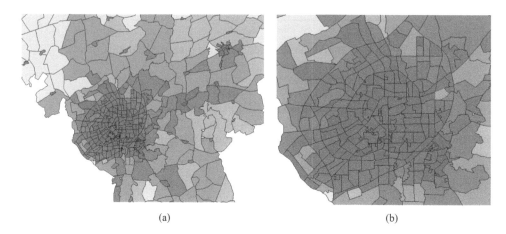

(a)   (b)

Figure 4. (a) Geographic information of LTE network deployment. (b) Partial enlarged view of downtown LTE network deployment.

Table 4. Example result of LTE deployment.

City	Site ID	Target Customers persons/Km²	Packet Traffic Load MBs/Km²	4G Smart Phones terminals/Km²	Streaming Traffic Load MBs/ Km²	Revenue RMB/Km²	LTE Deployment Value
A	site1	63350	2963792	36647	114315	58165	4.3
A	site2	23982	1925369	30459	55846	2422	3.1
A	site3	15567	1029365	19659	30937	51546	2.01
A	site4	1177	560583	1069	3968	1195	0.21
A	site5	–	–	–	–	–	–

system in in part 3.2. Given the sensitivity of the data, we anonymized the city name and the Site ID and only presented a typical example of final result in Table 4.

Then we revealed our results in the form of geographical visualization in Figure 4 to make LTE deployment value more accessible, understandable and usable. In Figure 4, the darker red region means that it is more urgent for LTE network deployment. The map indicates that a large fraction of the darker red region located in the populated downtown area as shown in Figure 4 (b).

Lastly, based on the deployment geographical visualization, budget for network construction and the situation of network resources, we concluded the LTE deployment strategy for this local network. We presented the priority site lists of LTE deployment with four levels, including high level, medium level, normal level and low level. The site deployment order will depend on the deployment assigned priority. The result is shown in Table 5.

Table 5. LTE deployment priority result.

Priority Level	Site Number	Percentage
High	1518	56.9%
Medium	870	32.6%
Normal	201	7.5%
Low	80	3%

From the result in Table 5, we can see that roughly 56.9% of total sites in this local network should place at the top priorities by using our proposed scheme. By following the above mentioned steps, the operator in this local network reduce the number of LTE deployment sites by 4%, comparing with the traditional ways. Meanwhile, the old method only concern the capacity of the network rather than the target LTE user's experience. To deploy these sites with urgent demand for LTE network can greatly improve customers' experience and increase their loyalty. On the other hand, the priority site lists enable the operator to apply right level of priority at different stages. These differentiated deployment strategy make the investment in LTE network more cost-effective and can achieve competitive advantage in local deployed areas.

5 CONCLUSIONS

The goal with this work is to solve the problems of conventional LTE deployment method in network planning. Our proposed scheme, multi-dimensional elements for evaluating LTE deployment areas can avoid the waste of capital expenditure and make the investment in LTE network more cost-effective and enhance customers' experience. The geographical visualization results as calculated from the real collected cellular network data show how it can be used to select the candidate deployment continuous coverage area.

In the future, we plan to investigate how cellular network data can be used in the course of the operators' routine business. As the burst of the mobile data, operators should speed up the process to develop new technologies, and to test new and sustainable business solutions to monetize big data.

REFERENCES

Becker, R.A., Caceres, R., Hanson, K., Loh, J.M., Urbanek, S. Varshavsky, A. & Volinsky, C. 2011. A tale of one city: Using cellular network data for urban planning. *IEEE Pervasive Computing* 10(4): 18–26.

Becker, R., Cáceres, R., Hanson, K., Isaacman, S., Loh, J.M., Martonosi, M., Rowland, J., Urbanek, S., Varshavsky, A., & Volinsky, C. 2013. Human mobility characterization from cellular network data. *Communications of the ACM* 56(1): 74–82.

Calabrese, F., Pereira, F.C., Di Lorenzo, G., Liu, L., & Ratti, C. 2010. The geography of taste: analyzing cell-phone mobility and social events. *Proc. Pervasive'10 Proceedings of the 8th International Conference on Pervasive Computing, Berlin, Heidelberg*, pp. 22–37.

Dean, J. & Ghemawat S. 2010. MapReduce: simplified data processing on large clusters. *Communications of the ACM* 51(1): 107–113.

Han, J., Haihong, E., Le, G. & Jian, Du. 2011. Survey on NoSQL database. *Proc. Pervasive Computing and Applications (ICPCA), 2011 6th International Conference on*, pp. 363–366. DOI: 10.1109/ICPCA.2011.6106531.

Kivi, Antero. 2007. Measuring mobile user behavior and service usage: methods, measurement points, and future outlook. *Proc. Proceedings of the 6th Global Mobility Roundtable*, pp. 1–2.

Lexi Xu, Xinzhou Cheng, Yue Chen, Kun Chao, Dantong Liu & Huanlai Xing. 2015. Self-optimised coordinated traffic shifting scheme for LTE cellular systems. *Proc. Conference on Self-Organizing Networks, Beijing, P.R.China.*

Lexi Xu, Yue Chen, KoK Keong Chai, Tiankui Zhang & John Schormans. 2012. Cooperative load balancing for OFDMA cellular networks. *Proc. European Wireless 2012, Poznan, Poland*, pp. 1–7.

Lexi Xu, Yue Chen, Kok Keong Chai, John Schormans & Laurie Cuthbert. 2015. Self-Organising Cluster-Based Cooperative Load Balancing in OFDMA Cellular Networks. *Wiley Wireless Communications and Mobile Computing* 15(7):1171–1187.

Mona Matti, & Tor Kvernvik. 2012. Applying big-data technologies to network architecture. *Ericsson Review*.

Onnela, J., Saramaki, J., Hyvonen, J., Szab'o, G., Lazer, D., Kaski, K., Kertesz, J. & Barabasi, A. 2007. Structure and tie strengths in mobile communication networks. *Proceedings of the National Academy of Sciences* 104(18): 7332–7336.

Rose, G. 2006. Mobile phones as traffic probes: practices, prospects and issues. *Transport Reviews* 26(3): 275–291.

Wang, H., Calabrese, F., Di Lorenzo, G. & Ratti, C. 2010. Transportation mode inference from anonymized and aggregated mobile phone call detail records. *Proc. Intelligent Transportation Systems (ITSC), 2010 13th International IEEE Conference on*, Madeira Island, Portugalpp.318–323.

Signal and Information Processing, Networking and Computers – Chen & Huang (Eds)
© 2016 Taylor & Francis Group, London, ISBN 978-1-138-02881-4

Five-dimension labeled 4G user immigration model based on big data analysis

Heng Zhang, Lei Zhang, Chuntao Song & Xinzhou Cheng
Department of Network Optimization and Management, China Unicom Network Technology
Research Institute, Beijing, P.R. China

ABSTRACT: For the purpose of 4G precision marketing, a novel prediction model named UIM-FDL is proposed in this paper to predict potential 4G customers from existing 2G/3G users. And five-dimension factors related to user immigration preference are derived from mining the massive business and operation data. Besides the traditional user character and consumption behavior, service experience and social circle are innovatively considered in immigration model. And six clever attributes are selected as the input label of each consumer based on evaluation of correlation matrix. We apply logistic regression and neural network respectively to identify the 2G/3G users who are most likely to immigrate to 4G network. The results demonstrate that the UIM-FDL model we proposed can achieve much greater performance than traditional marketing schemes, and UIM-FDL-NN outperforms UIM-FDL-LR scheme to some extent.

Keywords: 4G; user immigration; BP neural network; logistic regression; big data

1 INTRODUCTION

With explosive growth of mobile data traffic, building a high quality 4G/3G cooperative network offering excellent network coverage and speed is of great significance for China Unicom. However, 4th generation (4G) mobile communication systems have been deployed on a large scale in recent years, the 3G and 2G network still contribute a very large proportion of total data traffic. To improve the operating efficiency and user experience, it is in an urgent need to immigrate 2G/3G users to 4G network for telecomm operators. For the aim of precision marketing, it is of great importance to find out the potential 4G users who are most likely to immigrate from 2G/3G users.

Recently, the research related to telecomm marketing mainly focuses on customer churn management (Xu 2011), the study on how to identify the potential 4G users from existing 2G/3G networks have not been researched comprehensively. The traditional marketing methods adopt some common factors such as age, DOU (Discharge of Usage), MOU (Minutes of Usage) and ARPU (Average Revenue Per User) to select the high quality users, with setting fixed threshold artificially. However, it cannot satisfy the demand of internet+ age with diverse data services and applications, as we cannot describe user behavior efficiently with simple parameters. In literature (Hsu 2014), a framework based on CDR data is proposed to infer the potential users who are likely to join target services from other competitors. But the model based on the features of receiving and sending calls is not appropriate for mining potential 4G users who are fond of data traffic. Literature (Shi et al. 2014) develops a method based on K-Means clustering algorithm with canopy approach to identify 2G user groups that are most likely to immigrate to 3G/4G networks. However, only traffic and mobility dimensions are considered, the user character, consumption behavior and service quality which are important factors of user immigration have not been mentioned in this model. Therefore, the existing solutions cannot be used for the problem of predicting 4G potential precisely.

To address the problem, a five-dimension labeled 4G user immigration model (UIM-FDL) is proposed in this paper to predict potential 4G users from existing 2G/3G users based on logistic regression (UIM-FDL-LR) and neural network (UIM-FDL-NN) respectively. In recent years, LR and NN are widely applied in many fields, such as assess risk factors for various diseases, traffic and population prediction. A logistic growth curve model for mobile user growth is proposed in (Tao et al. 2012), and a multiple logistic regression model is used for hypothesis testing to identify the significant factors that contribute to accident severity in (MA 2009). And BP neural network model is also applied to predict network traffic forecasting strategy in (Li and Lei 2009, Li and Zhang 2009).

The rest of this paper is organized as follows: Section 2 presents the proposed system architecture of 2G/3G/4G immigration platform based on big data analysis. Section 3 mainly analyzes 4G consumer characters and behaviors. In section 4, the methodology of modeling is described in details. The results and performance evaluation are developed in section 5. And the conclusions are drawn in section 6.

2 SYSTEM ARCHITECTURE

The essence of 2G/3G/4G subscriber immigration is the best matching of subscriber, product and terminal. The whole work of user immigration can be divided into 3 parts, including potential 4G user mining, product adaption and terminal adaption. The first part is oriented to existing 2G/3G users holding terminals supporting LTE network, while the other two parts are targeted to non-4G terminal users. We mainly discuss the first part in this paper, the other parts including product and terminal marketing will be studied in the future.

As shown in Figure 1, the system architecture of immigration platform is composed of three parts, including data warehouse, prediction modeling and application layer. The data warehouse is the input of system, which is build based on 'B+O' original data such as user information, billing data, IUPS and Gn interface data. The prediction modeling is the decision center of 4G potential users, which is self-learning with six important features by logistic regression and neural network. In application layer, 4G potential subscriber list and marketing strategies will be output and deployed, and immigration performance will be monitored. And we mainly discuss the modeling part in this paper.

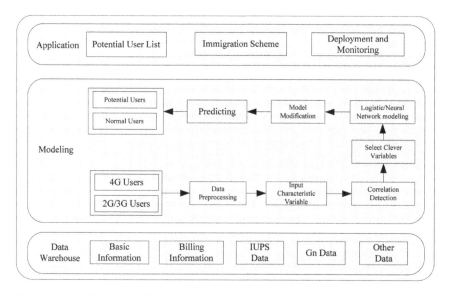

Figure 1.   System architecture of big data mining platform for 2G/3G/4G immigration.

3 PRELIMINARIES

Understanding 4G consumer features and behaviors is a vital step to help us build the mining model. The analysis results below are based on the data provided by China Unicom in a medium developed city.

As required, the whole 4G Users Group (WUG) can be divided into Original User Group (OUG) and Immigration User Group (IUG). The IUG are those who have immigrated to 4G from 2G/3G in an updating way, while OUG are newly developed 4G users who have never connected the operator' s network before. The statistical data show that about 74.2% of 4G users are belonged to OUG, and 25.8% are IUG. However, 84% of IUG come from 3G network, and only 16% are 2G users. In the view of terminal, only 18.9% 4G users' mobile phone can support LTE network, and 51% hold 3G devices.

In this section, the customer features and behaviors of WUG and IUG will be analyzed independently. As can be seen in Table 1, young and middle-aged people are the main 4G user groups. Male and female are favor of 4G almost equally, and the ratio of male to female is nearly equal to whole 2G/3G/4G networks. The most outstanding feature between IUG and WUG is the VIP ratio of IUG is much higher than WUG. About 76% of IUG users are VIP before immigrating, and the ratios of gold and diamond card are 49% and 13% respectively.

From Table 2, we can infer that DOU and ARPU of IUG are much higher than WUG, but MOU is a little smaller than WUG. The results indicate the 4G users immigrating from 2G/3G can consume more data traffic and tele-revenue. Compared to 3G users, the average DOU of 4G (709MB) is more than twice of 3G, and ARPU (116) is also 24.5 larger than 3G.

Table 1. Comparision of user features between WUG and IUG.

Attribute	Value	Proportion	
		WUG	IUG
AGE	0–30	66.0%	60.4%
	30–40	20.6%	27.5%
	>40	13.4%	12.1%
VIP Level	Diamond	8.0%	13.2%
	Goal	13.6%	48.8%
	Sliver	8.8%	13.8%
	Normal	69.6%	24.2%
GENDER	Male	72.2%	72.5%
	Female	27.8%	27.5%

Table 2. Comparision of user behaviors between WUG and IUG.

Attribute	Value	Proportion	
		WUG	IUG
DOU	0–500M	56.8%	40.4%
	500M-1G	22.5%	18.9%
	1G-4G	18.6%	35.9%
	>4G	2.2%	4.9%
MOU	0–600	75.0%	55.5%
	600–1000	13.6%	23.6%
	>1000	11.5%	20.9%
ARPU	0–100	56.1%	30.8%
	100–200	29.2%	32.3%
	>200	14.7%	36.9%

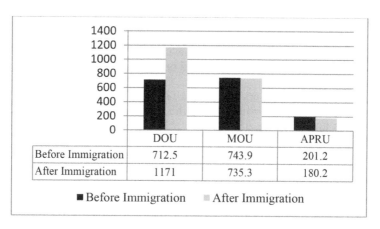

	DOU	MOU	APRU
Before Immigration	712.5	743.9	201.2
After Immigration	1171	735.3	180.2

■ Before Immigration   ▨ After Immigration

Figure 2.   Comparison of user behaviors before and after immigration.

The user behaviors of IUG before and after immigrating to 4G are compared in Figure 2. It is obvious that DOU increases from 712.5 MB to 1171 MB, but the fluctuation of MOU is very little. However, average ARPU decreases from 201.2 to 180.2, the main reason is 4G users with cheaper package can enjoy more data traffic and voice time with the same expense as 3G.

In conclusion, we can infer that young VIP users with high data traffic requirement are the main 4G user groups. And the 4G users immigrating from 2G/3G can consume more data traffic and tele-revenue than original 4G customers. Therefore, it is of great significance to identify the potential 4G users from existing customers.

4 METHODOLOGY

4.1 *Key factors of user immigration*

As seen in Table 3, the significant factors influencing user immigration are derived from five dimensions, including user attribute, terminal attribute, communication behavior, user experience and social circle. The details of selected attributes are described as below.

Firstly, the demographic variables such as age and VIP level are important basic factors in this model. Young people are more attractive to new things and diverse internet services. High level VIP users have larger probability to immigrate to 4G, and their traffic growth may be much larger than normal users. Secondly, the supportable network of terminal is considered as it will decide whether the customer can access LTE network without changing new 4G mobile phone. The user group holding 4G devices should be given high priority for marketing. Thirdly, the communication history indicates the users' consuming ability and demand for data and voice service. The higher DOU and ARPU are, the higher probability they are willing to enjoy better service experience using 4G network. However, MOU can just be taken as the assistant factor to be examined, as it have no significantly difference between 3G and 4G users. Fourthly, the service perception will decide their satisfaction of network quality, and affect their willingness to try high-speed 4G network. And integrated service perception is evaluated comprehensively by voice/http/download/instant message/streaming media service. Lastly, the social circle may make some effect on their impression for new things. If their relatives and friends are fond of 4G network, they are more likely to be influenced to join 4G network actively.

4.2 *Characteristic attributes*

The importance of listed variables in Table 3 is different, and some of them may have high correlation. Thus, the most significant attributes should be selected rationally. The procedure of selecting characteristic attributes is given in Figure 3.

Table 3. Main factors related to 4G user immigration.

Index	Classification	Main attributes
1	User attribute	Age, Gender, VIP Level, High Value User (HVU)
2	Terminal attribute	Terminal Price, Contract Phone, Supportable Network(2/3/4G)...
3	Communication behavior	DOU, MOU, ARPU...
4	User Experience	Integrated Service Perception (ISP), Voice/HTTP/Download/ Instant message/Streaming Media Service Perception...
5	Social circle	4G User Ratio of TOP 10 Social Circle (4GUR_TSC), VIP User Ratio of TOP 10 Social Circle...

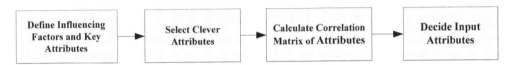

Figure 3. The procedure of selecting clever attributes.

In order to evaluate the similarity, the correlation levels between the above variables are calculated by Pearson Correlation Coefficient. And correlation matrix and decision criterion is proposed to measure the similarity. The correlation value ρ between input variable x_i and y_i can be defined as

$$\rho_{x_i,x_j} = \frac{\text{cov}(x_i,x_j)}{\sqrt{Dx_i}\sqrt{Dx_j}} \tag{1}$$

$$\text{cov}(x_i,y_i) = \sum_{i=1}^{n}(x_i - \bar{x})(y_i - \bar{y}) \tag{2}$$

$$D(x_i) = \sqrt{\sum_{i=1}^{n}(x_i - \bar{x})^2}, \ D(y_i) = \sqrt{\sum_{i=1}^{n}(y_i - \bar{y})^2} \tag{3}$$

The results of correlation matrix are listed in Table 4. As required, High Correlation (HC), Middle Correlation (MC) and Low Correlation (LC) are defined to evaluate the similarity degree. The correlation coefficient will be $0 \leq \rho < 0.5$ in LC, $0.5 \leq \rho < 0.8$ in MC and $0.8 \leq \rho \leq 1$ in HC.

It is obvious that correlation coefficients between VIP level and ARPU, HVU are higher than 0.7, which indicates the VIP level and ARPU, HVU have relatively high correlation. The main reason is VIP level is classified according to average ARPU in recent months, and ARPU also plays important parts in deciding high-value user. Thus, we select VIP level as the clever input attribute and drop the other two.

Base on the analysis above, six important attributes are selected as the input variable of predicting model. And the detailed description is illustrated in Table 5.

It is worth to mention that integrated service perception is the comprehensive evaluation of service experience, including voice, http, download, instant message, streaming media service. Each service experience is evaluated by perception matrix with many variables independently, and the decision results are excellent, good, average and bad which corresponding to 4,3,2,1. As ISP is the output of other modules, the detailed computational method will not be described in this paper. The ISP is averaged by all service perception and can be defined as

$$ISP = \frac{SP\_V + SP\_H + SP\_D + SP\_IM + SP\_SM}{5} \tag{4}$$

Table 4. Correlation matrix of different attributes.

	Age	VIP level	DOU	MOU	ARPU	HVU	ISP	4GUR_TSC
Age	1							
VIP Level	0.26	1						
DOU	0.23	0.49	1					
MOU	0.15	0.55	0.29	1				
ARPU	0.25	**0.79**	0.34	0.41	1			
HVU	0.19	**0.74**	0.32	0.38	**0.81**	1		
ISP	0.03	0.14	0.11	0.08	0.12	0.11	1	
4GUR_TSC	0.05	0.12	0.09	0.07	0.15	0.13	0.01	1

Table 5. Attribute specification of the predicting model.

Variable	Attribute name	Attribute values	Value sets
X1	Age	User age	Positive Integer
X2	VIP level	Diamond Member; Goal Member; Sliver Member; Normal Member	{3, 2, 1, 0}
X3	DOU	Average DOU in recent 3 months	Positive Real Number (MB)
X4	MOU	Average MOU in recent 3 months	Positive Real Number (Min)
X5	ISP	Excellent, Good, Average, Bad	{4, 3, 2, 1}
X6	4GUR_TSC	As defined in equation (5)	{0%, ..., 100%}

The 4G user ratio of TOP 10 social circle can be defined as

$$4GUR\_TSC = \frac{\sum_{i=1}^{10} is\_4G(U_i)}{10} \times 100\% \qquad (5)$$

where U_i is the user type of linkmen whose voice-contact frequency is in top ten. The function $is\_4G()$ is defined to judge whether the contact person is a 4G user.

$$is\_4G(U_i) = \begin{cases} 1, & U_i \text{ is 4G user} \\ 0, & U_i \text{ is not 4G user} \end{cases} \qquad (6)$$

4.3 Logistic regression model

Logistic regression and neural network model are proposed respectively to identify the 2G/3G users who are most likely to update 4G network. And binary logistic regression is applied in this paper since the dependent output variable Y can only take two values: Y = 1 indicates potential 4G users and Y = 0 indicates normal users with low priority. The probability of potential 4G user is modeled as logistic distribution in equation (7).

$$P(X) = \frac{\exp(\beta_0 + \beta_1 X_1 + \beta_2 X_2 + \beta_i X_i \ldots + \beta_s X_s)}{1 + \exp(\beta_0 + \beta_1 X_1 + \beta_2 X_2 + \beta_i X_i \ldots + \beta_s X_s)} \qquad (7)$$

The logit of the multiple logistic regression model is given as

$$logit(P(X)) = \ln \frac{P(X)}{1 - P(X)} = \beta_0 + \beta_1 X_1 + \beta_2 X_2 + \beta_i X_i \ldots + \beta_s X_s \qquad (8)$$

where P(X) is conditional probability of potential 4G user. And X_i is the value of the i_{th} independent variable, with β_i as the corresponding coefficient, for i = 1,2,...s. And s is the number of independence variables.

4.4 BP neural network

BP neural network model is also applied to describe the relationship between 4G potential metric and user behaviors of existing 2G/3G users. As in Figure 4, the typical BP neural network model is a full-connected neural network including input layer, hidden layer and output layer.

The input variables are the attributes described in Table 5. If the learning samples are applied to the model, the transferred value will be propagated from the input layer through middle layer to the output layer. The goal of the training process is used to find the weights that minimize some overall error measures. In order to reduce the error between actual output and expected output, connection weights will be adjusted form output layer to every middle layer, and ultimately to input layer. With the ongoing adjustment by the back-propagation, the correct rate for the network response to input also increases continuously.

The node number of input layer depends on the eigenvector of input significant attributes, and we set 6 in this paper. The output layer is the attribute which indicates the probability of customers toward 4G network. And two decision classes are output including decision class 0 (the consumer will not immigrate 4G) and decision class 1 (the consumer will immigrate 4G).

The BP neural network algorithm is described in Table 6. And the mathematical expression of neuron model can be described as:

$$y_j = f(S_j) \tag{9}$$

$$S_j = \sum_{i=0}^{n} w_{ji} x_i \tag{10}$$

where column vector X is input vector, row vector W_j is the connection weight vector for neuron j, S_j represents the input of neurons. And non-linear function $f()$ is the transfer function which must be differentiable and is a bounded S function. And Sigmoid function used in this paper can be defined as:

$$f(x) = \frac{1}{1 + e^{-x}} \tag{11}$$

Fitness function is the indicator to evaluate the ability of neural network, and we use Mean Square Error (MSE) to generate objective function. Considering there are P pairs of patterns, the p_{th} output layer error E_p can be defined as equation (12). The input of j_{th} unit in output layer is:

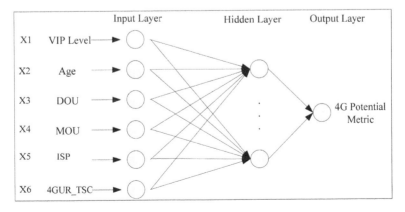

Figure 4. The framework of BP numeral network model.

277

Table 6. The BP neural network algorithm.

1	Initialize weight matrix V and W.
2	Calculate the output of neuron $Z = F[VX], Y = [WZ]$.
3	Given the input sample X_P, calculate accumulative error $E = \frac{1}{2}\sum_{p=1}^{P}\sum_{j=1}^{m}(t_{jp} - y_{jp})^2$.
4	As the definition of equation (14) and (15), obtain error signal σ_Y and σ_Z.
5	Adjust the weight of output layer and hidden layer respectively. $V = V + \eta \delta_Z X^T$ $W = W + \eta \delta_Y Z^T$
6	Return to step 2, add next sample to the network, until P pairs are all calculated then to step 7.
7	If $E < E_m$ (predefined threshold), stop training; otherwise go back to step 2.

$$E_p = \frac{1}{2}\sum_{j=1}^{m}(t_{jp} - y_{jp})^2 \tag{12}$$

$$S_j = \sum_{k=0}^{q} w_{jk} z_k \tag{13}$$

where t_j is the expected output and z_k is the output of hidden layer.

As $f()$ is differentiable, $f'_{yj}(S_j)$ can be obtained as the derivative of transfer function of j_{th} neuron in output layer to its clear input S_j. And the error signal σ_Y and σ_Z can be defined as

$$\delta_{yj} = \frac{\partial E_p}{\partial S_j} = (t_j - y_j)f'_{yj}(S_j) \tag{14}$$

$$W_k^T \delta_y f'_z = \begin{bmatrix} (\sum_{j=1}^{m}\delta_{yj}w_{j1})f'_{z1}(S_1) \\ \dots \\ (\sum_{j=1}^{m}\delta_{yj}w_{jk})f'_{zk}(S_k) \\ \dots \\ (\sum_{j=1}^{m}\delta_{yj}w_{jq})f'_{zq}(S_q) \end{bmatrix} = \begin{bmatrix} \delta_{z1} \\ \dots \\ \delta_{zk} \\ \dots \\ \delta_{zq} \end{bmatrix} = \delta_Z \tag{15}$$

5 EXPERIMENTAL RESULTS

5.1 *Importance of factors*

The dataset applied in this paper was provided by China Unicom in a medium-developed city. The billing data is averaged by continuous 3 months for each user. The dataset under study contains about 35490 users and can be divided into positive samples (19311) and negative samples (16179). The positive samples are those who have immigrated to 4G network from 2G/3G, and negative samples are those who are still 2G/3G users. And both logistic regression and neural network algorithms are applied to the proposed model to mine 4G potential users.

In Figure 5, the results indicate the importance analysis results based on UIM-FDL-LR and UIM-FDL-NN are nearly the same. The top 3 important factors are VIP level, age and DOU, and the significance of VIP level is much larger than others in user immigration. Similar to the pre-analysis results mentioned above, MOU has very limited influence and can only be taken as an assistant factor. However, the service experience should be considered in mining model,

278

as customer perception of 2G/3G network will affect their expectation for 4G network. The weight of 4G user ratio of TOP 10 social circle is about 0.01~0.02, which indicates the users with many relatives and friends using 4G network have relatively high probability to join 4G.

Based on the evaluation results, we can conclude that young VIP users with high data traffic demand are more likely to update 4G network. The service experience and social circle character should be taken into account in mining high quality marketing customers, and MOU can just be taken as a reference variable. More importantly, the conclusions drawn from simulation results can be matched well with 4G user character analysis results in section 3.

5.2 Performance evaluation

In order to estimate the performance of predicting models, evaluation matrix is proposed in Table 7. And three indexes are defined to compare the accuracy of three schemes, including Prediction Accuracy Rate (PAR), Prediction Hitting Rate (PHR) and Prediction Coverage Rate (PCR). The traditional scheme selects the VIP users with DOU larger than 500 MB as potential 4G users in this paper.

The evaluation indexes are defined as

$$PAR = \frac{A + D}{A + B + C + D} \tag{16}$$

$$PHR = \frac{D}{C + D} \tag{17}$$

$$PCR = \frac{D}{B + D} \tag{18}$$

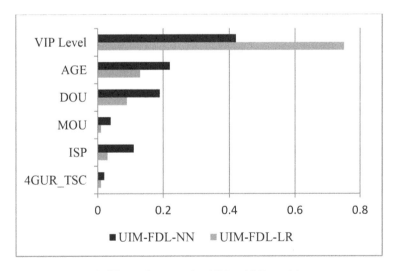

Figure 5.   The importance of different factors using NN and LR model.

Table 7.   The evaluation matrix of predicting model.

	Predict not immigrating	Predict immigrating	Total
Not immigrate actually	A	C	A + C
Immigrate actually	B	D	B + D
Total	A+B	C+D	A + B + C + D

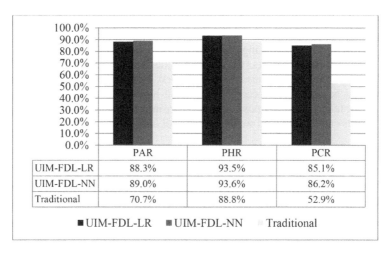

	PAR	PHR	PCR
UIM-FDL-LR	88.3%	93.5%	85.1%
UIM-FDL-NN	89.0%	93.6%	86.2%
Traditional	70.7%	88.8%	52.9%

■ UIM-FDL-LR   ■ UIM-FDL-NN   Traditional

Figure 6.   The comparison of prediction accuracy between different schemes.

As can be seen in Figure 6, the performance of NN and LR outperform the traditional scheme to a great extent, and the prediction accuracy of NN (89.0%) is better than LR (88.3%). The prediction hitting rate is relatively high (about 93.5%), which means the proposed UIM-FDL model can precisely identify the potential 4G users and achieve better marketing benefit. As some unpredictable factors can influence the potential users whether immigrate to 4G, the prediction coverage rate of NN and LR are not very high, with 86.2% and 85.1% respectively.

6   CONCLUSIONS

In this paper, a comprehensive 4G user immigration model named UIM-FDL is proposed to identify the valuable potential 4G customers from existing 2G/3G users. Base on the massive data mining and correlation evaluation, five dimension factors and six key attributes are innovatively considered as the input variable of logical regression and neural network model. The results demonstrate that the UIM-FDL immigration model can achieve greater performance than traditional schemes, and UIM-FDL-NN scheme outperforms the UIM-FDL-LR to some extent. Furthermore, the proposed scheme can help mobile operators to target potential 4G user groups in marketing promotion.

REFERENCES

Fang Xu. 2011. Research on a predictive model of the customer churn. *Proc. International Conference on Wireless Communications, Networking and Mobile Computing, Quintana-Roo, Mexico*: 1–3.
Haibin Shi, & Suyang Huang. 2014. Segmentation of mobile user groups based on traffic us age and mobility patterns. *Proc. International Conference on Computational Science and Engineering, Chengdu, China*: 224–230.
Jin Tao, & Deyong Gao. 2012. Study of logistic growth curve model for mobile user growth. *Proc. International Conference on System Science, Engineering Design and Manufacturing Informatization, Dalian, China:* 188–192.
Tsun-Hao Hsu. 2014. Inferring potential users in mobile social networks. *Proc. International Conference on Data Science and Advanced Analytics, Shanghai, China*: 347–353.
Yuanyuan Li, & Ming Zhang. 2009. Research on network traffic forecasting strategy based on BP neural network. *Proc. International Conference on Computational Intelligence and Software Engineering, Wuhan, China*: 1–4.
Zhu Li, & Qin Lei. 2009. A novel BP neural network model for traffic prediction of next generation. *Proc. International Conference on Natural Computation, Tianjing, China*: 32–38.
Zhuanglin M.A. 2009. Analysis of logistic model for accident severity on urban road environment. *Proc. IEEE Intelligent Vehicles Symposium, Xi'an, China*: 983–987.

Signal and Information Processing, Networking and Computers – Chen & Huang (Eds)
© *2016 Taylor & Francis Group, London, ISBN 978-1-138-02881-4*

Telecom big data based investment strategy of value areas

Mingjun Mu, Yongfeng Wang & Weiwei Chen
Network Technology Research Institute, China United Network Communications Corporation Limited, Beijing, China

ABSTRACT: In order to maintain a stable revenue growth, telecom operators must pay more attention to the investment income ratio, which requires a clear understanding of the characteristics of the investment region. This paper computes the relevant indicators of the investment region based on big data and mines the intrinsic link among the indicators by factor analysis to show the different characteristics of investment region. The results can provide technical support for the telecom operators' investment strategy.

Keywords: value areas; factor analysis; big data

1   INTRODUCTION

With the development of Internet, the mobile Internet business is growing in leaps and bounds which put forward higher requirement for mobile networks. In this context, telecom operators' investment on the network construction is increasing and the competition among operators' network quality is fiercer than ever. While at the same time, the customers' incremental bonus is disappearing, the business increment is slowing down and the social voice for the operators' price cuts is very high. As a result, large and random resource allocation will lead to the continuous decline of the output ratio which will seriously affect the economic benefits of telecom operators and put them at a disadvantage against their competitors in the telecommunications market.

To cope with the fierce competition in the telecommunications market and achieve the maximum investment benefit, operators need to have a clear understanding of their own network state, to accurately analyze the value of the coverage, to replace the blind investment with a more sophisticated fixed investment. The analysis of operators' regional value has become their key work.

The regional value is a comprehensive concept for the operators, which covers the economic value, social value, brand value, potential value and so on. Related indicators for the regional value include regional income, business volume, user number, user level, terminal price and terminal capacity in Qian et al. (2013) and Xiong and Chen (2014). In order to accurately calculate these indicators, it's necessary to deal with IUPS and other data through big data in Ye et al. (2015). Because these indicators cover different fields and have inconsistent data in the same area, researchers usually focus on one or two indicators and ignore others according to the difference of market development and the difference of cover scene in Zhang (2015), this scheme can produce hundreds of categories and can't form a visual understanding of the regional value. Some researchers weight all the indicators and get the comprehensive value of the region by the AHP (Analytic Hierarchy Process) algorithm in Wang et al. (2014), but the shortcoming of this scheme is that the weight of indicators is set subjectively by several experts and the results are lack of objectivity.

In this paper, we collect data as much as possible related to the regional value of the telecom operators, including the data from bill, IUPS, GN, and so on. To calculate the various indicators of regional value, we process big data by the database based on Hadoop architecture. After that, we mine the relationship among the various indicators based on factor analysis to reduce the dimension of indicators. Finally, we will calculate the regional value and propose investment advice.

2 INDICATOR CALCULATION BASED ON BIG DATA

Before the age of big data, telecom operators can only count the business volume of the region limited by data storage and data processing capabilities. The market revenue can only be estimated by multiplying the business volume by the unit price of standard business. There is no way to get the information about users and terminals.

In this paper, we collect all kinds of information of telecom operators, including bill, IUPS, GN, CDR, RNC call trace, MR and KPI. Because operators provide different unit price for different packages, we need the bill to calculate the user consumption accurately in the region which can provide the information of user packages. The information about business volume of the region in the bill cannot meet the requirement of accurate calculation, so we need the data from IUPS and GN to provide information about the time and traffic of data service and the CDR to provide information about the time and traffic of voice service. The RNC call trace and MR can correct the location of core network data. The KPI data can be used to verify the results of our model.

Considering user's mobility, packages, contracts and other factors, six value indicators are calculated through big data's modeling and processing. The involved indicators are shown in Table 1.

3 FACTOR ANALYSIS

3.1 *The principle of factor analysis*

The basic purpose of factor analysis is to use a small number of factors to describe the link among many indictors or factors. Several close variables are grouped in the same class, each type of variable becomes one factor, then most information of the raw data will be reflected with less number of factors. We can easily find out the main factors which affect the regional value and calculate their influence.

Assume that there are m primitive variables $x_1, x_2, \ldots, x_m$, k factors will be get by factor analysis $f_1, f_2, \ldots, f_k$, where k< m. We have that

$$\begin{cases} x_1 = a_{11}f_1 + a_{12}f_2 + \cdots + a_{1k}f_k + \varepsilon_1 \\ x_2 = a_{21}f_1 + a_{22}f_2 + \cdots + a_{2k}f_k + \varepsilon_2 \\ \cdots \\ x_m = a_{m1}f_1 + a_{m2}f_2 + \cdots + a_{mk}f_k + \varepsilon_m \end{cases} \tag{1}$$

Where a_{ij} is the load both on the i_{th} variable and on the j_{th} factor, ε_i is special factor which indicates some parts of primitive variables that cannot be explained by the public factor.

Table 1. Indicators of regional value.

Index name	Explanation
Income	Market revenue provided by region in the statistical period.
Business volume	Voice and data traffic generated in the region in the statistical period.
User number	Average user number serviced by region in unit time.
User level	Distribution of user level in the region.
Terminal price	Average price of terminals serviced by region in unit time.
Terminal capability	Distribution of the highest uplink and download speed of terminals.

We can get the formula

$$X = AF + \varepsilon \tag{2}$$

where

$$A = \begin{pmatrix} a_{11} & \cdots & a_{1k} \\ \vdots & \ddots & \vdots \\ a_{m1} & \cdots & a_{mk} \end{pmatrix} \tag{3}$$

If we can succeed in calculating the load matrix A and controlling ε in a certain range, then we can make the conclusion that public factor F can explain most information of primitive variables X.

3.2 *KMO Test and Bartlett's Test of Sphericity*

Before factor analysis, we must analyze the data of **KMO** (Kaiser-Meyer-Olkin) Test and Bartlett's Test of Sphericity to verify whether the indicators are suitable for the factor analysis or not. **KMO** is a parameter about the sampling suitability. The higher the value of **KMO** measure (close to 1), the more common factors there are among the variables, the more suitable for factor analysis the data is. The calculation method is given by

$$KMO = \frac{\sum\sum_{i \neq j} r_{ij}^2}{\sum\sum_{i \neq j} r_{ij}^2 + \sum\sum_{i \neq j} p_{ij}^2} \tag{4}$$

where r_{ij} is the correlation coefficient of a_{ij}, p_{ij} is the partial correlation coefficient of a_{ij}.

The purpose of the Bartlett's Test of Sphericity is to test whether the correlation matrix is an identity matrix or not. In general, it will indicate that there may be a meaningful relationship between the primitive variables if the value of significant level is less than 0.05.

We calculate the value indicators of more than 14,000 regions covered by wireless network in some city, and test the results by **KMO** and Bartlett's Test. Test results are shown in Table 2.

From the test result, the value of **KMO** Test is 0.737 and the value of significant level is 0.000. In the statistical sense, the results of value indicators are suitable for the factor analysis.

3.3 *Data analysis*

Because there are different dimensions among these indicators, we need to carry out the standardization of indicators before the analysis. In this paper, the normalization method is nondimensiongalization. Assume that z_{ij} is the result of standardization, which can be expressed as

$$z_{ij} = \frac{x_{ij} - \bar{x}_j}{s_j} \tag{5}$$

Table 2. Results of KMO and Bartlett's Test.

Kaiser-Meyer-Olkin Measure of Sampling Adequacy		0.737
Bartlett's Test of Sphericity	Approx.Chi-Square	54490.104
	df	15
	Sig.	0.000

Where x_{ij} is the jth indicator in the i_{th} area, $\bar{x}_j$ is the average of the jth indicator, S_j is the root-mean-square error of j_{th} indicator.

After normalization, we calculate matrix R which is the correlation coefficient matrix of Z and the front n eigenvalues of R arranged from big to small as $\lambda_1 \geq \lambda_2 \geq \cdots \geq \lambda_n$. The screening condition for the front n eigenvalues is that their cumulative variance is close to or more than 80%. The cumulative variance is given by

$$Q = \frac{\sum_{i=1}^{n} \lambda_i}{\sum_{i=1}^{m} \lambda_i} \tag{6}$$

The result is shown in Table 3.

The results show that the sum of the front two eigenvalues' variance is 79.258%, it indicates that the front two factors have retained most information of the indicators, so the front two factors are set as the public factors. When the public factors are determined, their component matrix can be calculated by formula (2). The results are shown in Table 4.

In order to make the public factors easier to explain, we deal with the component matrix of public factors by varimax orthogonal rotation. After three iterations, the component matrix comes to convergence. The results are shown in Table 5.

The final result of factor analysis shows that the two public factors can be used to characterize all the indicators of the regional value. There are three indicators putting their most load on factor 1, income's load is 0.945, user number's load is 0.933, business volume's load is 0.912, so the factor 1 can characterize the network's capacity. There are three indicators putting their most load on factor 2, terminal capability's load is 0.848, terminal price's load is 0.879, user level's load is 0.726, so the factor 2 can characterize the user potential.

Based on the component matrix after rotation, the score matrix of the public factors can be given as

$$\begin{cases} F_1 = w_{11}x_1 + w_{12}x_2 + \cdots + w_{1m}x_m \\ F_2 = w_{21}x_1 + w_{22}x_2 + \cdots + w_{2m}x_m \\ \cdots \\ F_n = w_{n1}x_1 + w_{n2}x_2 + \cdots + w_{nm}x_m \end{cases} \tag{7}$$

Table 3. Results of public factor extraction.

Component	Initial Eigenvalues			Extraction Sums of Squared Loadings		
	Total	% of Variance	Cumulative%	Total	% of Variance	Cumulative%
1	3.195	53.245	53.245	3.195	53.245	53.245
2	1.561	26.012	79.258	1.561	26.012	79.258
3	0.671	11.186	90.443			
4	0.248	4.126	94.570			
5	0.220	3.669	98.238			
6	0.106	1.762	100.000			

Table 4. Component matrix of public factors.

	Component	
	1	2
Income	0.87	-0.402
User number	0.858	-0.399
Business volume	0.82	-0.417
Terminal capability	0.594	0.621
Terminal price	0.692	0.592
User level	0.445	0.575

Table 5. Component matrix after rotation.

	Component	
	1	2
Income	0.945	0.16
User number	0.933	0.156
Business volume	0.912	0.12
Terminal capability	0.139	0.848
Terminal price	0.236	0.879
User level	0.042	0.726

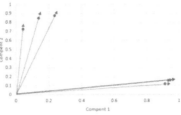

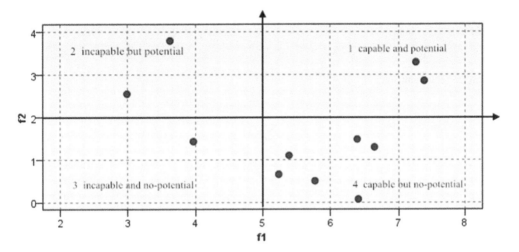

Figure 1. The value of sample areas.

where m is the number of primitive variables, n is the number of public factors after the factor analysis, w_{ij} is the j_{th} indicator's load on the i_{th} factor.

The result in Table 5 shows that the value of a region can be summed up in two public factors. we take some sample areas, calculate their value and show them on the frame of axes in Figure 1.

When the value falls in the first quadrant, it shows that both the network's capacity and the development potential of the region are very high. First of all, we must guarantee the quality of the current network. Secondly, business forecasting and expansion plan should be done to prepare the network's potential needs. When the value falls in the second quadrant, it shows that the region lacks service capacity while it has a very strong development potential. We need to focus on the optimization and problem investigation for the network to avoid the development of the region stopped by the network's problem. When the value falls in the third quadrant, it shows that the network capacity and development potential are weak in these regions. In the case of limited investment, these regions will not appear in the list of concerns. When the value falls in the fourth quadrant, it shows that the region's network capacity is very strong while the development potential is insufficient. What we need to do is just keeping the network's good service.

Based on the analysis results, we can build the value map (Figure 2) to help the telecom operators have a more intuitive understanding of the investment area. In contrast with the previous experience and other value except the telecommunication industry, operators can make the right decision on their investment.

Through the four quadrants analysis, we can know the key work and the investment direction for every region. At the same time, we can use the public factors' variance contribution rate to get the final score given by

285

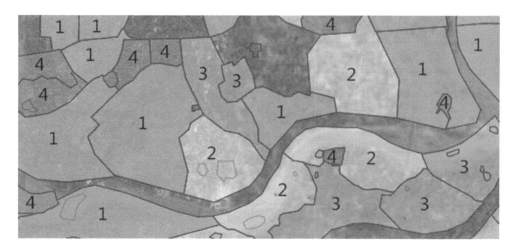

Figure 2. The value map.

$$Y = \left. \sum_{i=1}^{n} \lambda_i F_i \middle/ \sum_{i=1}^{n} \lambda_i \right. \tag{8}$$

When the telecom operators need sort all the region by their comprehensive value considering both the network capacity and the development potential, Y can help them as the objective data.

4 CONCLUSION

In this paper, a large amount of data is extracted from the telecom operators, through modeling and processing based on big data, we accurately calculate various indicators related to the value of region. After mining the intrinsic correlation of these indicators by factor analysis, we succeed in reducing numbers of indicators into two factors. Based on the two factors, we analyze the value characteristics of each region and propose the investment advice. At last, a comprehensive evaluation for all regions is given to have a good understanding of all the city's area.

It should be noted that the analysis results of the regional value only provide data support for the telecom operators' investment strategy. For investment with a strong purpose, it should be based on actual needs and local conditions.

REFERENCES

Jing Zhang. 2015. Research on the investment strategy of the network value region. *China New Telecommunications* 11: 43.

Kezheng Wang, Chuntao Song, & Xinzhou Cheng. 2014. Study of value-oriental region based on AHP. *WMC'14*: 219–223.

Lexi Xu, Yue Chen, KoK Keong Chai, Yuting Luan, & Dantong Liu. 2013. Cooperative Mobility Load Balancing in Relay Cellular Networks. *Proc. IEEE International Conference of Communications in China, Xi'An, P.R.China*: 141–146.

Lexi Xu, Yue Chen, Kok Keong Chai, John Schormans, & Laurie Cuthbert. 2015. Self-Organising Cluster-Based Cooperative Load Balancing in OFDMA Cellular Networks. *Wiley Wireless Communications and Mobile Computing* 15: 1171–1187.

Liling Qian, Chunpeng Song, & Fenghua Li. 2013. Discussion on the construction program of 3G network in valuable region. *Designing Techniques of Posts and Telecommunications* 6: 38–42.

Peigang Qiu. 2012. Analysis of Mobile Communication Network Value Model. *Journal of Beijing University of Posts and Telecommunications (Social Sciences Edition)* 6: 59–65.

Qing Ye, Kezheng Wang, & Meng Ran. 2015. The screening of value-oriental region and application research based on Iu-PS. *Designing Techniques of Posts and Telecommunications* 3: 35–38.

Qiong Huang. 2006. The application of the factor analysis method in the grades evaluation of higher vocational students. *Journal of Hubei Vocational-Technical College* 1: 35–38.

Raoting Zhang. 1999. An introduction to multivariate statistical analysis. Beijing: Science Press.

Shen, Y., C. Jiang, T. Quek, H. Zhang, & Y. Ren. 2015. Pricing Equilibrium for Data Redistribution Market in Wireless Networks. *Prco. IEEE International Conference of Communications, London, UK.*

Wei Xu. 2004. SPSS statistical analysis method and application. Beijing: Electronic Industry Press.

Zeshen Feng, Lan Gu, Qiong Chen, & Jiarong Gao. 2013. Responses of stream habitat factors to land use change. *Bulletin of Soil and Water Conservation* 1: 39–43.

Zheng Li, & Xingwang Guo. 2010. Factor analysis used in pulsed infrared thermographic NTD. *Journal of Beijing University of Aeronautics and Astronautics* 5: 622–626.

Zhuang Xiong, & Yanfen Chen. 2014. Analysis of multidimensional valued region model construction for china Unicom mobile network. *Mobile Communications* 14: 72–78.

Signal and Information Processing, Networking and Computers – Chen & Huang (Eds)
© *2016 Taylor & Francis Group, London, ISBN 978-1-138-02881-4*

A Coverage of Self-Optimization Algorithm using big data analytics in WCDMA cellular networks

Jie Gao, Xinzhou Cheng, Lexi Xu, Lijuan Cao & Kun Chao

Department of Network Optimization and Management, China Unicom Network Technology
Research Institute, Beijing, P.R. China

ABSTRACT: To address the abnormal coverage problems caused by unreasonable parameters settings or antenna issues more effectively, a coverage self-optimization algorithm using big data analytics is proposed. Recently, a series of approaches was developed to improve the performance of coverage for access network. However, because of the diversity deficiency of the data source these simulation approaches have their own limitations in supervised and unsupervised learning. Since big data analytics have become more and more popular in wireless optimization, evaluating and improving the coverage performance turns into a very effective approach by analyzing the huge amount of wireless network measurements and diagnostic data. By obtaining and analyzing the counters (records of the performance indicators of networks) from an existing network, a Coverage Self-Optimization Algorithm (CSOA) is proposed, based on big data analytics. The numerical results show that the coverage optimization process has a low-cost and high-efficiency for mobile operators.

Keywords: UTMS network; Big data; Coverage optimization; Downlink coverage

1 INTRODUCTION

As the rapid development of telegraphy and smart mobile phone, exhibits a significant increase in the complexity of operating cellular networks. In such a competitive market, with the purpose of continual revenue increase, the network operators have to provide high quality of network coverage and excellent traffic experience. To maintain the stability of network and good quality of the service, it will also cost operators a huge amount of Operating Expense (OPEX) and Capital Expenditure (CAPEX) (Lexi Xu et al. 2012) to maintain and optimize our daily work in order to increase the satisfaction and loyalty of the users. Facing a huge amount of network traffic that is generated, how to keep good network performance and quickly respond to the needs of users will be a great challenge for network operators.

With the rapid growth of information, a series of services have sprung up, including voice call, web traffic, download, and video traffic (Youtube, QQlive) etc. As a result, the Operation Support System (OSS) will collect a large amount of wireless network performance measurements and traffic data. These data can be used to improve the network performance, and find the correlative relationship of inhomogeneous data, as signaling and traffic data is the key to form an efficient approach. To achieve that objective, big data technology in O.F.Celebi et al. (2013) can provide nearly real-time solutions for processing a large amount of data, and it is useful for the coverage optimization. In this paper, a Coverage Self-Optimization Algorithm (CSOA) is proposed based on big data analytics, in order to improve the wireless coverage in WCDMA networks.

Current researches on the optimization of network coverage usually summarizes coverage problems into several aspects, such as coverage blind spots, extreme coverage, weak coverage, and mismatching of uplink and downlink coverage (3 GPP 2009). Some strategies in Lexi Xu et al. (2009) and Lexi Xu et al. (2015) can improve the coverage performance via reasonable resource allocation. The problems of downlink coverage usually appear as coverage blind spots and extreme coverage, which may be caused by unreasonable combinations of downtilt and transmission power of node base station (NodeB). Recently Tao Shu et al. (2010) and Ali.K.A et al. (2008) say that adjusting the transmission power is the major method for coverage of optimization. Actually, the transmission power has its own reasonable range, and the maximum value in UMTS (Universal Mobile Telecommunications System) network is 43 dBm. The coverage problem could not be optimized effectively by adjusting the power when the transmission power of NodeB reaches the adjustment limitation. But it can be addressed by adjusting the antenna downtilt effectively. For example, when the antenna downtilt installed is too large, it may result in a narrow vertical beam and weak coverage at the edge of the cell with a high Call Drop Rate (CDR) of downlink performance. On the contrary, if the antenna tilt adjusted is too small, it may cause extreme coverage and impact the donor cell and neighbor cell. For the donor cell, it will reduce the downlink coverage performance because of the high probability of weak Measurement Report (MR). For the neighbor cell, it will increase the interference level of downlink coverage. As a result, because of that extreme coverage could not be held very well by the donor cell, the wireless signal becomes instable and volatile. So call drops happen easily.

This paper is organized as follows. In section II, researches related to coverage optimization are discussed. Section III describes the scheme of proposed algorithm. Section IV displays the numerical results of the effective scheme by analyzing the network data. In the end, section V gives conclusions in this paper.

2 RELEVANT WORK

To solve the downlink coverage problems, many approaches have been proposed in the past. The most common strategy is to find out coverage problems by Drive Test (DT), but it needs a large number of resources and high input cost during the process. In addition, there are many automated schemes. For example, the authors in Fagen D et al. (2008) and Siomina I et al. (2006) propose automated coverage optimization schemes, but these schemes are unable to adapt to the variable and complex networks. As they have limited awareness to huge amounts of traffic data. With the adoption of big data analytics, the proposed algorithm can make the large scale measurement data collected by OSS be fully utilized for decision making.

The typical methods of data mining in big data analytics include neural network, genetic algorithm, decision tree, rough sets, clustering method, and so on. Clustering method is an important method in data mining, which divides the data set into several classes based on the similarity of data. The most important purpose of the clustering method is to find out the defective pixels among the huge amount of data samples. For example, it can be used to locate the problem cell, which impacts the whole coverage performance of the cellular networks, and clustering method is much more effective than the reverse order method.

In this paper, the proposed downlink coverage optimization algorithm is based on the technique of clustering method by collecting the counters such as Propagation Delay (PD), number of call setup success and number of call abnormal release from OSS in-time. In order to enhance the operational efficiency of the strategy, k-means technology (Han Hu et al. 2014) is applied in the proposed algorithm. Running on database, the algorithm can generate an appropriate suggestion for downlink coverage optimization by adjusting a certain tilt automatically until the Euclidean distances of each K sample space converge to a constant. The whole process and the target are accomplished without manual operation.

3 EFFICIENT OPTIMIZATION SCHEME BY USING BIG DATA ANALYTICS

In order to enhance the operational efficiency of the strategy, K-means technology is applied in the following algorithm. K-means technology is one of the most widely used clustering algorithm. It makes each subset mean of all data samples as the representative of each cluster. The main idea of the algorithm is to divide all sample data into different categories by iteration, and achieve the optimal value of the criterion function (E), which evaluates the performance of the clustering. As a result, each cluster becomes independent with each other and compact itself. The criterion function is described as follows:

$$E = \sum_{i=1}^{k} \sum_{p \in X_i} \| p - m_i \|^2 \tag{1a}$$

$$\Rightarrow E = \sum_{i=1}^{k} [\sum_{j=1}^{n_j} (p_j - m_j)^2]^{\frac{1}{2}} \tag{1b}$$

$$\Rightarrow E = \sum_{i=1}^{k} [(p_1 - m_j)^2 + (p_2 - m_j)^2 + ...(p_{nj} - m_j)^2]^{\frac{1}{2}} \tag{1c}$$

where X contains k cluster subsets, such as X_1, X_2...X_k and m_i represents the center for each cluster, n_i represents the sample number of each cluster subset.

The K-means algorithm is illustrated in Table 1:

The complexity of k-means is $O(n \times k \times t)$, and t is the time of iteration. As the k-means algorithm is sensitive with K and initial center, it is important to determine the initial center point of each cluster and K. The result will be susceptive with the existing extreme or isolated data points, so Data Cleaning (DC) work is necessary before the process of k-means.

Figure 1 shows the process of CSOA and the process can be divided into four steps: collection, calculation, clustering analysis, and optimization.

Step 1: Collection
In the process of collection, the collector will collect counters and engineering parameter from OSS periodically. Usually, the period could be set as 30 minutes, because the counters will be reported every 15 minutes from NodeB/RNC to OSS. Tables 2–4 describe the real-time data collected from OSS, such as engineering parameters, PD counters and handover counters. The engineering parameters include some static data, such as longitude and latitude. The generation procedure of PD counters is that NodeB will receive a preamble signal from UE before

Table 1.   K-means algorithm of big data analytics.

INPUT: the number of clusters K, and database contains n data sample.
OUTPUT: K cluster subsets.
1. To determine an initial cluster center for each cluster, the K initial clustering center is obtained;
2. Assign the data samples in the database to the nearest neighbor clustering according to the
   minimum distance principle;
3. Using the sample mean of each of the cluster as the new clustering center;
For i from 1 to n, repeat the procedure 2 and 3 until the Euclidean distances of each K sample space converge to constant.

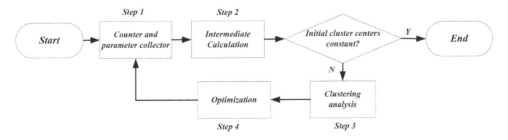

Figure 1.   The flowchart of CSOA.

291

Table 2. Engineering parameters.

City ID	RNC ID	Cell ID	Longitude	Latitude	Downtilt	Ncell ID
101	RNC3 A	2945B1	23.41	116.32	7°	2946 A1

Table 3. PD counters.

RNC ID	Cell ID	Date	Time	PD [1]	PD [2]	...... PD [40]
RNC3 A	2945B1	2015/6/1	8:00:00–8:30:00	10000	680	200

Table 4. Performance counters.

RNC ID	Cell ID	Ncell ID	Intra-frequency handover request	Intra-frequency handover request	RAB abnormal release	RAB normal release
RNC3 A	2945B1	2946 A1	50	48	3	47

any calls are made and measure the propagation delay according to the preamble known. Then NodeB calculates the distance between NodeB and UE by referring to the wireless propagation model. At last, NodeB will record the result into PD counters. The PD counters is a set of data, from PD [0] to PD [40]. The recorded values occur in a particular range of distances ([n%, (n+1)%]* CellRange), and the corresponding PD[n] will be added once. The CellRange represents the maximum rage that UE can access into the NodeB, default value is 3.5 km. The propagation distance of wireless signal can be expressed as (2):

$$D = (velocity\_of\_light * number\_of\_delay) / chip\_rate \qquad (2)$$

where CellRange is valued 3.5 km, the maximum number of delay is $(350003.84* 10^6)/3*10^8 = 450$, so that PD [1] records the times of RRC connection request from UEs, which have a distance between 0~350 m to the NodeB, and the other counters can be so on.

Step 2: Calculation

In the calculation step, several intermediate numbers need to be calculated for the clustering analysis step and all data samples have been ready in step 1. As mentioned above, coverage problems have two aspects: extreme coverage and weak coverage, and we assume the impact factors of them are ε and ω, which also meet the constraint of $\varepsilon+\omega = 1$.

A. Reasonable Coverage Radius (RCR)

Before the coverage optimization, the reasonable coverage distance is needed for judging whether the cell has extreme or weak coverage problem. In order to combine user's perception model and network static structure, we will calculate a weighted value for reasonable coverage radius (RCR) and set weight coefficient for the two aspects as α and β, also $\alpha+\beta = 1$, and as default, we set both of them as 0.5. The user's perception model radius is expressed as (3):

$$R_{user} = \frac{\sum_{i=1}^{n} D_i \times N_i}{\sum_{i=1}^{n} N_i} \qquad (3)$$

The static cell coverage can be expressed as (4):

292

$$R_{static} = \frac{2}{3} \times \frac{\sum_{i=1}^{n} D_i}{n} \tag{4}$$

where D_i represents the static site distance between each neighbor cell, and N_i represents the number of intra-frequency handover succeed in Table 4. Figure 2 shows the structure of cellular networks, and it is noted that the coverage radius is 2/3 site distance. Above all, we can get the RCR for each cell as (5):

$$R_{RCR} = \frac{\alpha \times R_{user} + \beta \times R_{static}}{\alpha + \beta} \tag{5}$$

B. Actual Coverage Radius (ACR)

After calculating the DCR, we need to evaluate the Actual Coverage Radius (ACR) of each cell. As PD records the different distance interval from UEs every 15 min as the format Table 2, we assume that 95% percent of the PD can represent the ACR of each cell, and ACR can be expressed as (6):

$$R_{ACR} = \frac{\sum_{i=1}^{40} (\frac{i\% \times CellRange}{2} \times N_i)}{\sum_{i=1}^{40} N_i} \times 95\% \tag{6}$$

where N_i is the number of records in PD[i], and CellRange is set 3.5 km as default value by equipment manufacturer.

C. Call Drop Rate (CDR)

We also need a combined network performance and service awareness to evaluate the reasonability of coverage, and CDR is the most direct and correlative index with the downlink coverage, which could be worse in the extreme or weak coverage environment. CDR can be described as (7):

$$CDR = \frac{rab\_abnor\_rel}{rab\_abnor\_rel + rab\_nor\_rel} \times 100\% \tag{7}$$

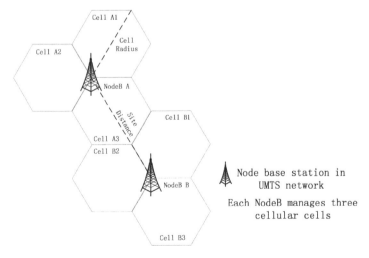

Figure 2.   The structure of cellular networks.

D. K and Initial Center

Clustering analysis needs K and initial center for each cluster to generate K cluster sets by iterative calculation. Because of that, we assume two aspects: coverage distance and service performance could duplicate coverage performance as much as possible. Then we set the structure of several cluster sets as M_n (CDR_n, D_{ACRn}/D_{RCRn}), where D_{ACR}/D_{RCR} can reflect the downlink coverage status as a quantitative index. And then, we set 3 as a default value for K. Also, we define an index P to represent the efficiency of CSOA, and the centers will be influenced by P at the end of CSOA. C1 is defined to represent the extreme coverage cluster, and C3 is weak coverage cluster and C2 is the normal coverage cluster. As a result, we can obtain the initial center for each cluster as follows. Figure 3 shows the structure of clustering analysis.

$$\begin{cases} C_1 & initialize: \{\overline{CDR} \times (1+P), \quad \frac{\overline{D_{ACR}}}{D_{RCR}} \times (1+P)\} \\ C_2 & initialize: \{\overline{CDR} \times (1-P), \quad \frac{\overline{D_{ACR}}}{D_{RCR}} \} \\ C3 & initialize: \{\overline{CDR} \times (1+P), \quad \frac{\overline{D_{ACR}}}{D_{RCR}} \times (1-P)\} \end{cases} \tag{8}$$

Step 3: Clustering analysis
In step 3, we will utilize the k-means algorithm as illustrated in table 5 to generate three cluster sets which represent for different coverage performance.

$$d(M_n, C_k) = \sqrt{(M_{nx} - C_{kx})^2 + (M_{ny} - C_{ky})^2} \tag{9}$$

where M_n is a sample of the database, and C_k is initial center of the kth cluster, and n is the number of database.

Step 4: Optimization
As mentioned above, we assume that the impact factor of extreme and weak coverage are ε and ω, which follow the constraint of $\varepsilon + \omega = 1$. So here, adaptive downtilt optimization (ADO) is defined for the step 4 to implement optimization for downlink coverage. For cluster 1–3, ADO is defined as follows:

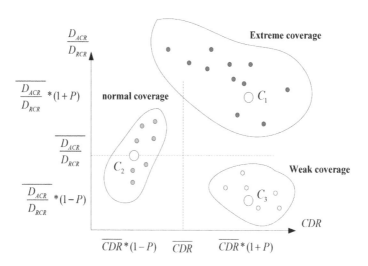

Figure 3.   The structure of clustering analysis.

Table 5. K-means algorithm of CSOA.

INPUT: K = 3, and database contains n data samples.
OUTPUT: three cluster subsets.
1. Determine each cluster center as (8);
2. Assign the data samples in the database to the nearest neighbor clustering according to
   the minimum distance principle by Euclidean distance model as (9);
3. Using the sample mean of each cluster as the new clustering center;
4. For i from 1 to n, repeat the procedure 2 and 3 until the Euclidean distances of each K samples
   space converge to a constant.

Table 6. The RSCP distribution comparison.

RSCP range	Before	After
>= -65dBm	34.06%	35.46%
[-75dBm,-65dBm)	16.27%	18.53%
[-85dBm,-75dBm)	20.82%	25.81%
[-95dBm,-85dBm)	14.93%	14.39%
[-100dBm,-95dBm)	8.24%	3.86%
-100dBm	5.68%	1.95%

$$\begin{cases} \theta_{t+1} = \theta_t \times (1+\varepsilon) & M_n \in C_1 \\ \theta_{t+1} = \theta_t & M_n \in C_2 \\ \theta_{t+1} = \theta_t \times \omega & M_n \in C_3 \end{cases} \qquad (10)$$

where M_n is a sample of the database, and θ_{t+1} is the downtilt of M_n after optimization.

After optimization, the process will return back to step 1 and this cyclic process will be finished until the center of each cluster set becomes constant.

4 NUMERICAL RESULT

In this paper, the case study of the wireless network is an urban area of China Unicom WCDMA network. The performance of the CSOA based on big data analytics is verified and the measurement data is processed by Modeler software. In the following numerical results, by comparing the probability distribution of RSCP (the downlink coverage indicator) and call drop rate of the whole network after CSOA, the results prove that the efficient coverage optimization scheme have effects on the coverage performance improvement for cellular networks. In addition, the efficiency of the CSOA is discussed in different distribution characteristics of the data.

The distribution range of the RSCP is illustrated in Table 6 and Figure 4, and the CDR distribution comparison is illustrated in Figure 5.

As illustrated in Table 6 and Figure 4, the distribution of RSCP has more than 5% below −100 dBm and doesn't satisfy the objective reference value of coverage before optimization. In this wireless environment, the wireless signal becomes instable and volatile. As a result, call drops will happen to the mobile users very easily. After CSOA, however, only 1.95% of the MR samples are below −100 dBm, and nearly 50% of the MR samples are above −85dBm. It is obvious that the coverage has a remarkable improvement and the total coverage level can reach an excellent level of coverage.

Besides from the coverage improvement, the user's experience can get a performance improvement as the call drop rate below 5% increases from 13.6% up to 58.4% and improves to 44.8%. Not only the cellular cells of high quality have performance improvement, but also the maximum call drop rate decreases from 25% down to 15%. Because CSOA takes coverage and key performance into consideration, the numerical results can demonstrate the proposed coverage optimization scheme has the capacity to address downlink coverage problem in cellular networks.

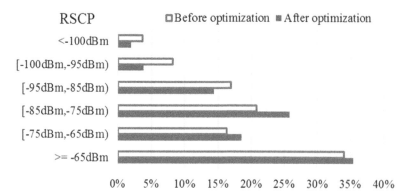

Figure 4.   Ratio of user number.

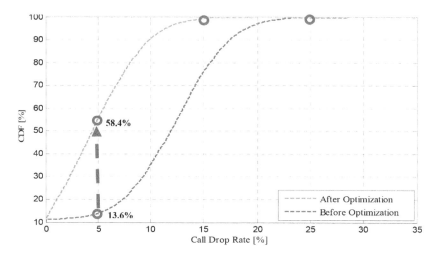

Figure 5.   The call drop rate distribution before & after optimization.

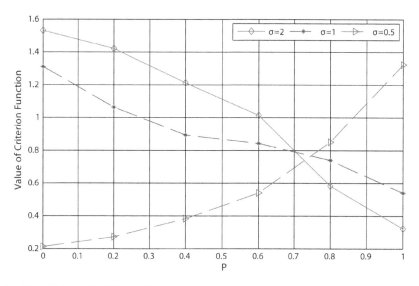

Figure 6.   The efficiency in different sigma scenarios.

296

The three networks in Figure 6 have the same $\overline{\frac{D_{ACR}}{D_{RCR}}}$, and $\overline{\frac{D_{ACR}}{D_{RCR}}} = 1$. Figure 6 shows the different efficiency in different sigma scenarios, as the k-means algorithm is strongly correlated with the distribution of data samples and initial centers. For example, when $\sigma = 2$, the value of the criterion function will go smaller as the value of P increases, because initial centers will get closer to data samples and $\sigma = 0.5$ is an opposite example of $\sigma = 2$. The scenario of $\sigma = 1$ is a little special, and the trend will become slower compared to $\sigma = 2$, as the distribution of data samples is more convergent to $\frac{D_{ACR}}{D_{RCR}}$. According to the result mentioned above, we can come to the conclusion that for different sigma scenarios (different cities or regions), it is necessary for network operators to set a reasonable P associated with σ of the data sample to achieve the best optimization consequence.

5 CONCLUSION

In this paper, an efficient coverage of optimization scheme using big data analytics is proposed for cellular networks. The proposed scheme can improve the coverage performance by locating and solving the problem cells with unreasonable antenna parameter settings without human intervention. Also, the proposed algorithm includes both coverage performance (network) and service performance (user) as analytical factors. It is more comprehensive than the algorithm in Fagen D et al. (2008), which considers coverage performance (overlap) as the only factor. We provide the consequence of CSOA as significant by practicing with the network data of China Unicom. The practice shows that the proposed scheme can solve the extreme and weak coverage problems effectively.

Although the analysis case in this paper is WCDMA networks, CSOA can also be applicable to a wide variety of wireless technologies, including other UMTS, WLANs and LTE networks, as the antenna propagation model is independent with the physical layer technology of the network system.

REFERENCES

3GPP R3–091927-SON. 2009. *Solution for coverage and capacity optimization.*

Ali K A, Hassanein H, & Mouftah H T. 2008. Power-controlled rate and coverage adaptation for WCDMA cellular networks. *Proc. IEEE Symposium on Computers and Communications*: 194–200.

Celebi, O.F., E. Zeydan, O.F. Kurt, O.F. Kurt, O.F. Kurt, & B. AykutSungur. 2013. On use of big data for enhancing network coverage analysis. *Proc. International Conference on Telecommunications, Casablanca*: 1–5.

Fagen D, Vicharelli P A, & Weitzen J. 2008. Automated wireless coverage optimization with controlled overlap. *IEEE Transactions on Vehicular Technology* 57(4): 2395–2403.

Han Hu, Yonggang Wen, Tat-Seng Chua, & Xuelong Li. 2014. Toward scalable systems for big data analytics: a technology tutorial. *IEEE Access* 21(2): 652–687.

Lexi Xu, & Yue Chen. 2009, Priority-based resource allocation to guarantee handover and mitigate interference for OFDMA systems. *Proc. IEEE International Symposium on Personal Indoor and Mobile Radio Communications, Tokyo, Japan*: 783–787.

Lexi Xu, Xinzhou Cheng, Yue Chen, Kun Chao, Dantong Liu, & Huanlai Xing. 2015. Self-optimised coordinated traffic shifting scheme for LTE cellular systems. *Proc. EAI International Conference on Self-Organizing Networks, Beijing, P.R.China*: 1–9.

Lexi Xu, Yue Chen, KoK Keong Chai, Tiankui Zhang, & John Schormans. 2012. Cooperative load balancing for OFDMA cellular networks. *Proc. European Wireless, Poznan, Poland*: 1–7.

Lexi Xu, Yue Chen, Kok Keong Chai, John Schormans, & Laurie Cuthbert. 2015. Self-organising cluster-based cooperative load balancing in OFDMA cellular networks. *Wiley Wireless Communications and Mobile Computing* 15(7): 1171–1187.

Siomina I, Varbrand P, & Yuan D. 2006. Automated optimization of service coverage and base station antenna configuration in UMTS networks. *IEEE Wireless Communications* 13(6): 16–25.

Tao Shu, & Krunz M. 2010. Coverage-time optimization for clustered wireless sensor networks: a power-balancing. *Proc. IEEE/ACM Transactions on Networking* 18(1): 202–215.

Signal and Information Processing, Networking and Computers – Chen & Huang (Eds)
© *2016 Taylor & Francis Group, London, ISBN 978-1-138-02881-4*

A user perception based value-added service strategy for 4G mobile networks

Lijuan Cao, Yinhe Zhou, Xinzhou Cheng, Kun Chao & Mingqiang Yuan
Department of Network Optimization and Management, China Unicom Network Technology
Research Institute, Beijing, P.R. China

ABSTRACT: The rapidly growing smartphones penetration, together with the fast development of Internet applications as well as the 4th Generation communication system accelerates the emergence of Big Data Era. The data growth has fostered interest of telecom operators in exploiting rich telecom data for value-added service while delivering positive user experience. In this paper, a novel value-added service strategy is proposed for 4G mobile communication, in which the dedicated bearer could be adaptively created based on user perception of the specific APP to provide guaranteed Quality of Service (QoS). Emphasis is placed on the criterion function by which user perception is evaluated and dedicated bearer is activated. The main objective of this service is to address the user satisfaction problem of specific APP, as well as make profits for both the APP providers and the operators. The experimental results reveal that the proposed strategy is reliable and can achieve good performance in improving user perception.

Keywords: Big Data; value-added service; dedicated bearer; QoS; user perception

1 INTRODUCTION

Big Data, which refers to datasets whose size is beyond the ability of typical database software tools to capture, store, manage and analyze, creates vast opportunities to transform raw data into knowledge, service and revenue potential. To counter these trends, telecom operators need new ways of monetizing mobile broadband data while increasing profits in the 4G markets by means value-added service (Nwanga, M.E. et al. (2015), Agarwal, V. et al. (2012), Šerval, D. et al. (2014) and Lexi Xu et al. (2015)). Traditional data service strategy provided by the telecom providers is to set a serial of volume-based data packages for all the end users, which brings operators limited profits due to their "data pipes" role. However, many operators have now realized that in order to better monetize their mobile broadband investments, they need to exploit much more sophisticated value-added service that effectively improve the customer satisfaction.

It should be noted that telecom operators have access to some important data sources which are not available to their competitors like mobile device manufacturers and Internet content providers, such as user information, user bill, real-time location information, Call Detail Record (CDR), resource usage and Key Performance Indicator (KPI) of network, data of S1 interface and so on. These data sources allow operators to create more value-added service to achieve incremental data revenues and provide better service.

Some deployed value-added data services (Kimbler, K. et al. (2012), Xiuli Yao et al. (2009), Salem, A.M. et al. (2013), Cheng Zhenyu et al. (2011), Smailovic, V. et al. (2013), Liu Yan et al. (2014), Hermet, G. et al. (2011) and Ekstrom, H.et al. (2009)) like Sponsored data (The content/App service provider pays the data charge instead of the end user, which makes extra revenue for the operators, like AT&T or Verizon), Time based service (It allows the operator to set price to period according to the resource usage of the network and allows subscribers to choose lower priced plans based on off-peak hours) and Shared wallet service (The family/group can share a data package and gets a single bill for the service) have been proved notable success among the world.

From the mentioned services above, it can be concluded that value-added service is now gradually becoming a new area for service innovation. However, to our knowledge, little attention has been devoted to priority service for specific applications or URLs. For this service, two problems have yet to be addressed are the means and criteria to guarantee the service priority. In this contribution, we propose an App priority service by means of setting dedicated bearer' QoS. And the emphasis of the strategy is placed on the criterion by which user perception is evaluated to activate dedicated bearer.

The remainder of this paper is organized as follows. Section 2 briefly presents the concept of dedicated bearer as well as the QoS control mechanisms in 4G Evolved Packet System (EPS). Section 3 describes the user perception based value-added service strategy proposed in this paper. In section 4, experimental results are presented in order to demonstrate the validity of our strategy. Section 5 gives the conclusion and future work of this strategy.

2 QOS CONTROL IN EVOLVED PACKET SYSTEM

The "bearer" is a central element of the EPS QoS concept and is the level of granularity for bearer-level QoS control. All packet flows mapped to the same bearer receive the same packet-forwarding treatment. The network-initiated QoS control paradigm specified in EPS is a set of signaling procedures for managing bearers and controlling their QoS assigned by the network. A bearer is either a default or a dedicated bearer. The default bearer is the bearer that is set up when the terminal attaches to the network. One default bearer exists per terminal IP address, and it is kept for as long as the terminal retains that IP address. The QoS level of the default bearer is assigned based on subscription data.

To provide different QoS in the network to two different packet flows for the same IP address of a terminal, one or more dedicated bearers are required. The dedicated bearer can be either a non—Guaranteed Bit Rate (non-GBR) or a GBR bearer. The operator can control which packet flows are mapped onto the dedicated bearer, as well as the QoS level of the dedicated bearer through policy and charging resource function (PCRF) or PDN GateWay(PGW).

In 3GPP TS. 23.203, Qos parameters for dedicated bearer are: QoS Class Identifier (QCI), Allocation and Retention Priority (ARP), Maximum Bit Rates (MBR), Guaranteed Bit Rate (GBR), as shown in Table 1.

Each standardized QCI is associated with standardized QCI characteristics. The characteristics describe the packet-forwarding treatment that the bearer traffic receives edge-to-edge between the terminal and the gateway in terms of bearer type (GBR or non-GBR), priority, packet delay budget, and packet-error-loss rate.

Table 1. Standardized QCI characteristics.

QCI	Resource Type	Priority Level	Packet Delay Budget	Packet Error Loss Rate	Example Services
1	GBR	2	100 ms	10^{-2}	Conversational Voice
2	GBR	4	150 ms	10^{-3}	Conversational Video (Live Streaming)
3	GBR	3	50 ms	10^{-3}	Real Time Gaming
4	GBR	5	300 ms	10^{-6}	Non-Conversational Video (Buffered Streaming)
5	Non-GBR	1	100 ms	10^{-6}	IMS Signalling
6	Non-GBR	6	300 ms	10^{-6}	Video (Buffered Streaming); TCP-based (e.g., www, e-mail, chat, ftp, p2p file sharing, progressive video, etc.)
7	Non-GBR	7	100 ms	10^{-3}	Voice; Video (Live Streaming); Interactive Gaming
8	Non-GBR	8	300 ms	10^{-6}	Video (Buffered Streaming);
9	Non-GBR	9	300 ms	10^{-6}	TCP-based (e.g., www, e-mail, chat, ftp, p2p file sharing; progressive video, etc.)

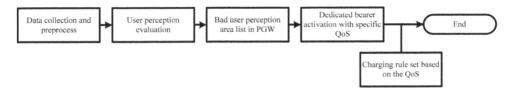

Figure 1. The executing process of the proposed strategy.

The ARP enables the EPS system to differentiate the control-plane treatment related to set-ting up and retaining bearers. That is, the ARP is used to decide whether a bearer establishment or modification request can be accepted or must be rejected due to resource limitations. In addi-tion, the ARP can be used to decide which bearer to release during exceptional resource limita-tions. The range of the ARP priority level is 1 to 15 with 1 as the highest level of priority.

The MBR and GBR are defined only for GBR bearers. These parameters define the MBR, that is, the bit rate that the traffic on the bearer may not exceed, and the GBR, that is, the bit rate that the network guarantees it can sustain for that bearer.

Based on the four parameters mentioned above, the service utilizing dedicated bearer can provide a guaranteed QoS, like delay, bandwidth and bearer priority.

3 USER PERCEPTION BASED APP PRIORITY SERVICE

The user experience of IP based services are dependent on bandwidth, latency and packet loss rate of the network. As Transmission Control Protocol (TCP) is the dominant transport layer protocol for the current IP networks and is used for 90% of the total traffic, the per-formance of TCP is vital to the user experience for IP services (Lexi Xu et al. (2015)). And the performance of Application Protocol, take Hyper Text Transfer Protocol (HTTP) as an example in this article, also contributes to the user perception on respond latency, bit rate and so on. Thus, we analyze the user perception of service using parsed TCP and HTTP packet data. Figure 1 shows the executing process of the proposed strategy (Lexi Xu et al. (2013)).

3.1 Data collection and preprocess

Collect the S1-U data for **D** days during 24-hour period, which contains traffic flow of vari-ous application generated by the mobile devices. Then filter out the traffic data of specific APP by means of Deep Packet Inspection (DPI), as the proposed strategy is based on the cooperation of specific APP provider and telecom operator. As three procedures (TCP link set up, TCP session and HTTP session) are considered in this section, Table 2 shows the key numeric columns which affect the user experience from three respective tables in the database after preprocess and data entry. And each record in the specific table represents a respective procedure, which means a TCP link set up, a TCP session and a HTTP session.

With these data resource, some important performance indicators can be calculated using the formulas below. The average success ratio of TCP link set up and HTTP session can be obtained using formula (1) and (2):

$$s_T = sum(TCP\_SUCCESS) / sum(SYN\_ATMPTS) \tag{1}$$

$$s_H = sum(HTTP\_SUCCESS) / sum(TOTAL\_ATMPTS) \tag{2}$$

The average delay for TCP link set up and the first successful HTTP packet can be reached by calculation following formula (3) and (4):

$$\sigma_T = sum(SUCCESS\_DELAY) / count(TCP\_SUCCESS = 1) \tag{3}$$

Table 2. Key numeric columns in the database.

Table	Column name	Explanation
TCP link set up	SYN_ATMPTS	Total SYN attempts from the user to the server during the TCP link set up
	TCP_SUCCESS	1 for TCP link set up success, while 0 for failure
	SUCCESS_DELAY	Delay of successful link set up, while NULL for failure
TCP session	TOTAL_LENGTH	Total length of packets transmitted during the TCP session, including retransmitted packets
	TOTAL_TIME	TCP session time, excluding link set up and unlinked period
	IDLE_TIME	Idle time during the TCP session
	RETRANS_LENGTH	Total length of retransmitted packets in the TCP session
HTTP session	TOTAL_ATMPTS	Total transmitted packets during the HTTP session, including the retransmitted packets
	HTTP_SUCCESS	Total successfully transmitted packets in the HTTP session
	FIRST_PACKET_DELAY	Delay of the first packet, NULL for failure
	FIRST_PACKET_STATUS	Transmission status of the first packet, 1 for success while 0 for failure

$$\sigma_H = sum(FIRST\_PACKET\_DELAY) / count(FIRST\_PACKET\_STATUS = 1) \quad (4)$$

And the average bit rate for the TCP session is calculated using formula (5):

$$v_T = average\left(\frac{TOTAL\_LENGTH - RETRANS\_LENGTH}{TOTAL\_TIME - IDLE\_TIME}\right) \quad (5)$$

3.2 User perception evaluation

Based on section 3.1, we can use other columns in these three tables to get the user perception. In this section, we define $E_{b,h,d,u}$ as the user perception for user 'u' in hour 'h' and day 'd', where 'b' indicate the e-NodeB id:

$$E_{b,h,d,u} = (s_{b,h,d,u,H} \times s_{b,h,d,u,T} \times 100) \times e^{-\frac{\sigma_{b,h,d,u,H} + \sigma_{b,h,d,u,I}}{\sigma_0}} \times E(v)_{b,h,d,u} \quad (6)$$

As $E(v)_{b,h,d,u}$ indicates the user perception effected by the bit rate in TCP session:

$$E(v)_{b,h,d,u} = \begin{cases} 1, & , v_{b,h,d,u,T} \geq V_0 \\ e^{1-\frac{v_{b,h,d,u,T}}{V_0}}, & , v_{b,h,d,u,T} < V_0 \end{cases} \quad (7)$$

σ_0 and V_0 are the reference delay and bit rate for the specific APP.

3.3 Bad user perception area list in PGW

We define E_0 as the threshold to indicate the acceptable user perception of analyzed APP. Then the bad user perception proportion for eNB 'b' and time 'h' can be calculated using formula (8):

$$Proportion_{b,h} = \sum_{d=1}^{D}\sum_{u=1}^{U_{b,h,d}} count(E_{b,h,d,u} < E_0) / \sum_{d=1}^{D}\sum_{u=1}^{U_{b,h,d}} count(E_{b,h,d,u} >= 0) \quad (8)$$

Based on bad user perception proportion, the indicator used to activate the dedicated bearer is measured as formula (9), where 1 indicates activating the dedicated bearer while 0 indicates not activating.

$$Indicator_{b,h} = \begin{cases} 0, & Proportion_{b,h} \geq P_{b,h} \\ 1, & Proportion_{b,h} < P_{b,h} \end{cases} \tag{9}$$

$P_{b,h}$ is the threshold of the bad user perception proportion for eNB 'b' and time 'h'.

$$P_{b,h} = \begin{cases} a, & U_{b,h} \in [0, 0.5 \times U_h] \\ b, & U_{b,h} \in (0.5 \times U_h, U_h] \\ c, & U_{b,h} > U_h \end{cases} \tag{10}$$

$U_{b,h}$ indicates the average APP user number for 'b' and 'h' each day, and U_h is the average APP user number for hour 'h'. (a,b,c) are the threshold that set for different load scenario of eNB. Save the four-tuple $(APP, b, h, indicator)$ in PGW, based on which, the operator can control when and which packet flows are mapped onto the dedicated bearer.

3.4 Dedicated bearer activation in PGW

Most PGW is equipped with Packet Inspection technology and Service Rule Matching technology, which are the two most vital technologies in content billing. The PGW identifys the specific traffic flow by means of Packet Inspection technology and creates dedicated bearer. Mainstream methods of this technology are SPI (Shallow Packet Inspection) and SA (Service Awareness). By the result of packet inspection, the PGW matches the three-tuple $(APP, eNB, hour)$ with the bad user perception area list and get the '$indicator$' value. If the indicator is 1, PGW execute the predefined QoS set, charging policy and action strategy to create dedicated bearer, as shown in Figure 2.

3.5 QoS and charging rule set

As mentioned above, QCI, ARP, GBR are the most important parameters to guarantee the QoS of the dedicated bearer, like latency, bandwidth, and packet error loss rate. The QoS

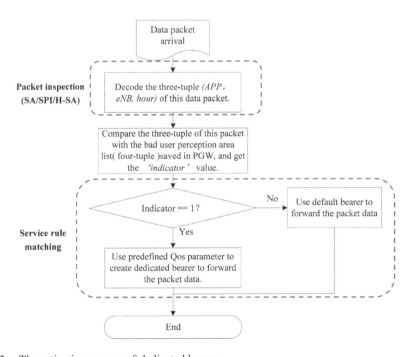

Figure 2. The activation process of dedicated bearer.

Table 3. QoS set and price rule for the dedicated bearer.

Number	(QCI, ARP, GBR)	Unit price per MB (*Pc*)
1	(QCI1, ARP1, GBR1)	P1
2	(QCI1, ARP1, GBR2)	P2
3	(QCI1, ARP2, GBR1)	P3
4	(QCI1, ARP2, GBR2)	P4
5	(QCI2, ARP1, GBR1)	P5
6	(QCI2, ARP1, GBR2)	P6
7	(QCI2, ARP2, GBR1)	P7
8	(QCI2, ARP2, GBR2)	P8

design needs negotiation of operator and APP provider, because QoS with low priority could not achieve the user experience improvements, while QoS with too high priority will cause the resource waste and resource load to the whole network. The telecom operator provides a serials of QoS set for the APP provider, and set unit price per MB '*Pc*' for the bill based on the QoS priority. So that the APP provider could choose a reasonable option based on their profit and pay monthly for the traffic generated by the user experiencing the APP.

$$Bill = P_c \times Traffic \qquad (11)$$

It should be noted that our goal in this article is to define a basic function, so to simply the performing we define QCI, ARP and GBR in a provided way in Table 3. With regard to implementation, APP-designed QoS should be set when different characteristics are considered by an application.

4   EXPERIMENTAL RESULTS

This section shows the experiential results that our strategy could achieve. It should be noted that as we couldn't value the business profit generated by this strategy for the telecom operators or the APP providers, the results mainly demonstrate the improvement of user perception. The dataset in this section contains 2.8 millions of data of S1-U interface generated by mobile devices in city **S** for application **H** during **D** =7 days period (6/1/2015–6/7/2015).

With regard to implementation, different results can be expected when different reference parameters (σ_0, V_0 and P_0) are considered by application **H**. We define $\sigma_0 = 100\,ms, V_0 = 1M$ based on the characteristic of the APP **H**. Figure 3 depicts the comparison of bad user perception cumulative distribution function between the scenarios of not using the strategy, and exploit the different QoS set for the dedicated bearer for the strategy at 20:00–21:00 period.

We define (a,b,c,E_0) = (80%,60%,40%,80) to denote the percentage criterion and the reference user perception. It should be noted that the performance gap between not-using and the proposed strategy is obvious: the proposed strategy could achieve much less bad user perception proportion, which can be explained that the dedicated bearer could guarantee the APP users the bandwidth, delay constraint, and the packet drop constraint. It also can be concluded that the stricter QoS set is exploited, the better user experience could be achieved, as shown by the bad user perception proportion comparison of the scenarios (QCI,ARP,GBR) = (2,5,1M), (QCI,ARP,GBR) = (3,5,1M) and (QCI,ARP,GBR) = (4,5,1M) during 6/8/2015–6/28/2015.

Additionally, for defined QoS set, various percentage criterions and the reference user perception will lead to different results. In Figure 4 we set (QCI,ARP,GBR) = (3,5,1M) to show the comparison of bad user perception cumulative distribution function of various the percentage criterions during 6/29/2015–7/19/2015. It seems that, the user experience will improve while the percentage criterion is more loose. Reasonable explanations for this phenomenon

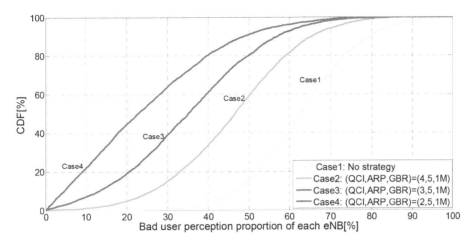

Figure 3. Bad user perception proportion comparison of different QoS sets.

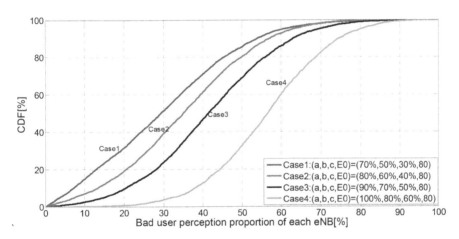

Figure 4. Bad user perception proportion comparison of different percentage criterion.

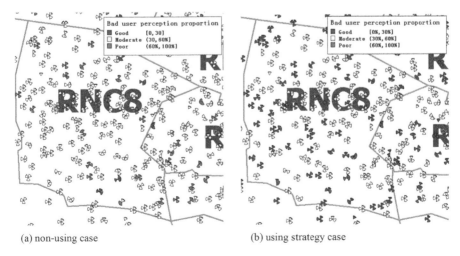

(a) non-using case   (b) using strategy case

Figure 5. Bad user perception proportion comparison of RNC8.

are that as the percentage criterion decrease, more user will have the access to dedicated bearer for this APP, and as a result the bad user perception proportion will also decline.

Figure 5(b) demonstrates the user experience improvement of RNC8 at 20:00–21:00 in 6/19/2015, when we set (QCI,ARP,GBR) = (3,5,1M) and (a,b,c,E0) = (80%,60%,40%,80). And the bad user perception decline by a large margin than the non-using case (a) as we expected.

5 CONCLUSION

We propose an adaptive strategy which could evaluate the user perception for the specific APP and improve the user experience by means of dedicated bearer when the user experience is relatively poor. Experimental result shows that the proposed strategy is reliable and could achieve good performance in improving the user experience.

REFERENCES

3GPP TS 23.203. 2010. *Policy and charging control architecture (Release 10)*.

Agarwal, V., Mittal, S., Mukherjea, S. & Dalal, P. 2012. Exploiting rich telecom data for increased monetization of telecom application stores. *Proc. IEEE Mobile Data Management, Bengaluru, Karnataka*: 63–68.

Cheng Zhenyu, Du Huiying, Lu Tingjie, & He Zhong. 2011. An empirical study of consumer adoption on 3G value-added services in China. *Proc. Management and Service Science, Wuhan, China*: 1–4.

Ekstrom, H. 2009. QoS control in the 3GPP evolved packet system. IEEE *Communications Magazine* 47(2): 76–83.

Hermet, G, & Combet, J. 2011. Mobile internet monetization: A methodology to monitor in real time the cellular subscriber transactional itinerary, from Mobile Advertising Exposure to Actual Purchase. *Proc. International Conference on Mobile Business, Como, Italy*: 307–312.

Kimbler, K. & Taylor, M. 2012. Value added mobile broadband services innovation driven transformation of the "Smart Pipe". *Proc. Intelligence in Next Generation Networks, London, UK*: 30–34.

Lexi Xu, Xinzhou Cheng, Yue Chen, Kun Chao, Dantong Liu, & Huanlai Xing. 2015. Self-optimised coordinated traffic shifting scheme for LTE cellular systems. *Proc. EAI First International Conference on Self-Organizing Networks, Beijing, China:* 1–9.

Lexi Xu, Yue Chen, KoK Keong Chai, Dantong Liu, Shaoshi Yang, & John Schormans. 2013. User relay assisted traffic shifting in LTE-Advanced systems. *Proc. IEEE Vehicular Technology Conference, Dresden, Germany*: 1–7.

Lexi Xu, Yue Chen, Kok Keong Chai, John Schormans, & Laurie Cuthbert. 2015. Self-organising cluster-based cooperative load balancing in OFDMA cellular networks. *Wiley Wireless Communications and Mobile Computing* 15(7): 1171–1187.

Liu Yan, Yu Ke, & Wu Xiaofei. 2014. Association analysis based on mobile traffic flow for correlation mining of mobile Apps. *Proc. Network Infrastructure and Digital Content, Beijing, China*: 420–424.

Nwanga, M.E., Onwuka, E.N., Aibinu, A.M. & Ubadike, O.C. 2015. Impact of big data analytics to Nigerian mobile phone industry. *Proc. Industrial Engineering and Operations Management, Dubai:* 1–6.

Salem, A.M., Elhingary, E.A. & Zerek, A.R. 2013. Value added service for mobile communications. *Proc. Power Engineering, Energy and Electrical Drives, Istanbul, Turkey*: 1784–1788.

Smailovic, V.., Galetic, V. & Podobnik, V. 2013. Implicit social networking for mobile users data monetization for telcos through context-aware services. *Proc. Telecommunications, Zagreb, Croatia*: 163–170.

Šerval, D., Marković, C. & Kovačević, S. 2014. 4G mobile internet, services, regulation and mobile operators in Bosnia and Herzegovina. *Proc.Information and Communication Technology, Electronics and Microelectronics, Opatija, Croatia:* 432–435.

Xiuli Yao, & Huaying Shu. 2009. Study on value-added service in mobile telecom based on association rules. *Proc. Software Engineering, Artificial Intelligences, Networking and Parallel/Distributed Computing, Daegu, Korea*: 116–119.

Signal and Information Processing, Networking and Computers – Chen & Huang (Eds)
© *2016 Taylor & Francis Group, London, ISBN 978-1-138-02881-4*

Big data assisted value areas of mobile internet

Chuntao Song, Xinzhou Cheng, Heng Zhang & Yuwei Jia
China Unicom Network Technology Research Institute, Beijing, P.R. China

ABSTRACT: Based on the massive network operating data, this paper introduces three basic factors for mobile network value, and the method for value calculation and mining of mobile internet. In this model, we can accurately locate the high value areas and draw out some kind of value maps, which can assist several branches of mobile operators to design their policy and deploy their investment.

Keywords: mobile Internet; value areas; big data

1 INTRODUCTION

With the popularity of smartphones and the prosperity of mobile applications, we are heading for a new age of mobile internet. The innovative Over The Top (OTT) services and the emerging of Mobile Virtual Network Operator (MVNO), have transferred traditional telecom operators to basic mobile internet channel providers. Nevertheless, the traditional operators are still acting as the host of the fundamental network infrastructure. Meanwhile, they are also in control of the real time operating data of the network, which is an unexploited goldmine and should be mined deeply using innovative methods.

The traditional methods to determine the value of a specific area are always based on a single factor, such as traffic, without showing concerns about the differences in the subscription information, user manner, and mobile terminal (Chunyan Fan et al. 2003). This paper proposes a novel model to determine value areas, which comprise of several quantitative factors, derived from telecom operating data and converged in several different levels using the AHP (Analytic Hierarchy Process) method. The value calculated above can be used in a series of application fields such as network planning, network optimization, marketing strategy launching, product development, and so on.

2 DEFINITION AND KEY FACTORS OF VALUE

In the mobile network, subscribers are the main body of value and mobile terminals are the medium to deliver value, while various kinds of services are the content of value. These three factors cover the typical contributions of mobile internet value, and they have different effects on the value. To ensure the comprehensiveness of the three effective factors as well as the correlation among them, we choose all these three factors as the basic factors for value analysis. Hence, the subscriber, mobile terminal, and service are the three main factors for value calculation in mobile internet. Furthermore, the subscriber factor contains information about user manner and user activities, such as user revenue, active user numbers, etc. The service factor is the illustration of different kinds of service, for example CS (Circuit Switch), and PS (Packet Switch) traffic. The mobile terminal factor reflects on the characteristics and network ability of the smartphones such as price, category (Radio Frequency capability), and so on.

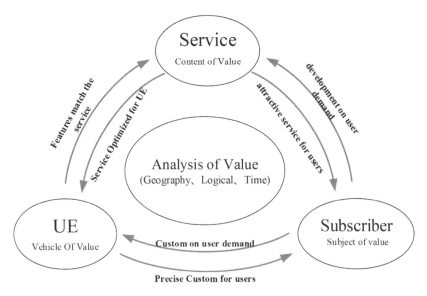

Figure 1. The internal relationship among three factors.

Figure 1 shows the tight internal relationship of the three factors mentioned above, joint analysis of the three factors can provide quantized value, which can be used for a series of application fields such as network planning, network optimization, marketing strategy launching and product development, and so on.

Data sources used for value analysis and value calculation are collected from different network entities, platforms, and interfaces. The collected data is an exclusive resource of mobile operators and it contains subscribers' ordering and consumption information, basic network resource information, PS and CS detailed records, call trace data, and so on. The value calculation method adopted in this paper is based on the joint analysis of all the above data sources.

The value of a mobile network can be presented in three different patterns, including the geographic pattern, the logical pattern, and the time domain pattern. The geographic pattern means the geographic distribution of value, which appears as different shapes of value areas such as a region, a line, or an island; the logical pattern means the logical distribution of value, such as commercial district, college, and stadium; the time domain pattern means the time-varying distribution of value, which can be categorized into different time distributions of value such as busy hour, festival, major activity etc.

3 MODELING AND CALCULATION

Value of mobile network is abstract, but it can be specified by a quantitative calculation. Then it can be used as productive factors in the mobile network operation. This paper proposes a method for the quantitative calculation of value in this section.

First, we collect the basic network data, which implies information of the three factors, including subscriber, service, and mobile terminal. Then we aggregate the data in the granularity of sector, which is the minimum unit for the value calculation in the next step. To be specific, the process above can be defined as sectorization. And sectorization here means clustering the cells which belong to one site and share the same antenna and azimuth but operate in different carrier frequencies. Sectorization can geometrically segment the target region through Tyson polygon, which is intuitive and persuasive for analyzers.

The value calculation based on the three factors is expressed in equation (1):

$$Value_{comp} = Value_{subscriber} + Value_{service} + Value_{UE} \qquad (1)$$

where $Value_{comp}$ stands for the comprehensive value, $Value_{subscriber}$, $Value_{service}$, and $Value_{UE}$ stand for the value component from the Subscriber, Service, and $Value_{UE}$, respectively.

The flow of quantitative calculation method for mobile network value is illustrated in Figure 2.

The calculation process of the sector level comprehensive value includes three steps. The first step is data mining and modeling.

There are five variables involved in this step, which are sector traffic volume (unit: MB), ratio of UE with the capability above the category 24 (%) in 3GPP TS 25.331 (2013) and 3GPP TS 25.306(2013), average price of UE (RMB), total revenue (RMB) and distinct numbers of subscribers. The variables mentioned above are calculated with granularity of sector and the time duration of the data source is one day.

Traffic is the RLC layer data volume numbering in MB, including data both from CS domain and PS domain. Cat24Ratio is the percentage of mobile terminals in a sector with network ability over category 24. Average Price is the mean price of all mobile terminals in this sector. Number of subscribers is the distinct active subscribers in this sector. Sum of Revenue is the total revenue from subscribers with different product types and activities in this sector.

The second step is zero-mean normalization (Z-score). The five variables of each sector defined above is standardized by z-score method in this step, the traffic standardization is shown in equation (2):

$$V_{trf\_Sector\_i} = \frac{T_{Sector\_i} - \mu}{\sigma} \qquad (2)$$

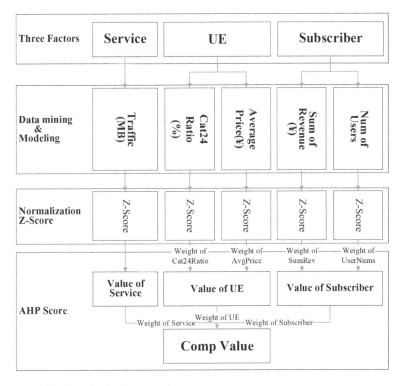

Figure 2. The derived method of sector value.

309

where $V_{trf\_sector\_i}$ is the standardization score of traffic volume for sector i, T_{Sector_i}, which is the real traffic volume of sector i, μ is the mean value of all sectors' traffic volume, and σ is the standard deviation of all sectors' traffic volume.

Using the same method as formula (2), we can calculate $V_{Cat24 Ratio\_sector\_i}$, $V_{AvgPrice\_sector\_i}$, $V_{UserNums\_sector\_i}$, and $V_{UserRev\_sector\_i}$.

The third step is AHP score. The comprehensive value of a sector is calculated by AHP method in Guohua Wang et al. (2004) amd Liling Qian et al. (2013). The seven weights used here are derived from the expert judgment matrix which is generated by survey from more than five experts.

The two-layer calculation for comprehensive value is stated as formula (3):

$$Value_{comp} = W_{subscriber} * V_{subscriber} + W_{service} * V_{service} + W_{UE} * V_{UE} \tag{3a}$$

$$= W_{subscriber} * (W_{UserNums-sector} * V_{UserNums\_sector} + W_{UserRev\_sector} * V_{UserRev\_sector}) + W_{service} * V_{service} | + W_{UE} * (W_{Cat24Ratio\_sector} * V_{Cat24Ratio\_sector} + W_{AvgPrice\_sector} * V_{AvgPrice\_sector}) \tag{3b}$$

where $Value_{comp}$ is the comprehensive value, $W_{subscriber}$ is the weight of the subscriber, $W_{service}$ is the weight of the service, and W_{UE} is the weight of the mobile.

In addition, $W_{UserNums\_sector}$ is the weight of the subscriber number, $W_{UserRev\_sector}$ is the weight of subscriber revenue, $W_{Cat24 Ratio\_sector}$ is the weight of mobile terminal percentage above category 24, and $W_{AvgPrice\_sector}$ is the weight of average price of mobile terminal.

The intermediate and final results provide both the single factor's value and the comprehensive value, which can be applied in different fields.

4 VALUE IN DIFFERENT PATTERNS

The value of a sector reflects its rank among all sectors in the network, and it should be distributed or integrated in different patterns for the comprehensive analysis. This paper presents three patterns for the value: the geographic pattern, the logical pattern, and the time domain pattern. The following section illustrates different value patterns in the form of value map or diagrams.

4.1 Value in geographic pattern

We can intuitively observe and locate the geographic value area in the value map, which is sectorized by Tyson polygon method and characterized with the sector value. The six maps in Figure 3 demonstrate the comprehensive value maps as well as maps of five basic value factors in the geographic pattern.

These maps can directly present the height of traffic, revenue, and mobile terminal. The map of service is the traditional value map based on traffic merely. By contract, through the new method it can provide more value maps from other perspectives. It can be seen from the maps above that these value results have obvious differences, which can be used for different purposes. They are useful for network planning, network optimization, marketing strategy, and product development.

4.2 Value in logical pattern

The physical sites are constructed for certain reasons and used for different purposes, they are often classified into different scenarios in the process of network planning.

Three typical classifications are region, line, and point, which are not adjacent in geographic areas. The detailed scenario list includes urban, country town, town, high railway, highway, campus, and scenic spot, etc. These should be defined as logical patterns of sector value, which are important references for resource allocation.

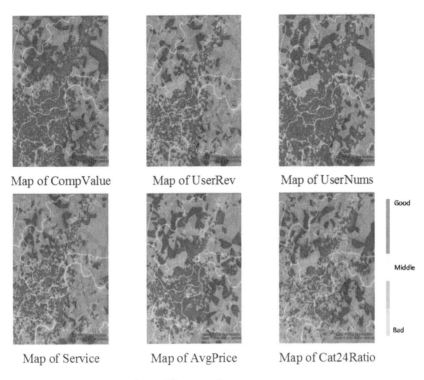

| Map of CompValue | Map of UserRev | Map of UserNums |
| Map of Service | Map of AvgPrice | Map of Cat24Ratio |

Figure 3. Sample of value maps for mobile network.

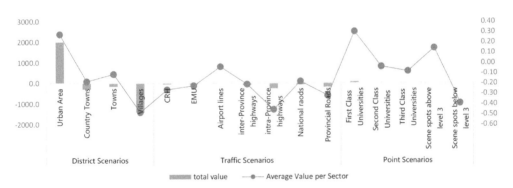

Figure 4. Sample of value comparisons among different scenarios.

Figure 4 shows an example of value comparison among different scenarios. Depending on the comprehensive and single factor values in logical pattern, we can discover the value of some special scenarios, for example in Figure 4, the railway scenario, a university scenario, and a famous scenic spot scenario etc. That makes it possible to accurately locate the outstanding scenarios or sub scenarios, which will give more in return for operators.

4.3 Value in time domain pattern

Subscribers are always moving; they have different habits especially during different times. This reflects the dynamic characteristic of the mobile network. The dynamic feature can be identified by the time varying value. For this purpose, we calculate the sector value with the interval of an hour to present the time varying value, as shown in Figure 5.

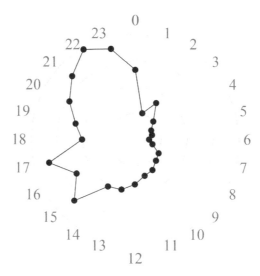

Figure 5. Sample of time varying value by hour.

Value of mining in the time domain is useful for understanding the value change with time. Figure 5 shows the time varying value of a residence community in a developed city of China. It is obvious that the most valuable time interval is 21, 22, and 23 o'clock. Hence, we get the conclusion that people in that city tend to use mobile network more frequently before midnight, while use is less during work time and after midnight. As we know the time domain value exactly, operators can provide some customized preferential product in busy hours for users in this city. Moreover, the operator can dynamically allocate network resources to guarantee user experience assurance after conjoint analysis with more network operation data.

5 VALUE APPLICATION

The ultimate purpose of the value mining is their application. The mass data from various sources, after being processed and integrated through big data method, presents lots of useful results with proper granularities and different patterns. This can provide quantization reference for a series of application fields, such as network planning, network optimization, marketing, product development etc. When the weights are adjusted for special purposes, the value results can be used for high-level strategic decision-making.

First, the operator can focus on value hot areas and put more investment there in order to get the profit as soon as possible, while put less investment into the plain and valley regions (Harri Holma et al. 2013). Then, the operators can determine resource allocation scheme for special regions with different values. For the region with a high user number and high traffic volume, the operator should pay more attention to the network capacity and resource utilization rate. For the region with high mobile terminal category ratio, advanced network up-gradation is prior. In addition, the operator should also develop more appropriate telecom products and mobile terminal promotion policies in the region with low terminal category ratios.

Figure 6 shows different shapes of comprehensive value regions in a southwest city of China. The result of the comprehensive value map and sub-factor value maps are verified and proved in accordance with local experts' cognition. These experts are telecom engineers concentrating on different research fields including network planning, network optimization, and marketing strategy and so on.

This local operator of the city has used the comprehensive value area to implement the LTE network planning and UMTS 900 M network planning step by step based on the value area research results. The quantified user number value and mobile terminal value have helped

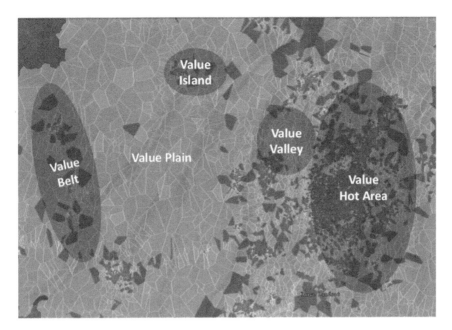

Figure 6. Sample of different comprehensive value regions in geographic pattern.

the local operator in terminal and product promotion as well as user immigration from 2/3G network to 4G network. As we know, all the implementations have achieved good results by now and the value analysis research is still in progress.

6 CONCLUSION

In the age of mobile network and information blooming, the content and pattern of network value is undergoing a profound change, even the operation mode is deeply transforming. Operators and service suppliers must reconsider the value of data resource in hand, explore and research deeply with advanced data mining methods to transfer the data into valuable assets. The proposed method of value areas in mobile network is just one single application of big data in mobile network to analyze, also it is superficial and simple. With the same data used for value calculation, we are doing more analysis such as user profile, trajectory, perception for traffic management, as well as network planning and optimization.

REFERENCES

3GPP TS 25.306. 2013. UE radio access capabilities (Release 11).
3GPP TS 25.331. 2013. Radio resource control (RRC) protocol specification (Release 11).
Chunyan Fan, Xiaoming Han, Weihua Tang. 2003. The extraction of expert judgment information in AHP and the comprehensive determination method of index weight. *Journal of air force engineering university*: 41–44.
Erik Dahlman, Stefan Parkvall, Johan Sköld, Per Beming. 2010. 3G Evolution, HSPA and LTE for Mobile Broadband. *Post & Telecom Press*: 23:335–367.
Guohua Wang, Li Xiong, Yu Dong, Zhiqiang Wang. 2004. The method of extracting and adjusting the expert judgment information in AHP. *Logistics and Management*: 65–67.
Harri Holma, Antti Toskala.2010. WCDMA FOR UMTS, HSPA Evolution and LTE. *John Wiley & Sons,* 8:173–218.
Liling Qian, Chunpeng Song, Fenghua Li. 2013. Discussion on construction scheme based on value area of 3G network. *Designing Techniques of Posts and Telecommunications*: 38–42.

Signal and Information Processing, Networking and Computers – Chen & Huang (Eds)
© 2016 Taylor & Francis Group, London, ISBN 978-1-138-02881-4

Spectrum allocation based on data mining in heterogeneous cognitive wireless networks

Chen Cheng, Xinzhou Cheng, Mingqiang Yuan, Lexi Xu,
Shiyu Zhou, Jian Guan & Tao Zhang
Department of Network Optimization and Management, China Unicom Network Technology
Research Institute, Beijing, China

ABSTRACT: Cognitive radio technology has been proposed to improve the spectrum efficiency by letting the cognitive radios act as cognitive users to opportunistically access underutilized frequency bands. Spectrum allocation is a direction of cognitive radio technology. In this paper, we propose a Single-objective Multivariable Quantum-inspired Particle Swarm Optimization (M-QPSO) algorithm and a multi-objective M-QPSO based on data mining for spectrum allocation to improve the network benefit in heterogeneous network. Simulation results show that the proposed Single-objective M-QPSO has advantages of both convergence rate and convergence accurate value compared with other intelligent algorithms. In addition, the proposed Multi-objective M-QPSO, which considers both utilization and fairness, can obtain the similar solutions to exhaustive search in much less time.

Keywords: data mining; cognitive radio; spectrum allocation; multi-objective optimization

1 INTRODUCTION

With the development of information technology and wireless communication technology, heterogeneous networks demand for radio spectrum resources increase rapidly (Xu et al. 2009). However, the scarceness of wireless spectrum will be the bottleneck for the continual development of wireless communication (Xu et al. 2011). Cognitive Radio (CR) is a kind of intelligent wireless communication technology with the ability to sense and adapt to the outside, which can alleviate the shortage of radio spectrum resources effectively and decrease operating expense (Xu et al. 2014).

Spectrum allocation, a part of the cognitive radio technology, aims at assigning spectrums to cognitive users appropriately. Some constraints should be taken into consideration in spectrum allocation, such as the space division multiplexing and the communication jamming (Peng et al. 2006) (Xu et al. 2015).

Spectrum allocation of cognitive radio can be regarded as a discrete optimization problem, it is difficult to obtain the optimal solution within reasonable computation time by the exact and analytical methods, therefore, intelligent algorithms are widely researched (Sun, 2009) (Xu, 2012). In Zhang et al. (2014) and Gao et al. (2011), Quantum Particle Swarm Optimization (QPSO), an intelligent algorithms based on Particle Swarm Optimization (PSO) was proposed to solve relay selection and spectrum allocation problem. Although the proposed QPSO has advantages of convergence rate and convergence accurate value compared with PSO, it cannot solve Multi-objective optimization and there is still room for improvement of convergence accurate. In Elhossini, Areibi and Dony (2010), the authors proposed pareto particle swarm optimization to solve Multi-objective optimization problem but it cannot be applied in spectrum allocation problem directly. In Gao and Cao (2011), a method for Multi-objective spectrum allocation is proposed, which is effective but complex to be applied and easy to becoming the optimal solution.

This paper proposes Multivariable Quantum-inspired Particle Swarm Optimization algorithm (M-QPSO), which has advantages of both convergence rate and convergence accurate value compared with QPSO and other algorithms. In addition, its Multi-objective optimization, which is based on data mining, is capable of finding near-optimal solutions within reasonable time when considers both Max-Sum-Reward network benefit and Max-Fair-Reward network benefit (Gao and Cao, 2011).

The rest of the paper is organized as follows. Section 2 introduces the system model of spectrum allocation in heterogeneous cognitive network. Section 3 and Section 4 propose the Single-objective M-QPSO and Multi-objective M-QPSO respectively to solve spectrum allocation problem. Finally, in Section 5, conclusions and future work are presented.

2 SYSTEM MODEL

2.1 *Cognitive heterogeneous network spectrum allocation model*

Figure 1 shows the model of cognitive heterogeneous networks (Leaves et al. 2004), where primary users and cognitive users are in the same space. We define cognitive users as wireless access points and primary users as macro base stations or home base stations. Overlay model is used here, which means if the channel is occupied by primary users, cognitive users are inaccessible.

Figure 2 shows the relationship between primary users, cognitive users and available channels M and N. Primary user x and cognitive user b, cognitive user a and d will interfere each other if they occupy the same channel.

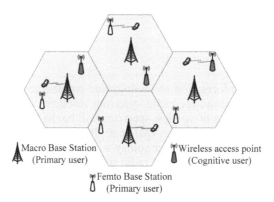

Figure 1. Cognitive heterogeneous networks model.

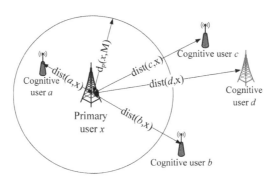

Figure 2. Distribution of users.

316

In Figure 3, channel M is occupied by primary user x, so cognitive user b and either c or d cannot occupy this channel. In Figure 4, there is no primary user occupying channel N, so only one of cognitive user c or b cannot occupy this channel.

2.2 Network benefit function and relative matrices

In this paper, we recommend two network benefit functions, Max-Sum-Reward and Max-Proportional-Fair, which means network utilization functions for the maximization of the utilization and fairness, as our object of spectrum allocation (Gao and Cao, 2011).

The benefit each user obtains can be expressed as:

$$r_n = \sum_{m=1}^{M} a_{n,m} \cdot b_{n,m} \tag{1}$$

The network benefit functions of Max-Sum Reward and Max-Proportional-Fair Reward are defined as (2) and (3) respectively

$$U_{\text{MSR}}(\mathbf{R}) = \sum_{n=1}^{N} r_n = \sum_{n=1}^{N} \sum_{m=1}^{M} a_{n,m} \cdot b_{n,m} \tag{2}$$

$$U_{\text{MPF}}(\mathbf{R}) = \sum_{n=1}^{N} \log(r_n) = \sum_{n=1}^{N} \log\left(\sum_{m=1}^{M} a_{n,m} \cdot b_{n,m} \right) \tag{3}$$

There are 4 matrixes in cognitive radio spectrum allocation model described as Table 1 (Leaves et al. 2004).

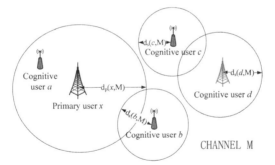

Figure 3.   Occupancy of *Channel M*.

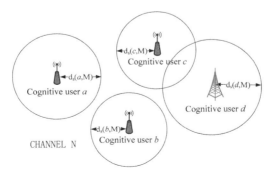

Figure 4.   Occupancy of *Channel N*.

Table 1. 4 matrices in cognitive radio spectrum allocation model.

Matrix name	Expression	Relationship
Available spectrum matrix	$L = \{l_{n,m} \mid l_{n,m} \in \{0,1\}\}_{N \times M}$	if $d_s(n,m) < d_{min}, l_{n,m} = 1$; otherwise, $l_{n,m} = 0, d_s(n,m) = Dist(n,x) - d_p(x,m)$
Network efficiency matrix	$B = \{b_{n,m}\}_{N \times M}$	$b_{n,m} = d_s(n,m)^2, d_{min} \le d_s(n,m) \le d_{max}$
Spectrum interference matrix	$C = \{c_{n,k,m} \mid c_{n,k,m} \in \{0,1\}\}_{N \times N \times M}$	$c_{n,k,m} \le l_{n,m} \times l_{k,m}$, When $n = k$, $c_{n,k,m} = 1 - c_{n,m}$
Spectrum allocation matrix	$A = \{a_{n,m} \mid a_{n,m} \in \{0,1\}\}_{N \times M}$	$a_{n,m} + a_{k,m} \le 1$

3 SINGLE-OBJECTIVE SPECTRUM ALLOCATION

3.1 Single-objective M-QPSO

The M-QPSO uses quantum code to represent the probabilistic of particles' position of "0" or "1". We call it quantum bit, which can be represented as a pair of composite numbers (α, β), where $|x_{ij}|^2 + |\beta_{ij}|^2 = 1, (j = 1,2,\cdots,M)$. α and β denote the status of "0" and "1" probability amplitude respectively. The quantum position of the ith particle is defined as:

$$\begin{pmatrix} \alpha_1 \alpha_2 \dots \alpha_l \\ \beta_1 \beta_2 \dots \beta_l \end{pmatrix} \tag{4}$$

where $0 \le \alpha_i \le 1, 0 \le \beta_i \le 1, \beta_i = \sqrt{1 - \alpha_i}$. So the quantum position of the ith particle can be simplified as (5), which is defined as the velocity of the particle:

$$v_i = (\alpha_{i1}, \alpha_{i2}, \dots, \alpha_{il}) = (v_{i1}, v_{i2}, \dots, v_{il}) \tag{5}$$

The quantum position of particles are updated by quantum rotation gate, which is expressed by quantum rotation angle θ to get higher fitness of objective functions in each evolution. It is expressed as:

$$U(\theta_{ij}^t) = \begin{pmatrix} \cos\theta_{ij}^t & -\sin\theta_{ij}^t \\ \sin\theta_{ij}^t & \cos\theta_{ij}^t \end{pmatrix} \tag{6}$$

The position of the jth quantum bit of the ith particle is updated as (7), which is simplified as (8).

$$v_{ij}^{t+1} = |U(\theta_{ij}^{t+1})v_{ij}^t| = \left| \begin{pmatrix} \cos\theta_{ij}^{t+1} & -\sin\theta_{ij}^{t+1} \\ \sin\theta_{ij}^{t+1} & \cos\theta_{ij}^{t+1} \end{pmatrix} v_{ij}^t \right| \tag{7}$$

$$v_{ij}^{t+1} = |v_{ij}^t \times \cos\theta_{ij}^{t+1} - \sqrt{1 - (v_{ij}^t)^2} \times \sin\theta_{ij}^{t+1}| \tag{8}$$

Assuming there are p particles in L-dimensional space. The velocity of ith particle is expressed as $v_i = (\alpha_{i1} \quad \alpha_{i2} \quad \cdots \quad \alpha_{il}) = (v_{i1} \quad v_{i2} \quad \cdots \quad v_{il})$. The position of ith particle is expressed as $x_i = (x_{i1}, x_{i2}, \cdots, x_{il})$, where $1 \le i \le p$, whose best position of all the evolutions, called individual optimal position, is expressed as $p_i = (p_{i1} \quad p_{i2} \quad \cdots \quad p_{il})$. The best position of all of p particles, called global optimal position, is expressed as $p_g = (p_{g1} \quad p_{g2} \quad \cdots \quad p_{gl})$. In the tth evolution, quantum rotation angle θ is computed as:

318

$$\theta_{id}^{t+1} = a_1(p_{id}^t - x_{id}^t) + a_2(p_{gd}^t - x_{id}^t) + \tilde{M} \tag{9}$$

where a_1 and a_2 are the coefficients of p_i and p_g respectively. M is a mutation operator, which is defined as:

$$M = \begin{cases} U(-\sigma_1,\sigma_1), & if\left(t \le \dfrac{T}{4}\right) \\ C(0,\sigma_2), & if\left(\dfrac{T}{4} < t \le \dfrac{3T}{4}\right) \\ N(0,\sigma_3), & if\left(\dfrac{3T}{4} < t \le T\right) \end{cases} \tag{10}$$

where $U(-\sigma_1,\sigma_1)$ is Uniform distribution function, whose global researching ability is strong but local researching ability is weak. $N(0,\sigma_3)$ is Gaussian distribution function, whose global researching ability is weak but local researching ability is strong. $C(0,\sigma_2)$ is Cauchy distribution function, whose global researching ability and local researching ability are between Uniform distribution and Gaussian distribution.

These designs enhance global researching ability in the preliminary stage of evolution to avoid it becoming locally optimal solutions, while enhance local researching ability in the later stage of evolution to accelerate its convergence rate.

We also introduce mutation probability η to avoid the solutions becoming locally optimal solutions instead of global solutions. η can be expressed as (11).

$$\eta = k\mu^{\frac{t}{T-t}} \tag{11}$$

where $0<\mu<1$, $0<k<1$. μ is the predefined attenuation factor and k is the mutation coefficient. The mutation probability η is close to k in the earlier stage of evolution, then decreasing with the increasing of evolution times, which is close to 0 in the end of evolution. The velocity of the jth quantum bit of the ith particle is updated as:

$$v_{id}^t = \begin{cases} \sqrt{1-(v_{id}^{t-1})^2}, & if \quad (p_{id}^{t-1} = x_{id}^{t-1} = p_{gd}^{t-1} \; \&\& \; r < \eta) \\ |v_{id}^{t-1} \times \cos\theta_{id}^t - \sqrt{1-(v_{id}^{t-1})^2} \times \sin\theta_{id}^t|, & else \end{cases} \tag{12}$$

The position of the jth quantum bit of the ith particle is updated as:

$$x_{id}^t = \begin{cases} 1, & r > (v_{id}^t)^2 \\ 0, & r \le (v_{id}^t)^2 \end{cases} \tag{13}$$

where r is a random number in the range (0, 1). Therefore, the quantum position of the particle will get a higher fitness value by this evolution.

3.2 Spectrum allocation using Single-objective M-QPSO

Assume a heterogeneous cognitive network of N cognitive users and M non-overlapping channels. The Available spectrum matrix $L = \{l_{n,m} \mid l_{n,m} \in \{0,1\}\}_{N \times M}$ is a binary matrix representing the channel available by cognitive users. Then assume the available spectrum matrix and spectrum allocation matrix as:

$$A = \begin{pmatrix} 0 & 0 & 0 & 1 & 0 \\ 0 & 0 & 0 & 0 & 0 \\ 1 & 0 & 0 & 0 & 0 \\ 1 & 0 & 0 & 0 & 1 \end{pmatrix}, L = \begin{pmatrix} 1 & 0 & 0 & 1 & 0 \\ 0 & 1 & 0 & 1 & 0 \\ 1 & 0 & 0 & 1 & 0 \\ 1 & 0 & 0 & 0 & 1 \end{pmatrix}.$$

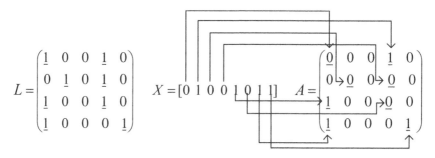

$$L = \begin{pmatrix} 1 & 0 & 0 & 1 & 0 \\ 0 & 1 & 0 & 1 & 0 \\ 1 & 0 & 0 & 1 & 0 \\ 1 & 0 & 0 & 0 & 1 \end{pmatrix} \qquad X = [0\ 1\ 0\ 0\ 1\ 0\ 1\ 1] \qquad A = \begin{pmatrix} 0 & 0 & 0 & 1 & 0 \\ 0 & 0 & 0 & 0 & 0 \\ 1 & 0 & 0 & 0 & 0 \\ 1 & 0 & 0 & 0 & 1 \end{pmatrix}$$

Figure 5. The mapping of quantum particles and spectrum allocation matrix.

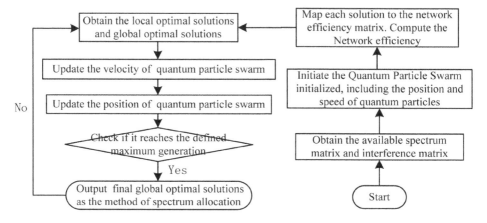

Figure 6. Flow chart of spectrum allocation using Single-objective M-QPSO.

We initiate a vector of 8 bits since the number of element "1" is 8 in the available spectrum matrix to avoid redundancy: $X = [0\ 1\ 0\ 0\ 1\ 0\ 1\ 1]$. The specific process is as Figure 5.

The steps of spectrum allocation using Single-objective M-QPSO is as Figure 6:

Step 1: Get the length of the velocity and position vector from given available spectrum matrix, which are equal to the number of "1" in available spectrum matrix.

Step 2: Assign "0" or "1" to the position of quantum particles randomly, and initiate the velocity of quantum particles as $1/\sqrt{2}$.

Step 3: Get the spectrum interference matrix according to Table 1. Compute the fitness of every particles according to the given network efficiency matrix.

Step 4: Compute the fitness of every particles according to the given network efficiency matrix, then get individual optimal position and global optimal position.

Step 5: Update the velocity of quantum particles according to (12).

Step 6: Update the position of quantum particles according to (13).

Step 7: If it reaches the defined maximum generation, output the outcome as the result of spectrum allocation; if not, go to step 4.

3.3 Simulation results

We compare M-QPSO with PSO (Leaves et al. 2004), QPSO (Gao, Cao and Diao, 2011), QGA (Zhao et al. 2009) and CSGC (Chang and Gong, 2014) in the background of spectrum allocation in heterogeneous cognitive wireless networks.

In the described model of cognitive heterogeneous network, we define cognitive users as wireless access points whose transmission power range from 1 to 4, and define primary users

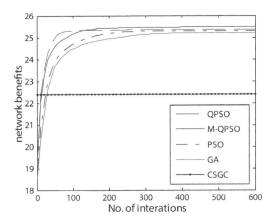

Figure 7. Max-Sum Reward (20 primary users, 10 cognitive users, 10 channels).

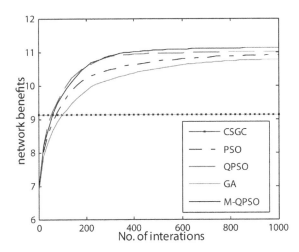

Figure 8. Max-Sum Reward (20 primary users, 30 cognitive users, 10 channels).

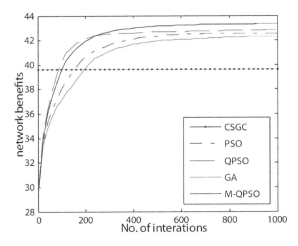

Figure 9. Max-Proportional-Fair Reward (20 primary users, 10 cognitive users, 20 channels).

as macro base stations whose transmission power are 4 or home base stations whose transmission power are 1. For M-QPSO, $a_1 = 0.06$, $a_2 = 0.06$, $\mu = 0.5$, $k = 0.2$. The parameters of other algorithms are set as references. Figures 7 and 8 are the results of network benefit of Max-Sum Reward, while Figures 9 and 10 are those of Max-Proportional-Fair Reward. These 4 Figures show that the convergence accurate of M-QPSO is higher than the other algorithms, and the convergence rate of M-QPSO is higher than GA and PSO. Especially, convergence rate of M-QPSO is lower than QPSO while the convergence accurate of M-QPSO is higher because M-QPSO has more variable quantities than QPSO, which enhance its global researching ability and prevent it from becoming local optimization.

4 MULTI-OBJECTIVE SPECTRUM ALLOCATION

4.1 Multi-objective M-QPSO

In the Single-objective M-QPSO presented above, the goal is to obtain the best solutions such as maximum or minimum according to one objective function, however, most of the real problems have more than one object, which can be solved by Multi-objective algorithms. In this paper, we propose Multi-objective M-QPSO based on data mining. Some conceptions are as follows (Gao and Cao, 2013):

Non-dominated solution: For solutions x and y in list S, if $f_i(x) \geq f_i(y), (i = 1, 2, \cdots m)$ for all i while $f_i(x) = f_i(y), (i = 1, 2, \cdots m)$ not for all i, x is a non-dominated solution, y is a dominated solution, and x dominates y. Otherwisie, there is no non-dominated relationship between this two solutions. We design a *data warehouse* to store the selected solutions with high non-dominated level and divergence. The *data warehouse*, which is used for computing the rotation angle θ to update solutions, is updated in each evolution and output the result in the end. Figure 11 is the interaction of *data warehouse* and evolutions.

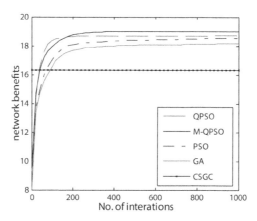

Figure 10.   Max-Proportional-Fair Reward (20 primary users, 10 cognitive users, 10 channels).

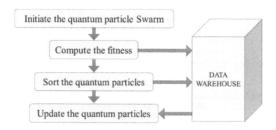

Figure 11.   The interaction of data warehouse and evolutions.

The steps of Multi-objective M-QPSO are as Figure 11:

Step 1: Assign "0" or "1" to the position of quantum particles randomly, and initiate the velocity of quantum particles as $1/\sqrt{2}$.

Step 2: Sort each quantum particle by its non-dominated level as follows:

    a. Traverse each solution a in S to obtain which solutions are dominated by a, then define them as S_a.

    b. Obtain the number of solutions which dominate a as n_a acoording to a). If $n_a = 0$, assign the non-dominated level of a as 1.

    c. For each solution a whose non-dominated level is 1, traverse each solution b in S_a, then let $n_b = n_b - 1$. Add solution b to list B if $n_b = 0$, then assign the non-dominated level of b in B as 1.

    d. Traverse each solution c in S_b as below then get the solutions list whose non-dominated level is 3.

    e. Repeat a)-d) till obtain the non-dominated level of every solutions in S.

Step 3: Compute the sum of the two Euclidean distance to both sides of the nearest solutions with the same non-dominated level as the divergence and sort them according to the divergence in descending order.

Step 4: Store the non-dominated solutions in the *data warehouse*. Assume there are s non-dominated solutions in total, then store $(M$-$s)$ solutions in the *data warehouse* to insure there are M solutions in it.

Step 5: Update S to S_{new}: In the tth evaluation, update rotation angle θ as:

$$\theta_{id}^{t+1} = \frac{a}{n_1 + n_2}[rand1 \sum_{k=1}^{n_1}(p_{(non-dom)kd}^t - x_{id}^t) + rand2 \sum_{k=1}^{n_2}(p_{(dom)kd}^t - x_{id}^t)] \tag{14}$$

where $p_{(non-dom)kd}$ represent the dth quantum of kth particle of which the non-dominated level is 1; $p_{(dom)kd}$ represents the dth quantum of kth particle of which the non-dominated level is not 1 (for the solutions of same non-dominated level, sort them by Euclidean distance in descending order). $rand1$ and $rand2$ are random numbers in the range (0, 1), which are the impact factors of solutions whose non-dominated level is 1 and the solutions whose non-dominated level is not 1 respectively. a is a constant between 0 and 1, representing amplitude of variation. Assume the number of solutions in *data warehouse* is N, let:

$$n_1 = \frac{N}{10}, n_2 = \frac{Nt}{10T} \tag{15}$$

where T is the defined maximum generation. In preliminary stage of evolution, the non-dominated level of most solutions is 1, which enhance the convergence rate; while in the later stage, the number of solutions whose non-dominated level is 1 is similar to number of solutions whose non-dominated level is not 1, which prevent it from becoming local optimization.

Step 6: Update the velocity and position of quantum particle according to (12) and (13).

Step 7: Mix S_{new} and S and then sort them as step 2 and 3, then store them in *data warehouse* as step 4.

Step 8: Let $S = S_{new}$ then go to step 2 if $t < T$. If not, stop the evolution, thus *data warehouse* is the list of solutions of Multi-objective M-QPSO.

4.2 *Performance of Multi-objective M-QPSO*

Two Benchmark functions called ZDT4 and SCH are used to show the performance of Multi-objective M-QPSO. We predefined $N = 100$, $a = 0.1$, $\mu = 0.5$.

1. ZDT4 function:
$$\begin{cases} \min f_1(x) = x_1 \\ \min f_2(x) = g(x) \times h(x) \end{cases} \tag{16}$$

where $g(x) = 11 + x_2^2 - 10\cos(4\pi x_2)$, $h(x) = \begin{cases} 1 - \sqrt{\dfrac{f_1(x)}{g(x)}}, & \text{if } f_1(x) \le g(x) \\ 0, & \text{if } f_1(x) > g(x) \end{cases} \tag{17}$

323

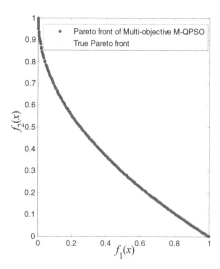

Figure 12.   ZDT4 solutions comparison of Multi-objective M-QPSO and real solutions.

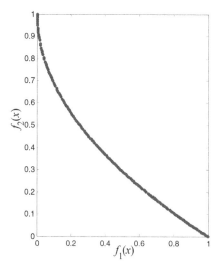

Figure 13.   ZDT4 solutions of multi-objective M-QPSO.

2. SCH function:
$$\begin{cases} \min f_1(x) = \begin{cases} -x & \text{if} \quad x \leq 1 \\ -2 + x & \text{if} \quad 1 < x \leq 3 \\ 4 - x & \text{if} \quad 3 < x \leq 4 \\ -4 + x & \text{if} \quad x > 4 \end{cases} \\ \min f_2(x) = (x - 5)^2 \end{cases} \tag{18}$$

Figures 12–17 are the simulation results of solutions obtained by Multi-objective M-QPSO and real solutions of ZDT4 and SCH functions. They show that the solutions obtained by Multi-objective M-QPSO are very similar with the real solutions.

4.3   Spectrum allocation using Multi-objective M-QPSO

The goal of Multi-objective spectrum allocation is to get solutions to take both network benefit of Max-Sum Reward and Max-Proportional-Fair Reward into account to balance the

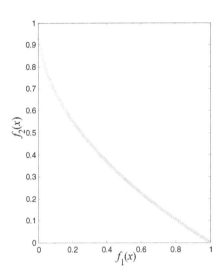

Figure 14.   ZDT4 real solutions.

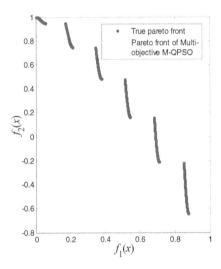

Figure 15.   SCH solutions comparison of Multi-objective M-QPSO and real solutions.

utilization and benefit as Figure 18. The steps 1–3 are same to the steps of Single-objective spectrum allocation in 3.2. After that, update quantum particles and *data warehouse* using Multi-objective M-QPSO described in 4.1, then get the final *data warehouse* as the solutions of spectrum allocation method. Figure 18 is the flow chart of spectrum allocation using Multi-objective M-QPSO.

The model of cognitive heterogeneous networks and parameters setting are same to Section 3.3. Simulation results are presented in Figures 19–21. In Figure 19, we compare Multi-objective M-QPSO with exhaustive search. The results show that the solutions of Multi-objective M-QPSO are same to those of exhaustive search, however, exhaustive search time costs 1020.5 s while Multi-objective M-QPSO only costs 41.543 s. So the proposed algorithm is much more efficient. Time costs grow exponentially with the increasing of users and channels so that we cannot get results by exhaustive search in complex spectrum allocation problems, like Figures 20 and 21. The results of Figures 20 and 21 show M-QSPO of Max-Sum Reward and M-QSPO of Max-Proportional-Fair Reward are two solutions of

325

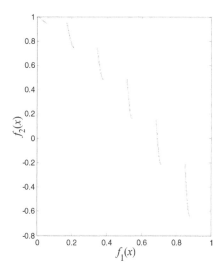

Figure 16.   SCH solutions of Multi-objective M-QPSO.

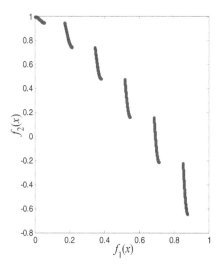

Figure 17.   SCH real solutions.

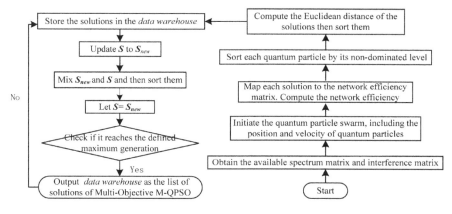

Figure 18.   Flowchart of spectrum allocation using Multi-objective M-QPSO.

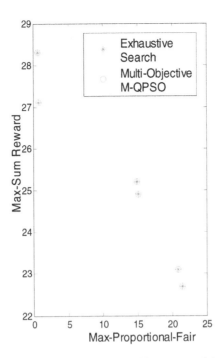

Figure 19. Network benefit of 10 primary users, 5 cognitive users and 6 channels.

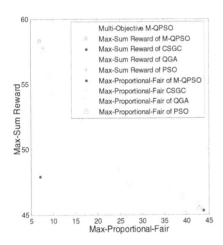

Figure 20. Network benefit of 20 primary users, 10 cognitive users and 10 channels.

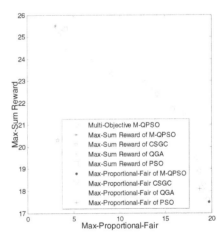

Figure 21. Network benefit of 20 primary users, 20 cognitive users and 10 channels.

Multi-objective M-QPSO, but Multi-objective M-QPSO can get solutions with more than one object taken into account, therefore, network utilization and fairness can be balanced.

5 CONCLUSIONS

This paper has proposed the Single-objective Multivariable Quantum-inspired Particle Swarm Optimization algorithm (M-QPSO) and the Multi-objective M-QPSO based on data mining, which can be used as spectrum allocation methods in heterogeneous cognitive networks.

Simulation results show that the proposed Single-objective M-QPSO has advantages of both convergence rate and convergence accurate value compared with other intelligent algorithms. In addition, the proposed Multi-objective M-QPSO, which takes both utilization and fairness into account, can obtain the similar solutions to exhaustive search in much less time, while in large dimensions, the M-QPSO can still obtain the efficient solutions which exhaustive search cannot obtain. Some interesting extensions of this work include studying dynamic spectrum allocation and other algorithms based on data mining to allocate spectrum in cognitive radio network.

REFERENCES

Chang Jun & Gong Wenlong. 2014. Structural modal parameter identification based on quantum-behaved particle swarm optimization combined with wavelet transformation. *Journal of Vibration and Shock* 33(23): 42–46.

Elhossini, A., Areibi, S. & Dony, R. 2010. Strength pareto particle swarm optimization and hybrid eapso for multi-objective optimization. *Evolutionary Computation* 18:182–197.

Hongyuan Gao & Jinlong Cao. 2013. Non-dominated sorting quantum particle swarm optimization and its application in cognitive radio spectrum allocation. *Journal of Central South University* 20:1878–1888.

Hongyuan Gao, Jinlong Cao & Ming Diao. 2011. A Simple Quantum-inspired Particle Swarm Optimization and its Application. *Information Technology Journal* 10:2315–2321.

Leaves, P, Grandblaise, D & Bourse, D. 2004. Dynamic Spectrum Allocation in Composite ReconFigureurable Wireless Network. *IEEE Communications Magazine* 15(4): 72–81.

Lexi Xu, Yue Chen, John Schormans, Laurie Cuthbert & Tiankui Zhang. 2011. User-Vote Assisted Self-Organizing Load Balancing for OFDMA Cellular Systems. *Proc. IEEE PIMRC, Toronto, Canada:* 217–221.

Lexi Xu, Yue Chen, KoK Keong Chai, Tiankui Zhang & John Schormans. 2012. Cooperative Load Balancing for OFDMA Cellular Networks. *Proc. European Wireless, Poznan, Poland:* 1–7.

Lexi Xu, Yue Chen, Kok Keong Chai, John Schormans & Laurie Cuthbert. 2015. Self-Organising Cluster-Based Cooperative Load Balancing in OFDMA Cellular Networks. *Wiley Wireless Communications and Mobile Computing* 15(7): 1171–1187.

Lexi Xu, Yue Chen. 2009. Priority-based Resource Allocation to Guarantee Handover and Mitigate Interference for OFDMA Systems. *IEEE PIMRC, Tokyo, Japan:* 783–787.

Lexi Xu, Yuting Luan, Kun Chao, Xinzhou Cheng, Heng Zhang & John Schormans. 2014. Channel-Aware Optimised Traffic Shifting in LTE-Advanced Relay Networks. *Proc. IEEE PIMRC, Washington, USA:* 1597–1602.

Peng, C., Zheng, H. & Zhao, B.Y. 2006. Utilization and fairness in spectrum assignment for opportunistic spectrum access. *ACM, Mobile Networks and Applications* 11(4): 555–576.

Sun jun. 2009. Quantum-behaved particle swarm optimization algorithm research. *Jiangnan University* 15(4): 22–28.

Tiankui Zhang, Jinlong Cao & Yue Chen. 2013. A Small World Network Model for Energy Efficient Wireless Networks. *IEEE Communications Letters* 4(16): 1–4.

Tiankui Zhang, Jinlong Cao, Zhimin Zeng & Dantong Liu. 2014. Joint Relay Selection and Spectrum Allocation Scheme in Cooperative Relay Networks. *Proc. IEEE VTC, Seoul, Korea:* 1–6.

Zhao, Z.J., Peng, Z., Zheng, S.L. & Shang, J.N. 2009. Cognitive radio spectrum allocation using evolutionary algorithm. *IEEE Transactions on Wireless Communications* 8(9): 4421–4425.

Signal and Information Processing, Networking and Computers – Chen & Huang (Eds)
© 2016 Taylor & Francis Group, London, ISBN 978-1-138-02881-4

Big data based mobile terminal performance evaluation

Chuntao Song, Yuwei Jia & Xinzhou Cheng
Department of Network Optimization and Management, China Unicom Network Technology Research Institute, Beijing, P.R. China

Xijuan Liu
Tianjin Nankai High School, Tianjin, P.R. China

ABSTRACT: With the rapid development of telecommunications, mobile terminals as bridges between users and networks have become vital entities in the communication process. However, mobile terminals of different brands or models are of varying quality, influencing their performance in practical networks. This paper proposes a novel mobile terminal evaluation scheme based on the call trace data, which truly records the operating conditions of all active terminals. The evaluation scheme investigates the mobile terminals from the perspective of both abnormal event ratio and limit EcIo, and measures terminal performance on both mean and extreme conditions. Moreover, the evaluation results are on a statistical basis, eliminating the possibility of disturbance or accidence. Thus, the results are objective and reliable, and can be used as a reference for a series of application fields, such as terminal purchasing, problem localizing, network optimizing, and so on.

Keywords: big data; mobile terminal; performance evaluation

1 INTRODUCTION

With the rapid development of telecommunications and mobile Internet, people are experiencing a society of data explosion. As the bridge between users and networks and mobile terminals is playing a more important role in the communication activity (Harri Holma et al. 2010, Erik Dahlman et al. 2010), benefiting from the prosperity are a large number of manufacturers that compete for producing mobile terminals with enhanced features and network ability in Han Hu et al. (2014). These terminals of different brands or models are of uneven quality and this may have vital influence on the network perception of users. Therefore, bad network experience is not always due to bad network quality. Moreover, this may mislead the work of network diagnosis and network optimization of telecom engineers. Under this circumstance, objective and comprehensive terminal performance evaluation is very important to telecom operators.

The conventional terminal evaluation is primarily carried out by terminal manufacturers and it mainly includes the stages of product design test, network access test, product quality test as well as the market launch test in Cukier K. et al. (2010). However, the evaluation process has some inherent disadvantages. First, it is based on individual terminal testing and the sample set is relatively small. Thus, it may introduce accidence in the evaluation process. Second, most evaluations are carried out based on the ideal inner field simulation, and doesn't account for the complicated and varying wireless environments (Han Hu et al. 2014).

To deal with the problem mentioned above, this paper proposes a novel mobile terminal evaluation scheme. It is based on the call trace data of active users in the practical networks, and evaluates the performance of mobile terminals from different perspectives. This can describe the network ability of mobile terminals more comprehensively under normal and extreme network conditions.

The rest of this paper is organized as follows. Section 2 gives the detailed illustration of the proposed scheme and section 3 presents the evaluation results of mainstream terminal models. Section 4 gives the significance and application fields of the proposed scheme. Finally, conclusions are drawn in section 5.

2 TERMINAL PERFORMANCE EVALUATION SCHEME

Mobile terminal is the media for users to interact with the cellular networks, and the performance of the mobile terminal is directly linked to the user perception of the network. Due to the difference in structural design and manufacturing level, mobile terminals of different brands or models can have different performances while operating in the cellular networks. Therefore, bad network experience is not always owing to bad network quality. There is need for network operators to develop a scheme to evaluate the terminal performance objectively and impartially. This will be beneficial to accurately address the network problem from the perspective of user perception. Moreover, the work of network optimization can get more effective and efficient.

This section starts from the introduction of operating data in practical cellular networks, and then proposes a novel scheme to evaluate the performance of mobile terminals comprehensively. The whole data set can be divided into three levels, which are message, event and KPI from the bottom to the top, respectively, in 3GPP TS25.331 (2013), and TS25.306 (2013). The examples of each level are listed in Figure 1.

Message: it's the minimum unit for interaction data between mobile terminals and the networks. It's produced by signaling data and aggregated into different events in the upper layer. It's the basis for deep analysis and problem localization.

Event: it's derived from message and aggregated into KPI in upper layer. It's divided into different categories, which describes different aspects of the network operating process. It can also be used to roughly localize and analyze network problems.

KPI: it's reflecting the main performance merits of the network from a general view. Typical merits can describe the network from the perspective of access ability, retain ability, mobility, and so on.

The three types of data mentioned above demonstrate the status of the network with different levels of specificity and different degrees of granularity. KPI is from the view of overall network condition, while the event and message are concerned about the underlying reasons. This paper concentrates on the event data in the medium layer, which can also be called as call trace data. It can be used to diagnose the network with relatively controllable data volume yet.

The event types in this scheme involve blocked call, dropped call, fast dormancy, handover from UTRAN as well as the call setup delay. These merits represent the performance of the mobile terminal from different aspects. To further determine the reasons for events mentioned above, another list of sub event types is shown in Table 1. The sub events which are related to the performance of mobile terminals are also labeled.

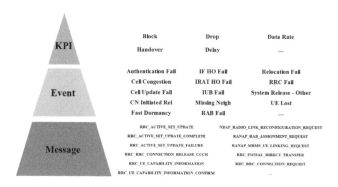

Figure 1.   Data set with three levels.

Table 1. The list of sub event types.

No	Sub Event	Related to UE
SubEvent1	Authentication Fail	
SubEvent2	Cell Congestion	
SubEvent3	CN Initiated Rel	
SubEvent4	Fast Dormancy	
SubEvent5	IUB Fail	
SubEvent6	Missing Neigh	
SubEvent7	RAB Fail - Code	
SubEvent8	RAB Fail - IUB Fail	
SubEvent9	RRC Fail - Code	
SubEvent10	RRC Fail - IUB Fail	
SubEvent11	System Release - Other	
SubEvent12	UE Lost - Low Coverage	
SubEvent13	UE Lost - No MR	
SubEvent14	Cell Update Fail	Yes
SubEvent15	IF HO Fail	Yes
SubEvent16	IRAT HO Fail	Yes
SubEvent17	RAB Fail - Other	Yes
SubEvent18	RAB Reconfig Fail	Yes
SubEvent19	Relocation Fail	Yes
SubEvent20	RRC Fail - DL Bad Quality	Yes
SubEvent21	RRC Fail - Other	Yes
SubEvent22	RRC Fail - Timeout	Yes
SubEvent23	UE Lost - DL Bad Quality	Yes
SubEvent24	UE Lost - Good RF	Yes
SubEvent25	UE Lost - UL Interference	Yes
SubEvent N	–	–

Note: the sub event index N depends on the event configuration of local call trace data.

By analyzing the percentage of abnormal events, which are related to mobile terminals, and calculating the limit EcIo when the mobile terminals are in the extreme network environment, the evaluation scheme is put forward and applied to investigate the mobile terminal performance in practical cellular networks.

Abnormal ratio: the percentage of all abnormal events, which are related to the performance of the mobile terminal. This is used to examine the general performance of terminals.

Limit EcIo: under the circumstance of bad signal quality or weak cell coverage, Limit EcIo can examine the terminal ability to retain the call.

Combination analysis of the two merits expressed above can be used to evaluate the performance of the terminal from the two different points of view. To be specific, the calculation of the two distinct merits are described as formulae (1) and (2).

$$AbnormalRatio = \frac{\sum_{i=14}^{i=25} SubEvent\_i\_times}{\sum_{i=i}^{i=25} SubEvent\_i\_times} \tag{1}$$

$$LimitEcIo = \frac{\sum_{i=12,20,23} SubEvent\_i\_EcIo}{\sum_{i=12,20,23} SubEvent\_i\_Times} \tag{2}$$

The Abnormal Ratio and Limit EcIo are calculated, respectively, for each individual terminal model. It's worth noting that, this scheme only considers terminal models with relatively large number of call trace records, and the threshold can be set as 10000 for instance. This is because the calculation results based on a relatively large number of samples are more

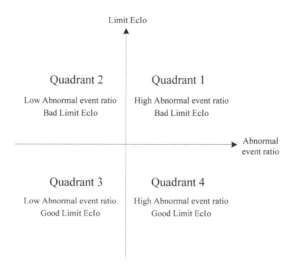

Figure 2. Terminal performance evaluation coordinate.

precise. Moreover, mobile terminals with abundant call trace data mean higher utilization rate, and the analysis is more meaningful for mainstream terminal models.

To clearly illustrate the pros and cons of various mobile terminal models, the two merits Abnormal Ratio and Limit EcIo are shown in a coordination system in Figure 2. The horizontal axis represents the Abnormal Ratio and the vertical axis represents the Limit EcIo.

Terminal models in quadrant 1 mean high Abnormal Sub event Ratio and bad Limit EcIo. Terminal Models in quadrant 2 mean low Abnormal Sub event Ratio and bad Limit EcIo. Terminal Models in quadrant 3 mean low Abnormal Sub event Ratio and good Limit EcIo. Terminal Models in quadrant 4 mean high Abnormal Sub event Ratio and good Limit EcIo. According to the analysis, it can arrive at the conclusion that terminal performance in quadrant 1 is worst, and that in quadrant 4 is best. Moreover, terminal performance in quadrant 2 and 3 falls in between.

3   EVALUATION RESULTS

In this section, it analyzes the performance of mainstream mobile terminal models according to the scheme proposed in the previous section. The evaluation results are based on the call trace data of a specific city's WCDMA network.

The abnormal ratio of mainstream models is shown in Figure 3. To be noted, only top 25 results are listed. The Limit EcIo of mainstream models is shown in Figure 4. Only top 25 results are listed as before.

The coordination system in Figure 5 shows the evaluation results of TOP 86 mainstream models from the perspective of multi dimensions.

Figure 6 is zoomed in from Figure 5, so that it is easy to figure out the clustering results of all terminal models, and find out the outlier models which need to be double checked. In the evaluation results shown above, it can be noted that the outlier models A and B fall in the first quadrant with high Abnormal Sub event Ratio and bad Limit EcIo. This means that the two models perform worse than most other models both in the normal and extreme network conditions.

The performance of the 86 terminal models in the coordination system is analyzed and the statistic results are shown in Table 2 as follows.

Brand A: the overall performance of Brand A terminals is good. Most models of this brand aggregate in the third quadrant, which means low Abnormal Sub event Ratio and good

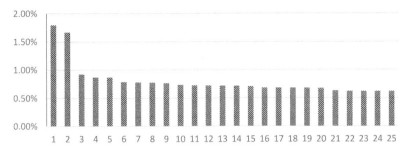

Figure 3.   Abnormal sub event ratio of TOP 25 terminal models.

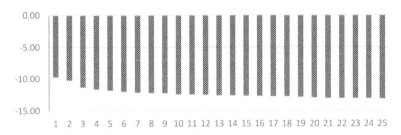

Figure 4.   Limit EcIo of TOP 25 terminal models.

Figure 5.   TOP 86 terminal models in the coordinate system.

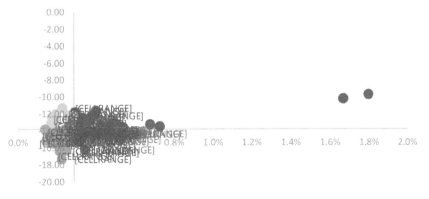

Figure 6.   Thumbnail view in the coordinate system.

Table 2. Summary of terminals in the coordinate system.

Brand	Number of models	Quadrant 1	Quadrant 2	Quadrant 3	Quadrant 4
A	27	2	8	15	2
B	18	13	3	1	1
C	11	4	2	3	2
D	7	5			2
E	6	2			4
F	6			6	
G	3			3	
H	2				2
I	1	1			
J	1	1			
K	1	1			
L	1	1			
M	1			1	
N	1			1	

Limit EcIo. There are also a few models falling on the second quadrant with relatively worse Limit EcIo.

Brand B: the overall performance of Brand B terminals is bad. Most models concentrate in the first quadrant, which means high Abnormal Sub event Ratio and bad Limit EcIo.

Brand C: the distribution of Brand C terminals in each quadrant is relatively uniform and the overall performance is fine. However, terminals modeled by M_1 and M_2 perform very badly in the coordinate system with high Abnormal Sub Event Ratio of over 1.7% and bad Limit EcIo of over −10.22dB.

Brand D: for the seven models evaluated in this brand, five models fall on the first quadrant, while the other two models lie in the fourth quadrant. The overall performance of Brand D terminals is not that good.

Brand E: similarly, 6 models of this brand mainly concentrate in the first and fourth quadrant with relatively high abnormal sub event ratio.

Others: six models of Brand F terminals and three models of Brand G terminals fall in the third quadrant, with a good performance both in Abnormal Sub Event Ratio and Limit EcIo.

4 APPLICATION FIELDS

Based on the massive call trace data, this paper analyzes the performance of mobile terminals in the perspective of both abnormal sub event ratio and limit EcIo. This evaluation system can go deep into the reasons for abnormal events and extreme conditions. The evaluation results can be applied in various fields like terminal purchasing, service developing, problem locating, network optimizing, and so on. The following section illustrates the details of some typical application cases.

Terminal purchasing: terminal performance in wireless networks can be an important merit for telecom operators in the process of terminal purchasing. The abnormal sub event ratio and limit EcIo demonstrate the statistical performance of a specific terminal model and reflect the terminal quality in mean and extreme network conditions. Thus, telecom operators can employ the evaluation results to assist the decision in terminal purchasing.

Problem localizing: based on the sub event types implied in the call trace data, telecom engineers can locate the reasons for high abnormal ratio problems. For example, the abnormal ratio of terminal model M_1 is very high in the previous evaluation, and it's caused by "UE lost – DL bad quality" and "UE lost—Good RF" as illustrated in Table 1. Then customized tests such as dialing tests can be employed to further locate and settle the problem.

Network optimizing: for terminal models with high penetration in the practical networks, telecom operators can carry out targeted network optimization strategy to address the problems. Take terminal model M_3 as an instance, it's of a high percentage in the networks. However, dropped call rate of this terminal model is high when it's in the condition of concurrent voice and data service. Potential reasons and corresponding optimization strategies can be carried out based on the signaling information.

5 CONCLUSION

This paper proposes a mobile terminal evaluation scheme based on the call trace data from the practical telecom networks. It's carried out on a statistical basis involving the whole sample data set, and evaluates terminal performance from the perspective of both the abnormal event ratio and limit EcIo. Thus, it can examine terminal network ability on both mean and extreme conditions objectively and reliably. This scheme is easy to undertake, and makes up for the shortcomings of the terminal field test. Moreover, it is applicable to both WCDMA and LTE networks, which is of great promotion value.

REFERENCES

3GPP TS 25.306. 2013. UE radio access capabilities (Release 11).
3GPP TS 25.331. 2013. Radio resource control (RRC) protocol specification (Release 11).
Cukier K. 2010. Data, data everywhere. *Economist*: 5–7.
Erik Dahlman, Stefan Parkvall, Johan Sköld, Per Beming. 2010. 3G evolution, HSPA and LTE for mobile broadband. *Post & Telecom Press*, 23: 335–367.
Han Hu, Yonggang Wen, Tat-Seng Chua, Xuelong Li. 2014. Toward scalable systems for big data analytics: A Technology Tutorial. *IEEE Access, vol. 2*.
Han Hu, Yonggang Wen, Tat-Seng Chua, Xuelong Li. 2014. Toward scalable systems for big data analytics: A Technology Tutorial. *IEEE Access, vol. 2*: 652–687.
Harri Holma, Antti Toskala. 2010. WCDMA for UMTS, HSPA evolution and LTE. *John Wiley & Sons,* 8: 173–218.

Signal and Information Processing, Networking and Computers – Chen & Huang (Eds)
© *2016 Taylor & Francis Group, London, ISBN 978-1-138-02881-4*

Human visual system based interference management method for video applications in LTE-A

Shiyu Zhou, Xinzhou Cheng, Mingqiang Yuan & Chen Cheng
Department of Network Optimization and Management, China Unicom Network Technology Research Institute, Beijing, China

ABSTRACT: In order to meet the urgent needs of rapid growth of wireless data traffic, especially video service data traffic, the LTE-A wireless network employs denser base stations. The frequent bursty interference caused by the dense base stations severely impacts the Quality of Experience (QoE) for video applications. This paper proposes a novel interference management method, which considers the characteristic of multiple services. The proposed method aims to improve the QoE for video applications by introducing the Rate Scaling Factor (RSF). Firstly, a novel QoE perception model based on characteristic of human visual system is designed. Then we develop the cellular interference traffic model. Combined the QoE perception models for video applications and Best-Effort (BE) applications, we utilize the proposed method to improve the QoE for video applications. Simulation results demonstrate the proposed method effectively improves the QoE for video users with the tolerated decrease in QoE for best effort users.

Keywords: video quality assessment; QoE; bursty interference; HVS

1 INTRODUCTION

As the rapid development of wireless network, multimedia services become more and more widely used in people's daily life (Xu et al. 2015). Meanwhile, mobile video traffic has been growing at an immense rate in recent years. Due to limited radio spectrum resources, the LTE-A wireless network (Xu et al. 2014) needs to deploy denser base stations to meet the growing demand. Studies show that by 2020, in order to deal with exponential growth of wireless data traffic, the density of base stations required to increase by about 20 times and the network will be heterogeneous network. Obviously, the densely deployed base stations will result in severe interference. The unpredictable, dynamic inter-cell interference becomes a major factor that affects the QoE of video service.

It is therefore crucial to address the interference issue and improve the QoE for video applications. Recently, many interference management algorithms have been proposed, such as interference decoding (Minero et al. 2012), interference alignment (Cadambe et al. 2008), interference mitigation (Xu and Chen 2009) etc. However, these methods do not consider interference characteristics of multiple services. In addition, some researchers take users' QoE into account in the interference management (Joseph et al. 2011) (Xu et al. 2013), however, they did not analyze the interference from other stations.

Above mentioned interference management methods pay little attention to the characteristic of different services. In this paper, we propose a novel interference management method based on Human Visual System (HVS) in dense deployment wireless networks. We consider the characteristic of HVS to develop the novel video quality assessment method.

Recently, video Quality Assessment (QA) is an important area of research. Popular HVS-based video quality indices include Perceptual Distortion Metric (PDM) (Winkler 1999), the Moving Pictures Quality Metric (MPQM) (Lambrecht and Verscheure 1996) etc. Under an assumption that the HVS is highly relevant to extracting structural information from a scene, the Structural Similarity (SSIM) (Wang et al. 2004) metric has been constructed. The SSIM method gained good performance in image quality assessment. In order to assess the video sequence quality, SSIM was extended to measure video distortion by considering motion information (Wang et al. 2002).

In this paper, we propose a novel interference management method, in which the drastic changes of video quality is decreased by spreading the power in temporal. In the proposed method, the characteristic of video applications and Best-effort applications are taken into consideration. The burstiness of interference is decreased by decreasing the peak rate of BE applications. Therefore, the QoE for video applications can be improved. In order to obtain the accurate video quality, we also propose a novel video quality assessment method, which employs a series of factors, including the human visual characteristics, the region of interest and temporal pooling. Considering the spatial and temporal characteristics, the saliency map can be obtained. And using the saliency map to weight the pixel quality which is produced by SSIM method, the final overall quality can be obtained. According to the perception of video quality, we decide whether employing the interference management method or not. Employing the proposed interference management algorithm, the QoE for video application can be improved and users' fairness can also be guaranteed.

The rest of this paper is organized as follows. Section 2 describes the novel QoE model for video based on HVS. Section 3 introduces the cellular interference and the QoE perception model for BE applications in the dense deployment wireless networks. In addition, Section 3 also introduces the rate scaling factor to reduce the burstiness of interference and improve the QoE for the video applications. The simulation results are given in Section 4. Finally, the conclusion is given in Section 5.

2 VIDEO QoE EVALUATION MODEL BASED ON HVS

The study on the Region Of Interest (ROI) mainly based on the human visual attention and eye movements. For the video applications, the human visual system tends to pay attention to the region of interest. In this paper, we use the saliency map to show the regions of interest. And we propose a model to generate saliency map, which is divided into spatial modeling and temporal modeling. The proposed video quality model has shown in Figure 1.

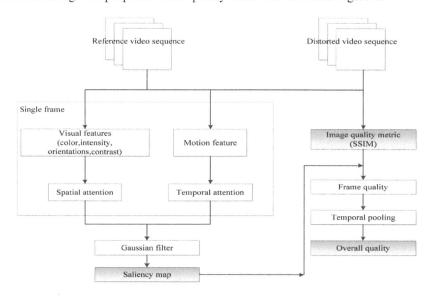

Figure 1. The proposed video quality model.

2.1 *The region of interest and saliency map generation*

The spatial attention take the color, brightness, contrast and direction features into account. In the field of image processing, color space, such as RGB, HSV and MTM, can be used to extract the color features in the image frames. In this section, we select HSV space for the extraction of color features. HVS space include hue (H), saturation (S) and value (V). H represents the basic color, S represents the depth of color with the value of 0 to 100%, and V represents the brightness, which shows the feeling of the human eye caused by the bright degree. Apparently, the human eye is more interested in the area of larger brightness contrast and pay more attention to this area. Whereas attention would be low.

First, we need to convert the color space from **RGB** to **HSV.**

When $R \neq G$ or $R \neq B$,

$$H = \arccos \frac{(R-G)+(R-B)}{2\sqrt{(R-G)^2 +(R-B)(G-B)}} \qquad (1)$$

When $B > G$,

$$H = 2\pi - \arccos \frac{(R-G)+(R-B)}{2\sqrt{(R-G)^2 +(R-B)(G-B)}} \qquad (2)$$

$$S = \frac{\max(R,G,B) - \min(R,G,B)}{\max(R,G,B)} \qquad (3)$$

$$V = \frac{\max(R,G,B)}{255} \qquad (4)$$

Then we need to quantitate the three component of color space (Xin et al. 2010), to obtain the final one-dimensional vector formula of color features.

$$C = \begin{cases} 0, & \text{if } V = 0 \text{ and } S = 0 \\ 6H + 3S + V, & \text{other} \end{cases} \qquad (5)$$

In formula, C represents the color features of visual interest region, which make full use of the image color information. Moreover, in the spatial attention model, the brightness, contrast and directivity are also the important elements. We use the method in (Itti et al. 1998) to process the brightness and directivity. A large number of experiments have shown that the response of the human visual system for brightness is mainly dependent on the relative changes of background luminance, not the absolute brightness value. So we adopt the way of brightness contrast to represent brightness characteristics.

Human eye vision is more sensitive to areas of high contrast, and distortion of this region has great impact on the result of the final video quality assessment. In this method, we introduced gaussian pyramid, by means of sampling and gaussian smoothing to highlight some features that cannot be seen in the original scale.

Finally, the saliency map is obtained by the weighted sum. The formula is as follows:

$$S_{am} = w_C * C + w_I * I + w_O * O + w_R * R \qquad (6)$$

where I denotes the intensity map, O denotes the orientation map and R denotes the contrast map. C, I, O and R are normalized results in the interval of [0, 1]. w_C, w_I, w_O and w_R denote the weighting value to the corresponding feature.

For the temporal attention, the motion is the main consideration. As we know, the impact from motion to the HVS is significant. HVS is more interested in the movement object in the image, and pay more attention to that. So the extraction of motion features pay an important role in the saliency map. In this paper, we use the method based on (Culibrk et al. 2009), which includes to part: the background model and the temporal filtering.

In Figure 2, for the background model, every frame of video sequence need to through Infinite Impulse Response (IIR), which is the low-pass filtering. We can obtain two reference frame from the IIR. And the processing as follows:

$$b_{t+1,l}(i,j) = (1-a_b)*b_{t,l}(i,j) + a_b*p_t(i,j), \quad l \in \{1,2\} \tag{7}$$

where a_b is the learning rate used to filter the background frame, $p_t(i)$ is the pixel value at location (i,j) in the current frame at time t, $b_{t,l}(i)$ is the pixel value at location (i,j) in the l-th background frame at time t.

According to (Culibrk et al. 2011), the learning rate is different for different background frame. And by a lot of experimental analysis, the relationship can be: $a_{b1} = a_{b2} \times 2$.

After the processing for the background model, the current frame and two background frames need to through the single-dimensional Mexican hat filter. The temporal filter is used to obtain the change of the current frame to the background and calculate the scale of the movement.

By the temporal filtering, the movement region of the video sequence can be obtained. However the area is not very good used in feature fusion processing, and it has singular point. So further processing needs to be normalized and eliminate singularities. In this method, the results from the temporal filtering will be normalized to the interval, then the value below the threshold value will be eliminated as shown in Figure 2. The validity and the accuracy of motion region extraction solution relies heavily on the setting of threshold. In the paper, we calculated the adaptive threshold by the results of the normalized, and the computation formula is as follows:

$$Th = \frac{1}{M*N}\sum_{i=1}^{M}\sum_{j=1}^{N} out\_norm(i,j) \tag{8}$$

Then based on the processing results of spatial and temporal attention, we can get saliency map. The weighted saliency map (A) is given by:

$$A = G*(\alpha*S_{am} + \beta*T_{am}) \tag{9}$$

where α and β are the weight values of each component, and should meet condition $\alpha+\beta=1$, G denotes a Gaussian filter.

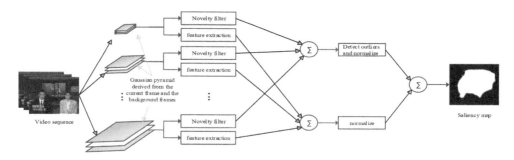

Figure 2.   Spatial-temporal attention model.

2.2 Image quality assessment and temporal pooling

Zhou Wang et al. [11] think that natural images are highly structured and the human vision system is more sensitive to structured information, so they proposed SSIM metric. And it becomes popular. In the paper, we choose SSIM metric as the frame assessment method, on account of its good performance and low complexity. SSIM include three components: luminance, contrast and structure, and the finally evaluate result is given by:

$$SSIM(x,y) = [l(x,y)]^\alpha \cdot [c(x,y)]^\beta \cdot [s(x,y)]^\gamma \tag{10}$$

where $l(x,y), c(x,y), s(x,y)$ denote luminance measurement, contrast measurement and structure measurement.

The 8×8 sliding window is selected to calculate the SSIM value of each pixel. And we can obtain the single frame quality as following:

$$SF_t = \sum_{i=1}^{M} \sum_{j=1}^{N} \frac{SSIM_t(i,j) \times A_t(i,j)}{\sum_{i=1}^{M} \sum_{j=1}^{N} A_t(i,j)} \tag{11}$$

where $[M,N]$ is the size of the video frame, $SSIM_t(i,j)$ denotes the SSIM value at location (x,y) at time t, and $A_t(i,j)$ denotes the attention value at location (x,y) at time t.

Then we have the single frame quality scores SF_t, which need to temporal pooling to obtain the final overall quality scores. The common schemes for pooling are Minkowski summation and direct average, which did not take the HVS characteristics for video sequence into consideration adequately. Some research have shown time-vary quality characteristics also can influence the video quality (Zink et al. 2003) and the human are more sensitive to degradation than to improvement for video quality (Masry et al. 2006). So in this paper, we consider the characteristics of HVS and utilize the low-pass function to adjust the single-frame quality scores:

$$SF_t' = \begin{cases} SF_{t-1}' + a_- \Delta_t, & if \ \Delta_t \leq 0 \\ SF_{t-1}' + a_+ \Delta_t, & if \ \Delta_t > 0 \end{cases} \tag{12}$$

where $\Delta_t = SF_t - SF_{t-1}'$. The values of a_- and a_+ are derived by training and it is different between them for the purpose of embodying the asymmetric tracking of human visual behavior. We use the parameter setting in (Masry et al. 2006), which present the values of a_- and a_+ are 0.04 and 0.5. SF_t' denotes the overall video quality.

3 INTERFERENCE MANAGEMENT METHOD FOR VIDEO APPLICATIONS

3.1 Cellular interference model

In this part, we focus on analysis of transmission scenario of video applications and the Best—effort applications in LTE-A (Xu et al. 2012).

As shown in Figure 3, we assume the BS_0 is serving the specific video client, UE_0, and there are J BS_s around, which are transmitting interfering data. The average Signal-to-Noise Ratio (SNR) is assumed to be γ at UE_0 from BS_0, and the interference to noise ratio (INR) from BS_j is denoted to be I_j, the other noise is n_0, ($n_0 << \sum_{j=1}^{J} I_j 1_j$), so an approximate expression of the signal to interference plus noise ratio (SINR) can be given by:

$$SINR = \frac{h_0 \gamma}{1 + \sum_{j=1}^{|J|} I_j 1_j} \tag{13}$$

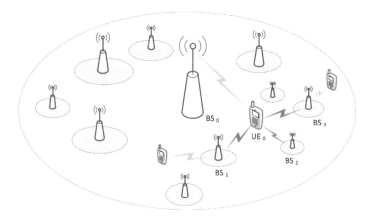

Figure 3.   The cellular interference model.

We know Shannon's capacity theorem shows the transmit rate is related to the transmit power. R_{oj} (at peak transmit power) is used to denote the average transmit rate at BS_j. Then we define the rate scaling factor α, which is a scalar values between 0 and 1. And the transmit power is scaled by $R_j = \alpha_j R_{oj}$ $(0 < \alpha < 1)$.

Then Wyner model (Wyner 1994) is used to simplify the analysis, $\gamma F_j(\alpha_j)$ would satisfy:

$$\alpha_j \log_2\left(1 + \gamma F_j(1)\right) = \log_2\left(1 + \gamma F_j(\alpha_j)\right) \tag{14}$$

So we can get the average received SINR at UE_0 as following:

$$SINR\left(\alpha_1, \cdots, \alpha_j\right) = \frac{h_0 \gamma}{1 + \sum_{j=1}^{|J|}\left(h_j\left(\left(1 + \gamma F_j(1)\right)^{\alpha_j} - 1\right)\right)1_j} \tag{15}$$

The interference probability depends on the traffic at BS_j, and the traffic can be modeled in two types. When a single user take web activity, his request time and think time are random. Some study have shown they obey exponential distribution. We set the request time is the duration of interference at BS_o, and the request probability can be given (Singh et al. 2012):

$$A_{f,j} = \mathrm{E}\left[1_j(\alpha_j)\right] = \frac{\Lambda_t}{\mu R(\alpha_j) + \Lambda_t} \tag{16}$$

For another situation, users' request frequency and service frequency is unrelated, such as download and mail check. The queue at BS_j is modeled as M/M/1 queue. The rate scaling factor α is considered in the model, the average service rate of this queue is R(α). The probability of the queue at BS_j can be given by,

$$A_{f,j} = \mathrm{E}\left[1_j(\alpha_j)\right] = \rho_j(\alpha_j) = \frac{\lambda}{\mu R_j(\alpha_j)} = \frac{\rho_j(1)}{\alpha_j} \tag{17}$$

$A_{f,j}$ denotes the interference probability. From formula we can see, by the interference shaping, the users 'request frequency is lower, thus the bursty interference are suppressed.

As we know, the QoE of video applications is negative correlation to the Packet Loss Rate (PLR). For the wireless video applications, the rate adaptive technology is adopted to adjust video bit rate. So to the simple analyze, we can use the average bit rate to be as the channel capacity, $B_n = \bar{C}(\vec{\alpha})$, $\vec{\alpha}$, is a vector, denoting the rate scaling factor.

Then the **PLR** of video applications is given as follows:

$$PLR(\vec{\alpha}) = E\left[\frac{\max(B_n - C_n, 0)}{B_n}\right] = \frac{E[\max(\bar{C}(\vec{\alpha}) - C_n, 0)]}{\bar{C}(\vec{\alpha})} \tag{18}$$

where C_n denotes the channel capacity at the time slot of nth frame. From the cellular interference model and the traffic model, we can get $\bar{C}(\vec{\alpha})$ and C_n.

For the single service, the **PLR** can be given as following,

$$PLR(\alpha_1) = \frac{A_{f,1}(1 - A_{f,1})\log\left(\dfrac{1 + h_0\gamma}{1 + \dfrac{h_0 r}{1 + (h_1((1 + rF_1)^{\alpha_1} - 1))}}\right)}{A_{f,1}\log\left(1 + \dfrac{h_0\gamma}{1 + (h_1((1 + rF_1)^{\alpha_1} - 1))}\right) + (1 - A_{f,1})\log(1 + h_0 r)} \tag{19}$$

3.2 QoE perception model for BE applications and video applications

We develop the QoE perception model using Weber-Fechner Law based the proposed framework in (Reichl et al. 2010). And we mainly consider web browsing and file download. In the model, we add the RSF, which can reduce the speed of the file download and increase the average page load time. The average page load time (or file download time) can be represented as:

$$E(D) = D_{back} + \frac{1}{\mu R(\alpha)} \tag{20}$$

where D_{back} denotes transmission delay and the server response delay, and $\frac{1}{\mu R(\alpha)}$ denotes the wireless link delay when the size of the page or file is $\frac{1}{\mu}$.

For web browsing service, there are logarithmic dependence between the user's QoE and page load time, namely:

$$MOS_{page} = \eta_1 \log\left(\frac{1}{\mu R(\alpha)}\right) + \eta_2 \tag{21}$$

For file download service, the user experience of file download and service completion time have strong logarithmic dependence relationship. So Weber-Fechner is utilized to represent the relationship:

$$MOS_{download} = \varepsilon_1 \log(\alpha) + \varepsilon_2 \tag{22}$$

Thus, we get the QoE perception model of BE application.

For the QoE of video applications, we use the model in the section2. The result calculated by the model are mapped to DMOS values (Sheikh et al. 2006), in which DMOS value is higher, the QoE of video service is lower. We define the parameters ξ be the improvement of video service QoE, which is proportional to with the adjustment factor α, shown by the following:

$$\xi = DMOS(RSF = 1) / DMOS(RSF = \vec{\alpha}) \tag{23}$$

We can obtain if interference shaping strategy improve the QoE of wireless video service.

4 SIMULATION ANALYSIS

In order to evaluate the effectiveness of the proposed method, we carry out the simulation experiment. In the simulation, we assume that there are 30 random users in the cell, including video users and Best-effort users. The reference video is uncompressed video sequences from video database EPFL–PoliMI, in YUV format, with the resolution of 352 * 288 and the frame rate is 30 frames per second. The video sequences use JM 13.1 for h. 264 / AVC to encode, and in the process of encoding, the way without B frame is used. The reference video coding rate is 1 MBPS. Video sequences are divided into multiple GOP (Group of Picture), and each GOP is 15 frames, including 14 P frame and 1 I frame.

In the process of simulation, we discuss two kinds of interference, one is the video user suffering one main interference base station in wireless network environment, and namely the interference of other stations can be neglected. The other kind of interference is the video user suffering the interference from multiple base stations around. In this situation, base stations need coordination in interference management, to ensure video service QoE.

First of all, considering two different types of interference in wireless network, comparative analysis is carried out. As shown in Figure 4, in two interference environment, we can see when the rate scaling factor $\alpha < 1$, the frame quality of the video sequence, Foreman, improved obviously, and the change of quality has remained roughly balanced compared with $\alpha = 1$.

According to the characteristics of HVS, the human eye is more sensitive to drastic changes of video quality, so we can learn that the method can effectively guarantee the overall QoE of video service. For the bursty interference caused by dense deployment of base stations in LTE-A wireless network, this method proposed could be effective for the video service.

Figure 5 shows that with the decrease of the rate scale factor, which means the scale is increasing, the QoE of video service enhances unceasingly. The improvement for QoE is defined to be ξ in the previous paper, and in the situation that there is a main interference base station, when $\alpha = 0.6$, the QoE for video sequence improve in two times. When $\alpha = 0.5$, the improvement can be three times. So the proposed method can effectively improve the QoE for video service in interference environment.

The improvement of QoE for video service is at the cost of reducing the Best effort service's experience quality, by lowering its transmission power. So, in order to ensure the users fairness, we also take the Best-effort service's QoE into account. Figure 6 shows that with reducing adjustment factor, which means the adjustment scale becoming bigger, the QoE for Best-effort service declines. However, the QoE for Best-effort service are not less than 3.5, and this is acceptable to Best-effort services.

Figure 7 shows the trade-off curve of Best-effort user and video user using the proposed QoE management methods. Figure 7 shows that even in the high value of rating scales factor, Best-effort service's QoE is in the tolerated decrease, however, the video service's QoE gets a significant improvement, improving more than 3 times.

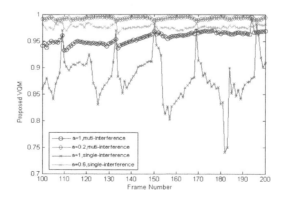

Figure 4.   The change of frame quality (foreman).

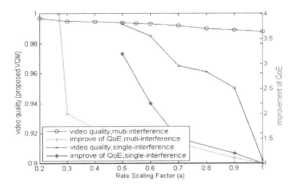

Figure 5.   The change of overall video quality.

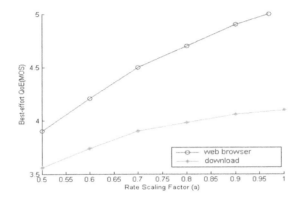

Figure 6.   The change of QoE for BE service.

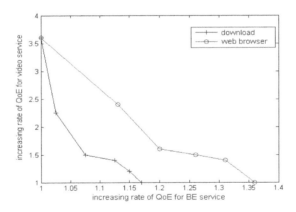

Figure 7.   The increase rate of QoE for video and BE applications.

Overall, simulation results show that the proposed interference management method can effectively guarantee and improve the QoE performance of the whole system.

5   CONCLUSION

In this paper, we have presented a novel interference management method based on HVS. In the method, we also proposed a video quality assessment model, which employs a series of factors, including the human visual characteristics, the region of interest and temporal pooling. Considering spatial attention and temporal attention, the saliency map was achieved. Then the SSIM index was used to assess each frame of the video sequences and a novel temporal pooling method was adopted to generate the overall video quality. In addition, we modeled the cellular interference, and developed the QoE perception models for video and BE applications. By introducing the rate scaling factor, the method proposed adjusted the peak rate of BE service to improve the QoE for video service. Simulation results demonstrated the proposed algorithm can effectively improve the QoE for video users with ensuring the user fairness.

REFERENCES

Cadambe, V.R. & Jafar, S.A. 2008. Interference alignment and spatial degrees of freedom for the K user interference channel. *Proc. IEEE International Conference on Communications*: 25–32.

Culibrk, D., Crnojevic, V. & Antic, B. 2009. Multiscale background modelling and segmentation. *Digital Signal Processing*: 1–6.

Culibrk, D., Mirkovic, M., Zlokolica, V. & Pokric, M. et al. 2011. Salient motion features for video quality assessment. *IEEE Trans. Image Process* 20: 948–958.

Itti, L., Koch, C. & Niebur, E. 1998. A model of saliency-based visual attention for rapid scene analysis. *IEEE Trans on Pattern Anal and Mach Intel* 20(11): 1254–1259.

Joseph, V. & de Veciana, G. 2011. Variability aware network utility maximization. *ArXiv Preprint ArXiv*: 1111–3728.

Lambrecht, C.J.B. & Verscheure, O. 1996. Perceptual quality measure using a spatiotemporal model of the human visual system. *Proc. The International Society for Optics and Photonics* 2668(1): 450–461.

Masry, M., Hemami, S.S. & Sermadevi, Y. 2006. A scalable wavelet-based video distortion metric and applications. *IEEE Trans. Circuits and Systems for Video Technology* 16: 260–273.

Minero, P., Franceschetti, M. & Tse, D.N.C. 2012. Random access: an information-theoretic perspective. *IEEE Transactions. Information Theory* 58(2): 909–930.

Reichl, P., Egger, S., Schatz, R. et al. 2010. The logarithmic nature of QoE and the role of the Weber-Fechner Law in QoE assessment. *IEEE International Conference* 2010: 1–5.

Sheikh, H., Sabir, M. & Bovik, A. 2006. A statistical evaluation of recent full reference image quality assessment algorithms. *IEEE Trans. Image Process* 15: 3440–3451.

Singh, S., Andrews, J.G. & de Veciana, G. 2012. Interference shaping for improved quality of experience for real-time video streaming. *IEEE J on Sel Areas in Commun.* 30(7): 1259–1269.

Wang, Z., Bovik, A.C., Sheikh, H.R. & Simoncelli, E.P. 2004. Image quality assessment: from error visibility to structural similarity. *IEEE Trans. Image Process* 13(4): 600–612.

Wang, Z., Lu, L. & Bovik, A.C. 2002. Video quality assessment using structural distortion measurement. *International Conference on Image Processing* 3: III-65–III-68.

Winkler, S. 1999. Perceptual distortion metric for digital color video. *Proc. The International Society for Optics and Photonics* 3644(1): 175–184.

Wyner, A. 1994. Shannon-theoretic approach to a Gaussian cellular multiple-access channel. *IEEE Trans. Inf. Theory* 40: 1713–1727.

Xin, L. & Shuaihua, D. 2010. A new animation image color-texture feature extraction method. *Proc. International Conference on Multimedia Technology*: 1–4.

Xu, L. & Chen, Y. 2009. Priority-based resource allocation to guarantee handover and mitigate interference for OFDMA systems. *Proc. IEEE Personal, Indoor and Mobile Radio Communications, Tokyo, Japan*: 783–787.

Xu, L., Chen, Y., Chai, K.K., Liu, D., Yang, S. & Schormans, J. 2013. User relay assisted traffic shifting in LTE-Advanced systems. *Proc. IEEE Vehicular Technology Conference, Dresden, Germany*: 1–7.

Xu, L., Chen, Y., Chai, K.K., Schormans, J. & Cuthbert, L. 2015. Self-organising cluster-based cooperative load balancing in OFDMA cellular networks. *Wiley Wireless Communications and Mobile Computing* 15(7): 1171–1187.

Xu, L., Chen, Y., Chai, K.K., Zhang, T. & Schormans, J. 2012. Cooperative load balancing for OFDMA cellular networks. *Proc. European Wireless, Poznan, Poland*: 1–7.

Xu, L., Luan, Y., Chao, K., Cheng, X., Zhang, H. & Schormans, J. 2014. Channel-aware optimised traffic shifting in LTE-Advanced relay networks. *Proc. IEEE Personal, Indoor and Mobile Radio Communications, Washington, USA*: 1597–1602.

Zink, M., Kunzel, O., Schmitt, J. & Steinmetz, R. 2003. Subjective impression of variations in layer encoded videos. *Lecture Notes in Computer Science* 2707: 137–154.

Signal and Information Processing, Networking and Computers – Chen & Huang (Eds)
© *2016 Taylor & Francis Group, London, ISBN 978-1-138-02881-4*

A novel complaint calls handle scheme using big data analytics in mobile networks

Yongfeng Wang, Xinzhou Cheng, Lexi Xu, Jian Guan, Tao Zhang & Mingjun Mu
Department of Network Optimization and Management, China Unicom Network Technology Research Institute, Beijing, China

ABSTRACT: The growth of data services has made it mandatory for telecom operators to identify factors which result in bad user experience. Most existing methods, especially key performance indicator based approaches, cannot precisely find the essential reason of the user complaint. In order to tackle these problems, this paper proposes a big data process scheme, via analyzing the large amounts of user data logged for telecom operators. In the proposed scheme, we process the whole information about complaints, including base stations, service type, user information, network Key Performance Indicators (KPIs) and user experience.

Keywords: telecom operator; big data; base station; user complaint

1   INTRODUCTION

Complaint call is vital for the network performance improvement. However, traditional analysis methods of complaint call data are inefficient, since it is hard to relate the mobile stations with user complaints. The telecom big data from core network to business support system (OSS) contain a large amount of user information (Chao et al. 2012). By associating user information with user behaviors in network, the big data analysis can trace the complaint users and reflect the drawbacks of networks. This can assist network operators in optimizing the network and improving construction efficiency (Yuxiao et al. 2011).

With the step of innovation of hardware and software, cell phones do play an important role in people's daily life. Wireless communication technologies help the development of mobile Internet and change user habits. However, telecom operators receive more complaints about bad experience of Apps or Internet surfing in recent years (Wojciech et al. 2012).

Customers dial complaint calls in order to express their dissatisfaction. However, these customers are not aware of the essential reason that disturbs themselves. From another perspective, when receiving a complaint call, telecom operators tend to regard it as a way to calm customers rather than a way to enhance network capability. Telecom operators usually check the network Key Performance Indicators (KPIs) to make sure whether any network device has encountered a major problem. The KPIs checking process is essential, however, operators are not able to find the exactly reason and records of both users and the network (Zhi et al. 2011).

Actually complaints stand for users suffering terrible telecom service experience. Since complaints have direct relation with critical customer value mining, further study is necessary. In most of traditional data analysis methods (Backstrom et al. 2010), the complaint call data based network quality improvement is inefficient. Since traditional data analysis methods ignore the relationship between business side and operation side of network, it is hard to pinpoint where user service starts. Big data, with strength of revealing the relation of massive data, is the best solution.

In order to effectively deal with the problem mentioned from the complaint call, we propose a complaint call handle scheme, where a big data system with several modules is established.

Every module plays a different role in the system. The system firstly collects users' history information and supplies the support staff with the portraits of complaint users. Furthermore it show user experience of different services after tracing the places where users have been. Then, by means of joint analysis of these modules, the system provides telecom operators with optimization plan.

The rest of this paper is organized as follows. Section 2 introduces objectives of the big data system as well as its logical structure. In section 3, we illustrate how support staff handle a complaint case with help of user portrait. Section 4 reveals how user trails are produced and how to utilize user trail to pinpoint network problems. Section 5 shows what composes of user experience and how the big data system relates network drawbacks with user experience. To illustrate how to improve network quality, in section 6, we show a demonstration of how the system improve network quality. Section 7 gives the conclusion.

2 SYSTEM OVERVIEW

The objective of the big data system is as follow:

– Suitable for providing user details for customer service staff.
  Big data provides on-demand customer provisioning, which dynamically provides user information during answering a complaint call.
– Increase the availability and efficiency.
  Big data increases the availability and efficiency in dealing with user complaint by detecting the underlying causes of that leads to bad user experience.
– Reduce the network investment.
  With complaints accumulate, drawbacks in network are highlighted in different areas and in various fields, which may precisely pinpoint where we invest money.
– Increase resiliency and robustness.
  Big data links user complaints system with network operation and maintenance. The interrelations between customer service and network management help telecom operators maintain more robust network.

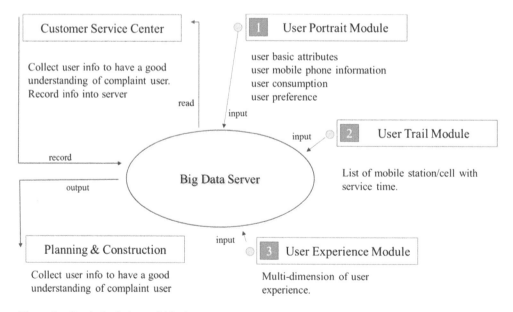

Figure 1.   Logical relations of big data system.

The big data sever aims to design and implement a large-scale platform, where support staff deal with user complaints, recording the key information to server. The overall goal is to detect what really affect the mobile network, so that detailed implementation programs can be carried out to enhance user experience.

The big data server is designed with three main functions. First of all, the big data server exchanges information with customer service center, so that support staff are able to fetch the whole user portrait when answer a complaint call. Meanwhile, the big data server helps staff transmit the real situation of the caller to the server, which may assist the server detecting the problem affecting the user experience. Secondly, the big data server has connections to several key modules that supply important data, including user portrait module, user trail module and user experience module. These modules separately draw a conclusion in different application fields through mathematic models. More information about users and mobile stations are available to server with the help of these modules. Thirdly, with big data analytics, the server outputs the planning and construction suggestions automatically. According to user complaints and big data analysis, the server outputs a solution as shown in Figure 1.

3 USER PORTRAIT

With the development of telecom big data, the value of subscriber information is widely appreciated. In the future, research on user relevant domain will be mainstreaming. Typical user information generally composes of many aspects, providing full profile of a mobile phone user. The user portrait extracts significant details from different data sources and aggregates them into several aspects. With user portrait, the following data is obtained:

- User basic attributes: phone number, IMSI, gender, age, service time etc.
- Mobile phone information: phone brand, phone type, phone operating system, price, capability etc.
- User consumption: average revenue per user, data of usage, minutes of usage, SMS usage, pricing package ID etc.
- User preference: main services, main phone applications, time preference, location preference etc.

Table 1 shows data source of user portrait. These data, including a large amount of both user basic information and user behavior information, are stored in the Business Support System (BSS). These data play a fundamental and necessary role in drawing user portrait.

As a significant connection between users and customer services, user portrait provide support staff with the most real and effective information about the user who are complaining on the other side of the phone. With user information, support staff are able to record more targeted details about what happened to the user, leading a more efficient conversation. After hanging up the phone, support staff submit key information during the conversation to the big data server, as shown in Figure 2. One complaint case generates an instance in the server, where all valuable information about the complaint user is recorded. It is worth

Table 1. Data source of user portrait.

Data	Source	Version
Monthly bill	Information department	1 month
Iu-ps/Gn	S-gateway	1 month
Detail record	P-gateway	1 month
User info	Information department	Latest version

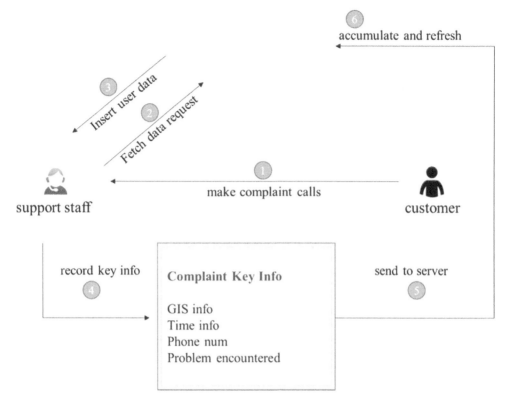

Figure 2.   The logical flow of user portrait module.

mentioning that the user information makes up an index of databases in the server, which links the other modules.

4   USER TRAIL

From the perspective of big data analysis, an effective solution is to find the correlation of various data sources. As shown in Figure 3, the complaint solution traces users' position to bond all the information together. Its essential character is that user trail lays the foundation of analysis basis. User trail can realize accurate distribution of users and user groups, as well as the order of position that user moves from one place to another.

With the help of user trail, telecom operators can rapidly seek out where user makes phone calls or surfing internet. Furthermore, the mobile stations where the phone call takes place are filtered out. Then the status and performance of the mobile stations can be checked to ensure if they function well. For more information, it is recommended to check the KPIs during a certain period.

The big data server aggregates all the modules together to be a more complex system. With maximum likelihood time matching algorithm, the big data server automatically calculates a relatively precise result that reveals the details about the bad experience of complaint user.

From the research perspective, this also implies that we can combine different source of data for different scenarios.

As shown in Figure 4, the user trail has three dimensions, including the service, the location and the time. When users send the service request message to network, their behaviors

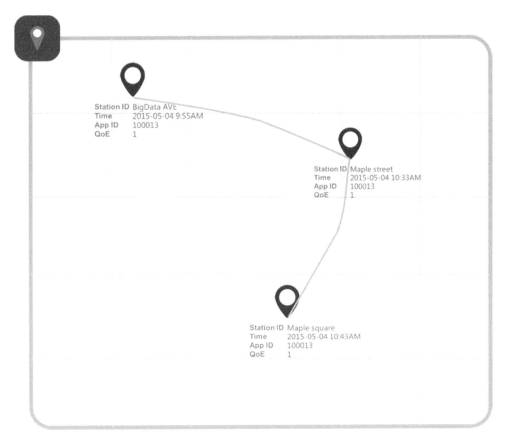

Figure 3.　A demonstration of user trail.

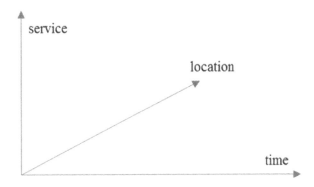

Figure 4.　The three dimensions of user trail.

are recorded in the user trail module. According to the time that complaint user mentioned on the phone, the system narrows the searching scope and matches a set of mobile stations, which have the large probability. The user trail module also helps staff retrospect the user's history footprint, which gives a global view of the complaint user.

The user trail module shown in Figure 5 is calculated on the basis of three categories of data. The Iu-ps/Gn/S1 data help revealing users' service feature of packet data service, such as webs and mobile applications. The call detail record data include user circuit data service information, while the mobile station data supply basic configuration of mobile stations.

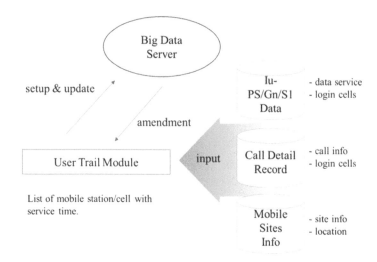

Figure 5.   Interaction of data and server in user trail module.

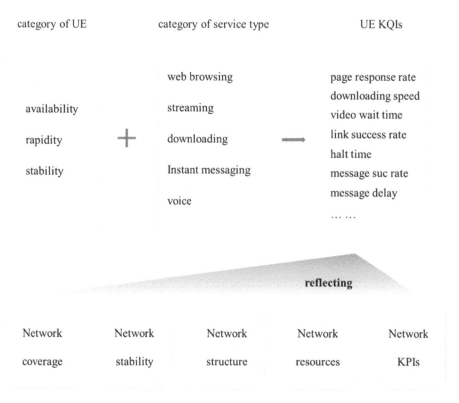

Figure 6.   Diagram of UE module affects network.

5   USER EXPERIENCE

However, the existing network optimization methods are facing the challenges of dramatic increasing in mobile application.

In order to calculate the user experience, we use Iu-PS/Gn data stream which is collected from core network. These data are an important input to the user experience algorithm. The user experience module calculates and generates a set of evaluation scores for every time user activates a service. Services are classified into five types, web browsing, streaming, downloading, instant messaging and voice. Each type of service has its own evaluation method. Through the module, UE KQIs are calculated, indicating the degree of user satisfaction with telecom services.

According to Figure 6, the big data server confirms what affect users by combine the user complaint data and UE module data together, fitting the network drawbacks.

6   NETWORK QUALITY IMPROVEMENT

The network suggestions are the visible outcome of big data server, which help telecom operators enact network construction or optimization plans.

A typical suggestion drawn from the server composes of many aspects of the network, including lack of resources, bad coverage, high interference, etc. To ensure the accuracy and efficiency, the big data server makes use of complaint data when its quantity reaches a relative large scale.

According to Figure 7, the big data server periodically output construction or optimization plans, after complaint data accumulate to a certain degree.

With user complaints accumulating, the network problems are more and more clear. Form Figure 8(a) to Figure 8(e), the network drawbacks are obvious in the timeline. In Figure 8(a), a few complaint calls are received. After multi-module reflection in server, complaint factors are transforms to networks drawback factors which are painted on the hot-map. While com-

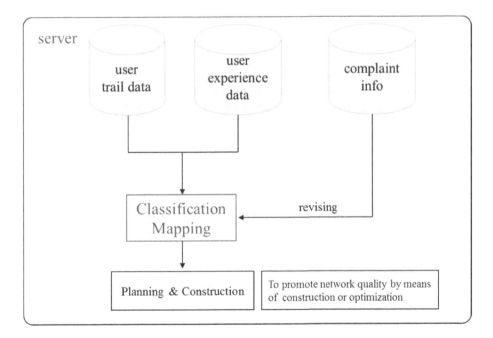

Figure 7.   The logical flow of network quality improvement.

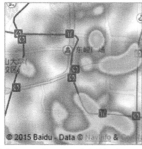 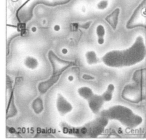 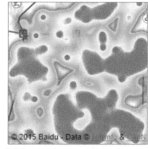

(a) 1 month complaint hotmap   (b) 2 months complaint hotmap   (c) 3 months complaint hotmap

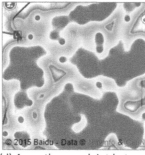 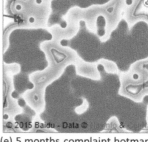

(d) 4 months complaint hotmap   (e) 5 months complaint hotmap

Figure 8.   User complaints distribution under different accumulation time.

plaints accumulating, network drawbacks are becoming more and more clear. After 5-months complaints collecting in Figure 8(e), hot-map is full of red regions, indicating emergency measures should be carried for network recovery.

7   CONCLUSION

This paper proposes a big data process scheme that help telecom operators optimize their network and meet users' demand. Based on the big data infrastructure, we establish an efficient and effective system, where a server with several powerful modules help solving complaint problems by outputting optimization and construction suggestions. We apply the big data analytical method to get the user complaint solution, which significantly reduces the time cost we spend on pinpointing network drawbacks and offers a macroscopic view of the whole network.

REFERENCES

Backstrom, L., E. Sun., & Marlow, C. 2010. Find me if you can: improving geographical prediction with social and spatial proximity, *Proc. World Wide Web Conf. New York, USA*: 1–7.
Chao Deng, Ling Qian, Meng Xu, Yujian Du, Zhiguo Luo, & Shaoling Sun. 2012. Federated cloud basedbig data platform in telecommunications, *Proc. Workshop on Cloud Services, Federation, and the 8th Open Cirrus Summit, New York, USA*: 44–48.
Do, T., & Gatica-Perez, D. 2012. Contextual conditional models for smartphone-based human mobility prediction, *Proc. ACM Int. Conf. on Ubiquitous Computing, Pittsburgh, USA*: 12–15.
Eagle, N., Pentland, A., & Lazer, D. 2009. Inferring friendship network structure by using mobile phone data, *Proc. the National Academy of Sciences*, 106(36):15274–15278.
Gonzalez, M.C., Hidalgo, C.A., & Barabasi, A.L. 2008. Understanding individual human mobility patterns, *Nature* 45(11): 779–882.

HDJ Jeong, Hyun, WS., Lim, J., & You, I. 2012. Anomaly teletraffic intrusion detection systems on hadoop-based platforms: A survey of some problems and solutions, *Proc. International Conference on Network-Based Information System, Vienna, Austria*: 766–770.

Lexi Xu, Yue Chen, KoK Keong Chai, Dantong Liu, Shaoshi Yang, & John Schormans. 2013. User relay assisted traffic shifting in LTE-advanced systems, *Proc. IEEE Vehicular Technology Conference, Dresden, Germany*: 1–7.

Lexi Xu, Yue Chen, KoK Keong Chai, Tiankui Zhang, & John Schormans. 2012. Cooperative load balancing for OFDMA cellular networks, *Proc. European Wireless, Poznan, Poland*: 1–7.

Lexi Xu, Yuting Luan, Kun Chao, Xinzhou Cheng, Heng Zhang, & John Schormans. 2014. Channel-Aware Optimised Traffic Shifting in LTE-Advanced Relay Networks, *Proc. IEEE International Symposium of Personal, Indoor and Mobile Radio Communications, Sept. 2014, Washington, USA*: 1597–1602.

Magnusson, J., & Kvernvik, T. 2012. Subscriber classification within telecom networks utilizing big data technologies and machine learning, *Proc. International Workshop on Big Data, Streams and Heterogeneous Source Mining: Algorithms, Systems, Programming Models and Applications, New York, USA*: 77–84.

Wojciech Indyk, Tomasz Kajdanowicz, & Przemysław Kazienko. 2012. Cooperative decision making algorithm for large networks using mapreduce programming model, *Cooperative Design, Visualization, and Engineering, Springer Berlin Heidelberg, Germany*: 53–56.

Yuxiao Dong, Qing Ke, Yanan Cai, Bin Wu, & Bai Wang. 2011. Teledata: data mining, social network analysis and statistics analysis system based on cloud computing in telecommunication industry, *Proc. International Workshop on Cloud Data Management, New York, USA*: 41–48.

Zhi Qiu, Zhaowen Lin, & Yan Ma. 2011. Research of hadoop-based data flow management system, *The Journal of China Universities of Posts and Telecommunications* 18(2): 164–168.

Signal and Information Processing, Networking and Computers – Chen & Huang (Eds)
© 2016 Taylor & Francis Group, London, ISBN 978-1-138-02881-4

A novel big data based problematical sectors detection algorithm in WCDMA networks

Pengfei Ren, Xinzhou Cheng, Lexi Xu, Mingqiang Yuan & Kun Chao
Department of Network Optimization and Management, China Unicom Network Technology Research Institute, Beijing, P.R. China

ABSTRACT: Recently, the optimization work for cellular networks becomes more and more complex, and the amount of cellular networks' data is increasing gradually. This paper proposes a novel big data based Problematical Sectors Detection Algorithm (PSDA) in WCDMA networks. Firstly, the architecture of the proposed algorithm contains three layers, including the target layer and the medium layer as well as the original data layer, which include some indicators separately. We can calculate the score of the target layer's indicator "Network Performance" for one sector by the medium layer and original data layer's indicators through the Z-Score algorithm and AHP algorithm. The higher the score is, the better the network performance is. Secondly, we sort the sectors by their network performance scores from high to low. We divide the sectors into three categories, such as the sectors with good, medium, or poor performance according to the sectors' network performance scores and the corresponding thresholds. Then network operators can detect the sectors with poor or medium network performance quickly and accurately, and then solve the network problems accordingly. In this way, network operators can improve their optimization work efficiency greatly. Because a large scale data analyzing is usually needed, we built a big data analyzing platform. The problematical sectors detection algorithm is deployed on the big data platform to enhance the computing efficiency.

Keywords: cellular networks; big data; problematical sectors; network performance

1 INTRODUCTION

In Liu Jun et al. (2013), it is described that telecom operators possess huge amounts of network data (CDR, MR, KPI, etc.) as Mobile Service Providers (MSPs). In Sahni et al. (2015), it is mentioned that the number of mobile subscribers are increasing gradually, and the optimization work for cellular networks becomes more and more complicated. Network operators need to respond to the changes of the networks to provide optimization work quickly.

Existing literatures mainly focus on some performance optimization of cellular networks which is described in Karatepe et al. (2014). A novel handover self-optimization scheme is proposed in Chien-Lung et al. (2014) which contains a self-configuration Neighbor Cell List (NCL) for new cells and a NCL self-optimization scheme of each cell in the network. In Pengfei et al. (2011) and Lexi et al. (2009), inter-cell interference coordination schemes are proposed to mitigate the inter-cell interference efficiently. And the access performance of cellular networks can be improved by the schemes proposed in Krishnamurthi et al. (2013) and Lexi et al. (2015).

In Celebi et al. (2013) and Lexi et al. (2015), it is mentioned that existing research and literatures mainly focus on detecting some network problems. And there are only few research and literatures concerning how to detect problematical sectors according to some comprehensive indicators which is described in Ramaprasath et al. (2015), such as accessibility, retainability, mobility and integrity indicators. However, detecting problematical sectors accurately and rapidly is the key for network operators responding to the changes of the networks quickly. Therefore, it is necessary and essential to develop an efficient scheme for network operators to detect problematical sectors accurately and improve their optimization work efficiency.

In this paper, a novel problematical sectors detection algorithm is proposed. The algorithm model includes three layers, including the target layer and the medium layer as well as the original data layer, which include some indicators respectively. Actually, in order to get the score of the target layer's indicator "Network Performance" for one sector, the algorithm includes the following steps. Firstly, we can calculate the scores of the original data layer's indicators by the network data from the network management system, using the Z-Score algorithm which is a mature algorithm and can be understood according to related research materials. Secondly, we can calculate the scores of the medium layer's indicators by the scores and weights of the original data layer's indicators. Thirdly, we can calculate the score of the target layer's indicator "Network Performance" by the scores and weights of the medium layer's indicators. The weights of these indicators can be calculated by the AHP algorithm which is also a mature algorithm and can be understood according to related research materials which is described in Chou et al. (2013). Actually, the score of the target layer's indicator "Network Performance" represents the sector's network performance level. The higher the score is, the better the network performance is. Fourthly, we sort the sectors by their network performance scores from high to low, and divide the sectors into three categories as the sectors with good, medium, or poor network performance according to their network performance scores and corresponding thresholds. Finally, network operators can detect the sectors with poor or medium network performance quickly and accurately, and then solve the network problems accordingly. Because of needing a large scale data analyzing, we built a big data analyzing platform to enhance the computing efficiency. And the proposed scheme is deployed on the big data platform.

The rest of this paper is organized as follows. In Section 2, our big data analyzing platform is introduced. And the algorithm model of the scheme proposed in the paper is described briefly in Section 3. And the following Section 4 describes the flow of the scheme in detail. Then, Section 5 provides the implementation effect of the scheme in a practical network optimization project. Finally, Section 6 summarizes our work.

2 BIG DATA ANALYZING PLATFORM

Figure 1 shows the architecture of our big data analyzing platform. The big data platform provides the execution environment of the problematical sectors detection algorithm proposed in this paper.

As illustrated in Figure 1, the big data analyzing platform contains Data Collection Layer, Data Processing Layer and Application Layer.

Data Collection Layer is responsible for collecting network data, such as KPIs, MR, engineering parameters, etc. Before the network data is analyzed, Data Processing Layer will pretreat the network data. After the pretreatment, the format of the network data is convenient for us to store and analyze the data. In Data Processing Layer, Hadoop, Hive and Spark SQL are responsible for storing and processing the network data. Through the big data modeling and mining, lots of novel applications become realizable or precise. In our big data analyzing platform, Application Layer includes Network planning, Network Optimizing and Precise Marketing. The problematical sectors detection algorithm proposed in this paper is a part of Network Optimizing. Because of deployed on the big data analyzing platform, the proposed algorithm works more efficiently.

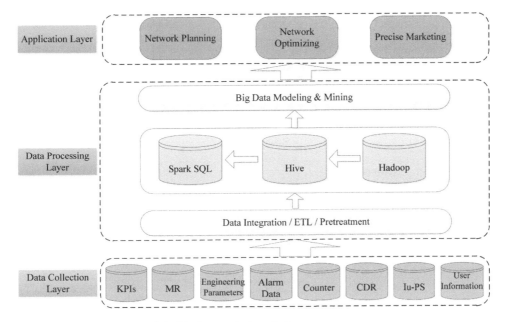

Figure 1.   The architecture of the big data analyzing platform.

3   SYSTEM MODEL OF PROBLEMATICAL SECTORS DETECTION ALGORITHM

3.1   *The model of the problematical sectors detection algorithm*

In this paper, the model of the proposed algorithm includes three layers such as the target layer, medium layer and original data layer. As illustrated in Figure 2, the original data layer contains some indicators such as "Access Success Probability", "Drop Call Probability", "Handover Success Probability" and "Data Rate". The medium layer includes some indicators such as "Accessibility", "Retainability", "Mobility" and "Integrity". The target layer contains one indicator that is "Network Performance". In fact, one sector's score of the indicator "Network Performance" represents the sector's network performance level. The higher the score is, the better the network performance is.

Every indicator has its attributes such as Zn and Wn (n is an Arabic number as illustrated in Figure 2) which represent its score and weight. We can calculate one indicator's attribute Zn using the Z-Score algorithm and calculate its attribute Wn through the AHP algorithm. The Z-Score algorithm and AHP algorithm are both mature algorithms and can be understood according to related research materials.

3.2   *Classification standard of network performance*

As mentioned above, we divide the sectors into three categories as the sectors with good, medium, or poor network performance according to their network performance scores and the corresponding thresholds as illustrated in Figure 3.

After calculating the score of the target layer's indicator "Network Performance" for every sector, we can calculate the total average score of all the sectors' indicators "Network Performance". And the total average score is denoted by $\overline{S}$. The parameter a% is used for threshold setting as illustrated in Figure 3, and its value can be adjusted according to our needs.

As illustrated in Figure 3, if the score of one sector's indicator "Network Performance" is between ($\overline{S}$ (1+a%) +∞), the sector's network performance is good. Similarly, if the score is

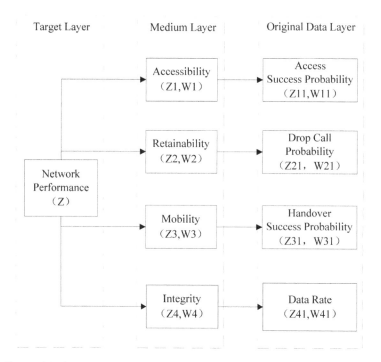

Figure 2.   The model of problematical sectors detection algorithm.

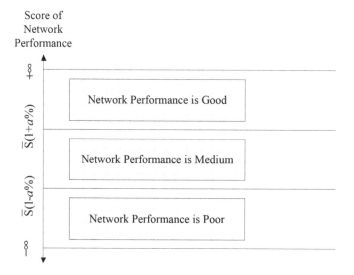

Figure 3.   Classification standard of network performance.

between $(\bar{S}\,(1-a\%)\,\bar{S}\,(1+a\%))$, the sector's network performance is medium. And if the score is between $(-\infty,\,\bar{S}\,(1-a\%))$, the sector's network performance is poor.

3.3   Calculations of basic parameters

In the following Section 4, some equations and parameters are needed. The original data layer's indicator "Access Success Probability" denoted by P_{AS} can be calculated by

$$P_{AS} = P_{RCES} \times P_{RSS} \tag{1}$$

where P_{RCES} is the indicator RRC connection establishment success probability, and P_{RSS} is the indicator RAB setup success probability. Actually, the indicators P_{RCES} and P_{RSS} can be calculated by the following equation (2) and equation (3) separately.

The indicator P_{RCES} can be calculated by

$$P_{RCES} = N_{RCES} / N_{RCER} \times 100\% \tag{2}$$

where N_{RCES} is the total number of RRC connection establishment success, and N_{RCER} is the total number of RRC connection establishment requests.

The indicator P_{RSS} can be calculated by

$$P_{RSS} = \frac{N_{CRSS} + N_{RPRSS} + N_{HRSS}}{N_{CRSR} + N_{RPRSR} + N_{HRSR}} \times 100\% \tag{3}$$

where N_{CRSS} is the total number of CS RAB setup success, N_{RPRSS} is the total number of R99 PS RAB setup success, N_{HRSS} is the total number of HSDPA RAB setup success, N_{CRSR} is the total number of CS RAB setup requests, N_{RPRSR} is the total number of R99 PS RAB setup requests, and N_{HRSR} is the total number of HSDPA RAB setup requests.

In equation (2) and equation (3), the indicators, such as N_{RCES}, N_{RCER}, N_{CRSS}, N_{RPRSS}, N_{HRSS}, N_{CRSR}, N_{RPRSR} and N_{HRSR}, are the KPI indicators of the cellular networks and can be gotten from the network management system which is mentioned in Brunner et al. (2006).

Table 1 lists system parameters and definitions which will be used in this paper.

The original data layer's indicator "Drop Call Probability" denoted by P_{DC} is

$$P_{DC} = \frac{N_{CRAR} + N_{RPRAR} + N_{HRAR}}{N_{CRR} + N_{RPRR} + N_{HRR}} \times 100\% \tag{4}$$

where N_{CRAR} is the total number of CS RAB abnormal release, N_{RPRAR} is the total number of R99 PS RAB abnormal release, N_{HRAR} is total number of HSDPA RAB abnormal release, N_{CRR} is total number of CS RAB release, N_{RPRR} is total number of R99 PS RAB release, and N_{HRR} is total number of HSDPA RAB release. The indicators, such as N_{CRAR}, N_{RPRAR}, N_{HRAR}, N_{CRR}, N_{RPRR}, and N_{HRR}, are the KPI indicators of the cellular networks and can be gotten from the network management system.

The original data layer's indicator "Handover Success Probability" denoted by P_{HS} is

$$P_{HS} = \frac{N_{SHS} + N_{IFHS}}{N_{SHR} + N_{IFHR}} \times 100\% \tag{5}$$

where N_{SHS} is the total number of soft handover success, N_{IFHS} is the total number of inter-frequency handover success, N_{SHR} is the total number of soft handover requests, and N_{IFHR} is the total number of inter-frequency handover requests. The indicators, such as N_{SHS}, N_{IFHS}, N_{SHR}, and N_{IFHR}, are the KPI indicators of the cellular networks and can be gotten from the network management system.

The original data layer's indicator "Data Rate" denoted by D_{DR} is

$$D_{DR} = \frac{T_{HRDT} + T_{RPRDT} + T_{CVCT}}{T} \tag{6}$$

where T_{HRDT} is the downlink traffic of the HSDPA RLC layer, T_{RPRDT} is the downlink traffic of the R99 PS RLC layer, T_{CVCT} (unit: Kbyte) is the converted traffic of the CS voice and T is the lasting time of the service. The indicators, such as T_{HRDT} and T_{RPRDT}, are the KPI indicators of the cellular networks and can be gotten from the network management system. In equation (6), the indicator T_{CVCT} can be calculated by equation (7).

Table 1. System parameters and definitions.

Parameter	Definition
W1, W2, W3, W4,W11, W21, W31, W41	The weights of the medium layer and original data layer's indicators as illustrated in Figure 2.
Z, Z1, Z2, Z3, Z4, Z11, Z21, Z31, Z41	The scores of the target layer, medium layer and original data layer's indicators as illustrated in Figure 2.
P_{AS}	The original data layer's indicator "Access Success Probability".
P_{RCES}	RRC connection establishment success probability.
P_{RSS}	RAB setup success probability.
N_{RCES}	The total number of RRC connection establishment success.
N_{RCER}	The total number of RRC connection establishment requests.
N_{CRSS}	The total number of CS RAB setup success.
N_{RPRSS}	The total number of R99 PS RAB setup success.
N_{HRSS}	The total number of HSDPA RAB setup success.
N_{CRSR}	The total number of CS RAB setup requests.
N_{RPRSR}	The total number of R99 PS RAB setup requests.
N_{HRSR}	The total number of HSDPA RAB setup requests.
P_{DC}	The original data layer's indicator "Drop Call Probability".
N_{CRAR}	The total number of CS RAB abnormal release.
N_{RPRAR}	The total number of R99 PS RAB abnormal release.
N_{HRAR}	The total number of HSDPA RAB abnormal release.
N_{CRR}	The total number of CS RAB release.
N_{RPRR}	The total number of R99 PS RAB release.
N_{HRR}	The total number of HSDPA RAB release.
P_{HS}	The original data layer's indicator "Handover Success Probability".
N_{SHS}	The total number of soft handover success.
N_{IFHS}	The total number of inter-frequency handover success.
N_{SHR}	The total number of soft handover requests.
N_{IFHR}	The total number of inter-frequency handover requests.
D_{DR}	The original data layer's indicator "Data Rate".
T_{HRDT}	The downlink traffic of the HSDPA RLC layer.
T_{RPRDT}	The downlink traffic of the R99 PS RLC layer.
T_{CVCT}	The converted traffic of the CS voice.
T	The lasting time of the service.
T_{CVT}	The CS voice traffic (unit: Erl).

$$T_{CVCT} = \frac{T_{CVT} \times 50 \times 3600}{8} \tag{7}$$

where T_{CVT} is the CS voice traffic (unit: Erl). The indicator T_{CVT} is the KPI indicator of the cellular networks and can be gotten from the network management system.

4 FLOW AND ANALYSIS OF PROBLEMATICAL SECTORS DETECTION ALGORITHM

4.1 *Calculating the scores of the original data layer's indicators*

Actually, we can calculate the values of the original data layer's indicators according to equations (1) to (7). And then we can calculate the scores of the indicators, such as Z11, Z21, Z31, Z41 as illustrated in Figure 2, according to the indicators' values through Z-Score algorithm.

The indicators of different dimensions become comparable through the scores calculated by Z-Score algorithm.

4.2 Calculating the weights of the original data layer and medium layer's indicators

Every indicator of the original data layer and medium layer has a weight, such as W1, W2, W3, W4, W11, W21, W31 and W41 in Figure 2. The weights can be calculated by AHP algorithm. From the medium layer to the original data layer, the indicator "Accessibility" owns only one branch which is the indicator "Access Success Probability", and so the weight W11 equals 1. The weights W21, W31 and W41 are similar to W11, and also equal 1.

4.3 Calculating the scores of the medium layer's indicators

After calculating the scores and weights of the original data layer's indicators, we can calculate the scores of the medium layer's indicators, such as Z1, Z2, Z3, Z4 in Figure 2, by the following equations.

$$Z1 = W11 \times Z11 \tag{8}$$
$$Z2 = W21 \times Z21 \tag{9}$$
$$Z3 = W31 \times Z31 \tag{10}$$
$$Z4 = W41 \times Z41 \tag{11}$$

where W11, Z11, W21, Z21, W31, Z31, W41 and Z41 are the scores and weights of the original data layer's indicators as illustrated in Table 1.

4.4 Calculating the score of the target layer's indicator

After calculating the scores and weights of the medium layer's indicators, we can calculate the score of the target layer's indicator "Network Performance" by the following equation.

$$Z = Z1 \times W1 + Z2 \times W2 + Z3 \times W3 \tag{12}$$

where Z1, W1, Z2, W2, Z3 and W3 are the scores and weights of the medium layer's indicators as illustrated in Table 1. In equation (12), the score Z is used for evaluating the network performance. And the higher the score Z is, the better the network performance is.

4.5 Evaluating the network performance

First, we sort the sectors from high to low by each sector's score Z which can be calculated by the equations (1) to (12).

Second, we calculate the total average score $\bar{S}$ of all the sectors' scores Z. As illustrated in Figure 3, the parameter a% is used for threshold setting, and can be adjusted according to our needs. If one sector's score Z is between $(\bar{S}(1+a\%), +\infty)$, the sector's network performance is good. And if the score Z is between $(\bar{S}(1-a\%), \bar{S}(1+a\%))$, the sector's network performance is medium. And if the score Z is between $(-\infty, \bar{S}(1-a\%))$, the sector's network performance is poor.

Finally, network operators should focus on the sectors which have poor or medium network performance. Because the user perception of these sectors is likely poor, we should firstly optimize these sectors and improve the network performance efficiently.

5 IMPLEMENTATION EFFECT OF PROBLEMATICAL SECTORS DETECTION ALGORITHM

In this section, the proposed problematical sectors detection algorithm is applied in a practical network optimization project as illustrated in Figure 4.

The area in Figure 4 is the downtown of some big city in China. In this area, the densities of the WCDMA sites and mobile users are high relatively and the wireless environment is complicated. The main network configuration and PSDA parameters are shown in Table 2.

As illustrated in Figure 4, the area of different colors is the coverage area of WCDMA micro sites, and the dots are WCDMA indoor distribution systems. The area in Figure 4 is a part of the network optimization area of this project. Take the micro sites' coverage area for an example. The area's different colors represent different network performance levels. For example, the red area's network performance is best and the blue area's network performance is worst. The numbers in the legend of Figure 4 illustrate different ranges of the score Z.

In this network optimization project, the total average score $\bar{S}$ of all the sectors' scores Z is 3.35 according to the scheme proposed in this paper. As illustrated in Table 2, the parameter a% is set as 50%.

According to the classification standard of the network performance mentioned in Section 3, if one sector's score Z is between $(5.02, +\infty)$, the sector's network performance is considered good and its coverage area is red as illustrated in Figure 4. And if one sector's score Z is between $(1.67, 5.02)$, the sector's network performance is medium. As the number of the sectors with medium network performance is very large relatively, the scores Z of the sectors with medium network performance are further divided into three ranges. And these

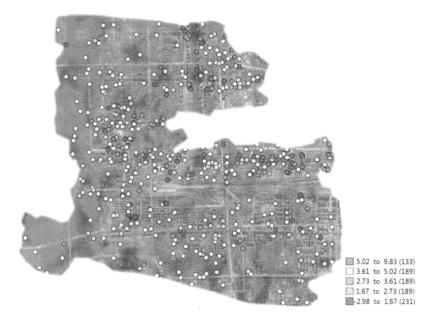

Figure 4.   Implementation effect of problematical sectors detection algorithm.

Table 2.   The main network configuration and PSDA parameters.

Parameter	Value
Network Type	WCDMA
Carrier Frequency	2.1 GHz
System Bandwidth	15 MHz
Inter-site Distance	435 m
Number of Antennas	$1T \times 2R$
Cell Total Transmission Power	20 W
a% in PSDA Threshold	50%

sectors' coverage area is in three colors as illustrated in Figure 4. In this way, after optimizing the sectors with poor network performance, we can focus on the sectors with relatively low scores Z among the sectors with medium network performance. And if one sector's score Z is between $(-\infty, 1.67)$, the sector's network performance is poor and its coverage area is blue as illustrated in Figure 4.

According to the proposed algorithm, we can accurately find the problematical sectors which are in the blue area of Figure 4. After optimizing the sectors in the blue area, we further optimize the other problematical sectors. The scores Z of these problematical sectors are low relatively among the sectors with medium performance. In this project, our optimization work is performed according to the scheme proposed in this paper. As a result, our optimization work efficiency is enhanced dramatically.

6 CONCLUSIONS

A novel big data based problematical sectors detection algorithm is proposed in this paper. In order to enhance the computing efficiency, we build a big data analyzing platform on which the proposed scheme is deployed. We implement the proposed scheme into our practical network optimization work. Through the scheme, we can objectively evaluate one sector's network performance according to the sector's score Z. In the proposed scheme, all the sectors are divided into three categories which are the sectors with good, medium, or poor network performance according to the sectors' scores Z and the corresponding thresholds. And network operators should focus on the sectors which have poor or medium network performance. In all, we can find the problematical sectors accurately by the novel scheme, and then solve the network problems. In this way, we can dramatically enhance our optimization work efficiency.

REFERENCES

Brunner, C., Garavaglia, A., Mittal, M., Narang, M. & Bautista, J.V. 2006. Inter-system handover parameter optimization. *Proc. Vehicular Technology Conference, VTC-2006 Fall. 2006 IEEE 64th*: 1–6.

Celebi, O.F., Zeydan, E., Kurt, O.F., Dedeoglu, O., Iieri, O., AykutSungur, B., Akan, A. & Ergut, S. 2013. On use of big data for enhancing network coverage analysis. *Proc. Telecommunications (ICT), 2013 20th International Conference on*: 1–5.

Chien-Lung Lee, Wen-Shu Su, Kai-An Tang, & Wei-I Chao. 2014. Design of handover self-optimization using big data analytics. *Proc. Network Operations and Management Symposium (APNOMS), 2014 16th Asia-Pacific*: 1–5.

Chou, Y.C., Yen, H.Y., Sun, C.C. & Hon, J.S. 2013. Comparison of AHP and fuzzy AHP methods for human resources in science technology (HRST) performance index selection. *Proc. Industrial Engineering and Engineering Management (IEEM), 2013 IEEE International Conference on*: 792–796.

Karatepe, Ilyas Alper, & Zeydan, Engin. 2014. Anomaly detection in cellular network data using big data analytics. *Proc. European Wireless 2014; 20th European Wireless Conference*: 1–5.

Krishnamurthi, R. & Gopal, K. 2013. Effective optimization and fault tolerance using multi-homed link assignment in cellular access network. *Proc. Contemporary Computing, 2013 Sixth International Conference*: 244–249.

Lexi Xu, & Yue Chen. 2009. Priority-based resource allocation to guarantee handover and mitigate interference for OFDMA systems. *Proc. IEEE International Symposium of Personal, Indoor and Mobile Radio Communications, Sept. 2009, Tokyo, Japan*: 783–787.

Lexi Xu, Xinzhou Cheng, Yue Chen, Kun Chao, Dantong Liu, & Huanlai Xing. 2015. Self-optimised coordinated traffic shifting scheme for LTE cellular systems. *Proc. EAI ICSON, Jan 2015, Beijing, P.R.China*: 1–9.

Lexi Xu, Yue Chen, Kok Keong Chai, John Schormans, & Laurie Cuthbert. 2015. Self-Organising cluster-based cooperative load balancing in OFDMA cellular networks. *Wiley Wireless Communications and Mobile Computing* 15: 1171–1187.

Liu Jun, Li Tingting, Cheng Gang, Yu Hua, & Lei Zhenming. 2013. Mining and modelling the dynamic patterns of service providers in cellular data network based on big data analysis. *Communications, China* 10:25–36.

Pengfei Ren, Xiaogang Li, Chengkang Pan, Xiaodong Shen, Jianming Zhang, Lin Sang, & Dacheng Yang. 2011. A novel inter-cell interference coordination scheme for relay enhanced cellular networks. *Proc. Vehicular Technology Conference (VTC Fall), 2011 IEEE:* 1–5.

Ramaprasath, Abhinandan, Srinivasan, Anand, & Lung, Chung-Horng. 2015. Performance optimization of big data in mobile networks. *Proc. Canadian Conference on Electrical and Computer Engineering (CCECE), 2015 IEEE 28th Canadian Conference on:* 1364–1368.

Sahni, A., Marwah, D. & Chadha, R. 2015. Real time monitoring and analysis of available bandwidth in cellular network-using big data analytics. *Proc. Computing for Sustainable Global Development (INDIACom), 2015 2nd International Conference on:* 1743–1747.

Signal and Information Processing, Networking and Computers – Chen & Huang (Eds)
© *2016 Taylor & Francis Group, London, ISBN 978-1-138-02881-4*

Big data assisted human traffic forewarning in hot spot areas

Yuwei Jia, Kun Chao, Xinzhou Cheng, Tao Zhang & Weiwei Chen
China Unicom Network Technology Research Institute, Beijing, China

ABSTRACT: The rapid development of mobile Internet contributes to the success of various value added services. As the carrier to deliver upper layer services to end users, telecom operators hold great advantage in understanding a user's behavior. Under this condition, this paper proposes a telecom's big data assisting human traffic forewarning scheme in three steps, which can eliminate potential danger or accidents, caused by the people overcrowding in hot spot areas. Moreover, some other applications, which can be developed based on the user location information, are listed further to illustrate the importance of big data to telecom operators.

Keywords: telecom operator; big data; user location

1 INTRODUCTION

Nowadays, the telecommunication industry is in the stage of rapid progress, 3GPP Long Term Evolution (LTE) networks fully reveals the inherent strength in the network capacity and transmission rate compared to conventional 2G/3G networks (Karatepe et al. 2014, Lexi Xu et al. 2013). Moreover, nationwide deployment of LTE networks by telecom operators brings the advantage of LTE into reality. Thus more users are able to get access to the network with a guaranteed Quality of Service (QoS) (H. Zhang et al. 2014).

The evolution of cellular networks also contributes to the vigorous development of mobile Internet in turn. Recently, as one of the hottest topics, the architecture of "terminal-channel-cloud" illustrates the substantial relationship among users, telecom operators, and the Internet Service Providers (ISPs) (Becker, R.A. et al. 2011). As shown in Figure 1, telecom operators play the role of "channel", which provides carriers for ISPs on the "cloud" side to deliver various kinds of upper layer services to users on the "terminal" side (Han Hu et al. 2014, Lexi Xu et al.2015).

Because of the architecture mentioned above, information of users as well as upper layer services is getting more and more transparent for telecom operators. In this paper, the massive information carried in telecom networks is interpreted as telecom big data. Typical features of telecom big data can be summarized as "4V", which is described in details as follows (Lexi Xu et al. 2014, Chien–Lung Lee et al. 2014).

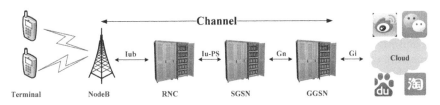

Figure 1. Illustration of the "terminal-channel-cloud" architecture.

a. Volume

    The volume of telecom big data can be extremely high (Sanqing Hu et al. 2014). Take China Unicom as an example, it serves hundred millions of mobile users in China. The networks are deployed in 31 provinces nationwide and the network types include Global System for Mobile Communication (GSM), Wideband Code Division Multiple Access (WCDMA), and LTE. Telecom big data is being produced and saved in entities of both wireless networks and core networks. Take a China's downtown statistical test as an example, call detail data can reach the size of nearly 10 TB (Terabyte) per day with over 100 billion records, and the billing detail data averages 1 TB per day with 17 billion records. The storage and processing of the massive data is up to PB (Peta-byte) level.

b. Variety

    Telecom big data is multi-dimensional, featured by varied sources, distinguished formats and diverse application directions. The whole data set can be generally classified into 23 categories, including information of telecom subscriber, network entity, network link, user perception, user location etc. It's a great challenge to manage these complicated forms of data.

c. Velocity

    Telecom big data can record all the information of the users and the networks periodically. Take measurement report (MR) as an example, each user equipment (UE) reports its MR to the base station periodically. Therefore, MR is an effective way for the telecom operator to track the real time operating status of the user. To guarantee the precise measurement, MR is periodically reported with time interval of 480 ms for GSM networks and 2 s~64 s for WCDMA networks in China Unicom (Y. Shen et al. 2015). Such fast paced data streaming requests high processing ability of the big data platform.

d. Veracity

    Telecom big data implies sufficient content about user identity, user preference, user trajectory, user network perception etc. The conjoint analysis of all these kinds of information is a significant way to portray a user comprehensively and solidly. The comprehensive information of users is the precondition and basis to apply telecom big data in diverse industry fields.

    From the "4V" characteristics of telecom big data mentioned above, it's obvious that telecom operators are aware of the behavior and interests of their users. Compared with big data in other industries, telecom big data have dominant advantage in large user base and wide geographical scope. Moreover, as for the data quality, it's more comprehensive and precise.

    To make full use of the strength mentioned above, telecom operators have attempted to employ telecom big data to assist telecom operation, such as network investment, network optimization, network construction, marketing strategy, and so on. However, as for the application of telecom big data in social services or other industries, it still has a long way to go. Considering the Shanghai trampling event in early 2015, this disaster could have been prevented by local government through telecom big data analysis.

    Under this condition, this paper studies the telecom's big data application in social services and proposes the data assisting human traffic forewarning scheme. It is aimed at eliminating the potential danger or accident caused by human overcrowding in hot spot areas. This scheme is understood in three steps. First, citywide human traffic is under real time monitoring, thus, hot spots with sustained high traffic density is identified. Second, as the core of this scheme, a User Trajectory Prediction Algorithm (UTPA) is designed to estimate the possibility of users travelling to the hot spots. Third, based on the prediction results, target areas are obtained which are still in the tendency of increasing human traffic. Relevant public management departments are informed to take measures to ensure the safety and stability of these target areas.

The rest of this paper is organized as follows. Section 2 illustrates the detailed three steps of the scheme mentioned above. Section 3 introduces the application in other fields based on telecom big data. Finally, conclusions are drawn in Section 4.

2 TELECOM BIG DATA ASSISTING HUMAN TRAFFIC FOREWARNING

For a city with a large population, such as Beijing, London, and New York, etc., is prone to danger and accidents that is induced by people overcrowding, especially on holidays. This will bring about unimaginably serious impact on people's life and safety of property. With the assist of telecom big data, not only the current status of human traffic can be monitored, but also the tendency of that can be predicted ahead of time. Thus, relevant public management departments can take measures to prevent the potential accident in advance. The following section gives the detailed description of this scheme.

2.1 Citywide human traffic monitoring

As mentioned previously, telecom operators hold great potential for providing necessary information to identify people's travelling behavior. The growing utilization of cellular phones implies that a large percentage of people keep their phone with them most of the time. Here, we make use of the user location information implied in telecom big data to analyze the geographic distribution of human traffic.

As shown in Table 1, it's a sample record of the user location information. *IMSI* (International Mobile Subscriber Identification) is the user network identifier. *Timestamp* denotes the moment when the user initiates a call or data service. *Longitude* and *Latitude* represents the location where the voice or data service is initiated. For example, $\langle Lon_i, Lat_i \rangle$ reflects the location of the ith record. In section 2, the record is expressed as $R_{\langle IMSI, Timestamp, Longitude, Latitude \rangle}$ and the set of all records is expressed as Φ.

This paper defines the variable $Traffic_{\langle Lon_i, Lat_i \rangle}$ as the human traffic around the location of $\langle Lon_i, Lat_i \rangle$.

$$Traffic_{\langle Lon_i, Lat_i \rangle} = distinct\left(IMSI, \langle Lon_i, Lat_i \rangle, \Phi\right) \quad (i \geq 1) \tag{1}$$

Equation (1) counts duplicate-removal *IMSIs* for all records in Φ at $\langle Lon_i, Lat_i \rangle$. The effect of the function *distinct*( ) is similar to that of the keyword **distinct** in Structured Query Language (SQL) (Ben Forta. 2013).

Then human traffic in the citywide can be quantified and visually presented on the heat map. It's worth noting that the human traffic is being monitored the whole day and the heat map is varying with the calculated result, in real time. Thus, citywide hot spots can be recognized, which means that the area in the heat map appears as red or orange in most of its instances. Figure 2 shows the example of the human traffic distribution map.

The target of the proposed scheme is the identification of hot spots. As shown in Figure 3, the histogram shows the real time status of the hot spots which are marked as A–J. There is a warning threshold set for each spot, which indicates the maximum allowed traffic load for the particular area. When human traffic at the statistical moment exceeds the warning threshold, the bar representing this spot is highlighted for notice.

Table 1. Sample record of user location information.

Timestamp	Longitude	Latitude	IMSI
2014-11-06 08:34:55	103.8226	36.05404	460011342602664

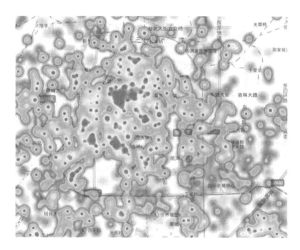

Figure 2.   Human traffic distribution of city X at 16:00.

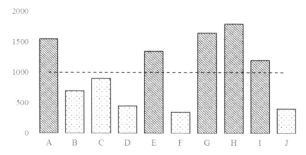

Figure 3.   Hot spots human traffic monitoring of city X at 16:00.

2.2   *Hot spots human traffic prediction*

For an area where the human traffic has exceeded the warning threshold by a certain margin, there is a high probability of overcrowding. Under this circumstance, potential arriving rate of the human traffic should be studied to find out the tendency of the human flow.

2.2.1   *Proposed user trajectory prediction algorithm*

To address the problem mentioned above, this paper proposes a User Trajectory Prediction Algorithm (UTPA). It is based on the conjoint analysis of the target user as well as the users surrounded within the same area are located. Then, the possibility of the target user travelling to a specific geographic point is calculated. Table 2 shows the flow of the UTPA.

Step 1:   The data source of this algorithm is the records $R_{\langle IMSI,Timestamp,Longitude,Latitude\rangle}$ in Φ.

Step 2:   In UTPR, the concept of grid is introduced. The grids are a series of adjacent rectangle arrays with uniform size, which are denoted by $G_{\langle grid\_x,grid\_y,a\rangle}$. **grid_x** and **grid_y** are the horizontal and vertical coordinates of the grid, respectively. **a** represents the side length of the grid.

For $R_{\langle IMSI,Timestamp,Longitude,Latitude\rangle}$, which is located in the area of $G_{\langle grid\_x,grid\_y,a\rangle}$, can be rewritten as reference point $D_{\langle grid\_x,grid\_y,m\rangle}$. $\langle grid\_x, grid\_y\rangle$ is the grid coordinate of the record at $\langle Longitude, Latitude\rangle$. **m** represents the number of records in grid $G_{\langle grid\_x,grid\_y,a\rangle}$ at **<Timestamp>**.

Step 3:   For the time interval $[t_i,t_j]$, the aggregation cluster of reference points $D_{\langle grid\_x,grid\_y,m\rangle}$ can be expressed as:

370

Table 2.   Flow of the UTPA.

Step 1	User location records are represented as R
Step 2	User location records R are transformed into reference point D
Step 3	Reference points D are aggregated into cluster C_{mov}
Step 4	Calculate the average weighted point of cluster C_{mov}
Step 5	Calculate the possibility P at the weighted point

$$C_{mov} = \{\langle t_1, grid\_x_1, grid\_y_1, m_1 \rangle, \cdots, \langle t_n, grid\_x_n, grid\_y_n, m_n \rangle\} \, (t_i \le t_1, \cdots, t_n \le t_j) \qquad (2)$$

Step 4: This paper defines $O_{\overline{\langle grid\_x, grid\_y, m, k \rangle}}$ as the weighted average point of all reference points in cluster C_{mov}. $\langle grid\_x, grid\_y \rangle$ is the grid coordinate of the weighted point. m is the number of all records in cluster C_{mov}. And last but not least, k means the radius of influence for point $O_{\overline{\langle grid\_x, grid\_y, m, k \rangle}}$.

For any $\langle grid\_x_k, grid\_y_k, m_k \rangle \, ((1 \le k \le n) \in C_{mov})$, there are the following equations:

$$\overline{grid\_x} = \sum_{k=1}^{n} grid\_x_k \times \frac{m_k}{\sum\limits_{r=1}^{n} m_r} \qquad (3)$$

$$\overline{grid\_y} = \sum_{k=1}^{n} grid\_y_k \times \frac{m_k}{\sum\limits_{r=1}^{n} m_r} \qquad (4)$$

$$k = \begin{cases} k_{max} & \sum\limits_{r=1}^{n} m_r \ge m_{max} \\ \sum\limits_{r=1}^{n} m_r & m_{min} < \sum\limits_{r=1}^{n} m_r < m_{max} \\ k_{min} & \sum\limits_{r=1}^{n} m_r \le m_{min} \end{cases} \qquad (5)$$

In Equation (5), m_{max} and m_{min} are the upper bound and the lower bound of record number in cluster C_{mov}, respectively. k_{max} and k_{min} are the upper bound and the lower bound of influence radius of the point $O_{\overline{\langle grid\_x, grid\_y, m, k \rangle}}$, respectively.

Step 5: Assuming the set of all weighted points $O_{\overline{\langle grid\_x, grid\_y, m, k \rangle}}$ in the time interval $\left[t_i, t_j \right]$ is S_o, which is expressed as:

$$S_o = \left\{ O_1{}_{<\overline{grid\_x_1}, \overline{grid\_y_1}, m_1, k_1>}, \cdots, O_n{}_{<\overline{grid\_x_n}, \overline{grid\_y_n}, m_n, k_n>} \right\} \qquad (6)$$

Then, the possibility of user occurrence at the weighted point $O_i{}_{<\overline{grid\_x_i}, \overline{grid\_y_i}, m_i, k_i>}$ during $\left[t_i, t_j \right]$ is marked as:

$$P_{o_i} = \frac{m_i}{\sum\limits_{i=j}^{n} m_j} \, (1 \le i \le n) \qquad (7)$$

2.2.2 *Algorithm verification*

The prediction accuracy of the UTPR is evaluated by comparing the predicted arriving rate and the actual condition of a specific spot. The involved parameters of the UTPA algorithm are listed in Table 3.

371

Table 3. Algorithm parameters.

Parameters	Values
City	Changchun
Area	Hongqi Street
Network type	WCDMA network
Equipment manufacturer	Huawei
Data Source	Gn interface data
Grid size	50 m, 100 m
Prediction lead time	0.5 h, 1.0 h, 1.5 h, 2 h, 2.5 h, 3 h, 3.5 h

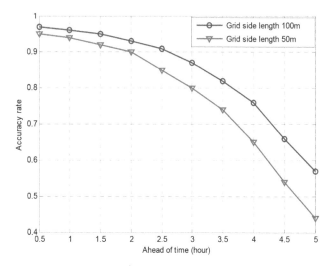

Figure 4.   Prediction accuracy of UTPA.

Figure 4 shows the prediction accuracy of UTPA under different prediction lead time and different grid sizes. The performance of this algorithm deteriorates with the extension of lead time. The prediction accuracy (comparison of the *actual human traffic 0.5 hour later* and the *predicted arriving rate*) exceeds 95% in this case. Besides, for a specific prediction lead time, UTPA with a larger grid side length can be more error tolerant. This means that the user base in a larger grid is higher, and then the error rate which is expressed by the ratio of the predicted error to the user base is lowered. Thus, it can present more accurate prediction results with larger grid side length.

2.3   *Management and control measures*

Based on the prediction results in the previous section, relevant management and control measures are put forward to eliminate the potential danger in section 2.3.

On the one hand, telecom operators can forward short messages to remind subscribers of the real time traffic condition at the hot spot. On receiving the message, a portion of people may choose alternative destinations instead. This may reduce human traffic to some extent at the source. On the other hand, security measures and police force should be strengthened to maintain the order at the hot spot. This is a significant method to keep the stability of the hot spot.

3 APPLICATIONS IN OTHER FIELDS

The comprehensive user information is the basis of various value-added applications or services. Previously, user information is collected in the form of online or offline surveys, which is a time-consuming and expensive process. Sometimes, large-scale investigations may take several years to get complete, and by that time many results may get out of date even before being released. This has been a tough job for a long time.

Nowadays, as intelligent terminals have become indispensable for many people, telecom operators get the opportunity to collect user information from telecom big data. Thus, it can provide a means to study user behavior frequently, cheaply, and at an unprecedented scale.

Section 3 lists a series of example applications, which can be developed based on the user information implied in telecom big data.

3.1 *Personalized recommending*

The location information implied by telecom big data is the basis of a series of Location Based Services (LBSs). Take the user trajectory as an example; it records the activities of users in daily life. These activities to some extent are the reflection of user intention and user preference. If a user often appears at sports venues, it indicates that this user may like sports activities. Frequent appearance at scenic spots may characterize the user's preference for travelling or outdoor activities. The combination of user trajectory and other characteristics identified through Deep Packet Inspection (DPI) is an effective way to explore the implicit needs of the users. Thus, it can provide practical suggestions or recommendations for the users before they figure out what they really want.

3.2 *Transportation scheduling*

Nowadays, bad urban traffic condition has become a tough problem. It's essential for transportation departments to acquire the travelling habit of the general public, thus transportation resource can be allocated reasonably and effectively. By means of telecom big data, the geographical and temporal distribution of human traffic can be identified more precisely. Thus, public transport network can be designed to be more aligned with the practical population migration situation. More public buses can be scheduled on the busy line, and congested intersections can be predicted more accurately.

4 CONCLUSION

This paper concentrates on the application of telecom big data in people's daily life. It proposes a human traffic forewarning scheme, which is aimed to eliminate the potential accident caused by human overcrowding in hot spot areas. This scheme was understood in three steps. First, citywide human traffic is under real time monitoring, thus, hot spots with sustained high traffic density is identified. Second, as the core of this scheme, a user trajectory prediction algorithm is introduced to estimate the possibility of the users travelling to hot spots. Third, based on the prediction results, target areas which are still in the tendency of increasing human traffic are identified. Relevant public management departments are informed to take measures to guarantee the safety and stability of the target areas.

REFERENCES

Becker, R.A., Caceres, R., & Hanson, K. 2011. A tale of one city: using cellular network data for urban planning. *IEEE Pervasive Computing* 10(4): 18–26.
Ben Forta. 2013. *SQL in 10 minutes*. Beijing: China Post and Telecommunications Press.

Chien-Lung Lee, Wen-Shu Su, Kai-An Tang, & Wei-I Chao. 2014. Design of handover self-optimization using big data analytics. *Proc. IEEE Asia-Pacific Network Operations and Management Symposium, Hsinchu, China:* 1–5.

Han Hu, Yonggang Wen, Tat-Seng Chua, & Xuelong Li. 2014. Toward scalable systems for big data analytics: a technology tutorial. *IEEE Access* 2: 652–687.

Karatepe, Ilyas Alper, Zeydan, & Engin. 2014. Anomaly detection in cellular network data using big data analytics. *Proc. Verband Deutscher Elektrotechniker European Wireless, Barcelona, Spain:* 1–5.

Lexi Xu, Yue Chen, Kok Keong Chai, John Schormans, & Laurie Cuthbert. 2015. Self-organising cluster-based cooperative load balancing in OFDMA cellular networks. *Wiley Wireless Communications and Mobile Computing* 15(7): 1171–1187.

Lexi Xu, Yue Chen, KoK Keong Chai, Dantong Liu, Shaoshi Yang, & John Schormans. 2013. User relay assisted traffic shifting in LTE-Advanced systems. *Proc. IEEE Vehicular Technology Conference, Dresden, Germany:* 1–7.

Lexi Xu, Yuting Luan, Kun Chao, Xinzhou Cheng, Heng Zhang, & John Schormans. 2014. Channel-aware optimised traffic shifting in LTE-Advanced relay networks. *Proc. IEEE International Symposium on Personal Indoor and Mobile Radio Communications, Washington, USA:* 1597–1602.

Sanqing Hu, Ye Ouyang, Yu Dong Yao, Fallah, M.H, & Wenyuan Lu. 2014. A study of LTE network performance based on data analytics and statistical modeling. *Proc. IEEE Wireless and Optical Communication Conference, Newark, USA:* 1–6.

Shen, Y., C. Jiang, T. Quek, H. Zhang, & Y. Ren. 2015. Pricing equilibrium for data redistribution market in wireless networks. *Proc. IEEE International Conference on Communications, London, UK:* 1–5.

Zhang, H., H. Liu, C. Jiang, X. Chu, A. Nallanathan, & X. Wen. 2014. A practical semi-dynamic clustering scheme using affinity propagation in cooperative picocells. *IEEE Transactions on Vehicular Technology* 2014.2361931. DOI: 10.1109.

Signal and Information Processing, Networking and Computers – Chen & Huang (Eds)
© 2016 Taylor & Francis Group, London, ISBN 978-1-138-02881-4

A novel big data based Telecom User Value Evaluation method

Kun Chao
China Unicom Network Technology Research Institute, Beijing, China

Pengfei Wang
Beijing Branch of China Unicom, Beijing, China

Lexi Xu
China Unicom Network Technology Research Institute, Beijing, China

Danyang Wu
Beijing University of Posts and Communication, Beijing, China

Xinzhou Cheng & Mingjun Mu
China Unicom Network Technology Research Institute, Beijing, China

ABSTRACT: Customers are essential for telecom operators. Tracking the trace and the distribution of users with different values are the vital determinate factors for the telecom operations; this includes the precise network planning, the diverse investments in different regions, the priority maintenance, the optimization ordering etc. It is supposed to assist telecom operators in maximizing both the network resources utilization and the investment benefit. Meanwhile, precise marketing and diverse services for different users are also a future development trend of telecom operators. This paper proposes a Telecom User Value Evaluation (TUVE) method for the mobile Internet era. In the TUVE method, we analyze telecom big data in order to obtain comprehensive information from the users. Based on these users' information, we construct the user value evaluation system composed of the current value and the potential value. Furthermore, we analyze the weights of key factors for the current value and the potential value. Finally, we calculate each user's comprehensive value. The TUVE method can assist the efficient network planning, maintenance, and optimization. In addition, the propose TUVE method can help telecom operators set a more precise market strategy.

Keywords: telecom user value evaluation; telecom operator; user current value; user potential value

1 INTRODUCTION

In the mobile Internet era, telecom operators envisage a series of challenges, which includes the gradually reduced investment, and the improved requirements of the users' experience in Shu Liu (2014). In order to address these challenges, the marketing related departments have already tried relevant strategies to provide different users with different services. Meanwhile, the network operation related departments have also attempted to do relevant adjustments for the network planning and investment (Wenbiao Zhang, 2012 and Lexi Xu et al. 2013). More specifically, the network construction method based on the signal coverage is being converted to the method based on the high-value users' distribution. Employing the differential investment, telecom operators can maximize the network resources utilization and guarantee the good service perception for high-value users. Therefore, it is very important to estimate each user's value effectively to the telecom operators.

This paper proposes a Telecom User Value Evaluation (TUVE) method. The TUVE method analyzes the current value and the potential value. Then, the TUVE method analyzes the weights of key factors for the current value and the potential value. Based on the weights, the TUVE method calculates each user's comprehensive value. In addition, based on the current value and the potential value clustering, we can get several user groups. Each group has a corresponding characteristic which is similar. It can assist the telecom operators to provide suitable services for these groups.

The rest of this paper is organized as follows: Section II introduces the existing user value evaluation methods. Section III presents the TUVE scheme, including the TUVE system construction, the index weights computation of the TUVE method and the comprehensive value calculation of each user. Section IV shows the analysis and the application of the calculation result using the TUVE method. Section V gives a conclusion of the whole paper.

2 EXISTING USER VALUE EVALUATION METHODS

The user value definition is different according to different industries. Many user value analyses are designed in the existing works. Among these works, RFM model and CLV models are widely used (Jiandong Du and Zhiqiang Zhao, 2011). RFM is a value analysis model proposed by Hughes in 1994 (Jiandong Du and Zhiqiang Zhao, 2011). This model uses three factors, which includes recently purchased day (Recency) and purchase frequency (Frequency) as well as purchase monetary (Monetary), to estimate the customer value and customer profitability. Relevant organizations can obtain these three factors of RFM model from customer transaction database, so the RFM model becomes one of the most popular methods of customer analysis. CLV (customer lifetime value) refers to the customer lifetime (or a long time) for the enterprise's profit value. Customer lifetime value is also known as the customer life value, which means the sum of revenue that every buyer may bring to the enterprise in future. The contribution of the customer to enterprise's profit can be divided into introduction period, rapid growth period, mature period and the decline period, which can then analyze the customer in different value circles. Recently, the user value analysis is also carried out with the combination of the RFM model and the CLV model (Jiandong Du and Zhiqiang Zhao, 2011). However, the user value estimation methods (e.g. RFM model, CLV model), which is considered only from the purchase behavior and returns, have some limitations. It is because those users have multiple dimensions of characteristics in the mobile Internet era, including the business consumption, the serving network, the terminal and the complex service. Similarly, these characteristics reflect the user value.

At present, telecom operators usually estimate the users' value on the basis of their monthly ARPU (Qiuqing Huang et al. 2005 and Haiming Fu et al. 2007), which represents the average revenue per user. These ARPU based methods have only some limitations, since they can only reflect the maximum current earnings to telecom operators but failed to reveal the intrinsic value of the customer. Therefore, these ARPU based methods cannot effectively measure the customer quality of telecom enterprises. Meanwhile, these ARPU based methods cannot meet the needs of customer segmentation and differentiation marketing for telecom enterprises. In addition, the users' telecom consumption gradually inclines to the traffic consumption. Hence, telecom operators should not only consider the monthly ARPU for the user value evaluation. Instead, user value evaluation should also consider the user's package information, traffic, application behavior features, and cellphone information. Especially on the diverse allocation of mobile network resources, not only the user's current value but also the potential value should be considered.

3 THE NOVEL TELECOM USER VALUE EVALUATION METHOD

3.1 *Construct user value evaluation index system*

Based on telecom operators' customer basic information, customer monthly billing and customer Internet behavior feature data analysis, we can decompose the customer value evaluation index system into two sub index systems: the user's current value and potential value. Then, we select the evaluation indexes for current value and potential value. Finally, we get the index system of the TUVE method, and the index system is as shown in Figure 1.

In the index system, each index can be divided into several intervals, and each interval corresponds to an evaluation score. Take the user grade as an example, the users with average ARPU≥400 RMB can be identified into one class, the users with average ARPU between 200 RMB and 400 RMB can be in another class, and the remaining with average ARPU≤200 RMB belong to the third class. Certainly, the division standard can be justified flexibly by our analysis requirement.

3.1.1 *Current value*

3.1.1.1 User grade
User grade reflects a certain past value of users. Take one telecom operator as an example, users can be divided into VIP users and non-VIP users by the consumption level in the past six months or the special social identity. VIP users are further subdivided into the Diamond user, the Gold user and the Silver user. Usually some important clients to telecom operators may be identified directly as the VIPs. Table 1 exemplifies the VIP division rules based on the consumption level in two cities. In the two cities, VIP user division standards are different since they have different economic conditions and service behavior.

3.1.1.2 Package level
With the current domestic telecom operators employing the consumption of packages in 3/4G network, the user's package largely determines the extent of his mobile communication consumption. By the population of contract terminals (contract with a telecom operator), the user can get an iphone by a very low price with a one-year contract or more. The longer the contract is, the lower price for the terminal itself will be. So the user's package reflects his means to continue spending budget on the communication in a period of time (1 years or 2 years or more). Therefore, the package is an important factor to measure the current value of users.

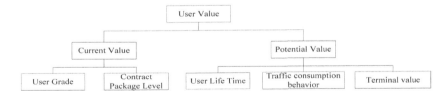

Figure 1.   User value evaluation index system.

Table 1.   Example of user grade division method in some cities.

| City | Consumption Level in 6 months | | | Additional requirements |
	Diamond	Gold	Silver	
A	ARPU ≥ 600 RMB	400 RMB ≤ ARPU < 600 RMB	200 RMB ≤ ARPU < 400 RMB	Only for after-paying users
B	APRU ≥ 350 RMB	350 RMB > APRU ≥ 200 RMB	200 RMB > APRU ≥ 100 RMB	The VIP client is valid from the rating that month to a year, monthly calculation

3.1.2 *Potential value*

3.1.2.1 User life time

User life time refers to the time duration, between the initial time of the user joining the telecom operator and the current time. Specifically, the user life time is calculated from the month of user joining in to the current month, and it is accumulated 12 months to 1 year. A user with longer time duration reflects this user has more stable relation and higher user stickiness with this telecom operator. In addition, the time duration is direct embodiment of user loyalty. Take a city's data as an example, the relationship between the amount users of different life time distribution and ARPU is shown in Figure 2.

Note: User loyalty always consists of the behavior loyalty and the emotional loyalty (Xiaofei Tang, 2008). User emotional loyalty refers to some subjective factors, and it is usually obtained from the user survey instead of the telecom big data. User behavior loyalty generally contains the user life time, the user consumption and the user service behavior. The user consumption and the user service behavior are concerned in other indicators. This paper primarily considers the behavior loyalty since we can periodically analyze the behavior loyalty via telecom big data. To avoid the overlapping of the indicators, we mainly consider the user life time to reflect the user loyalty.

From Figure 2, the curve shows the ARPU with different life time users. The ARPU of the users whose life time duration are within 2 years is the lowest. With the increasing of the users' life time over 2 years, the ARPU is generally stable over 100 RMB. Thus, one of the benchmarks may be identified as 2 years for the index.

3.1.2.2 Traffic consumption behavior

With rapid change of the mobile Internet business, more and more of the telecom users are becoming mobile Internet users. Meanwhile, from 3G to 4G network, users' communication consumption is increasingly inclined to traffic consumption (Ying Xia, 2014). Therefore, the development of user traffic consumption habits is very important for telecom operators.

Although, the user's traffic consumption has a close relationship with the contract package's traffic flow size, there are also some exceptions. Some users with large data packages may use only a small amount (50% or even less) of the traffic, while some other users may consume more traffic than the package contained and need to upgrade the package amount or purchase an extra traffic package. It notes the user's traffic consumption behavior, and it indicates the potential value of a user to telecom operators. Therefore, we can use the index-traffic/month over the package's traffic to reflect the user's traffic consumption behavior.

Then, we can make a benchmark. For example, the users, who have external traffic flow consumption (except the traffic flow in certain packages for the 6 consecutive months) or have promoted the package level in 6 months, form the mobile Internet habit partly and these users are willing to spend more on mobile Internet services. This indicates more income for the telecom operators.

3.1.2.3 Terminal value

It is generally known that each user is bounded to some kind of terminal together; therefore, each user has the terminal value in mobile networks. Users with different brands and different

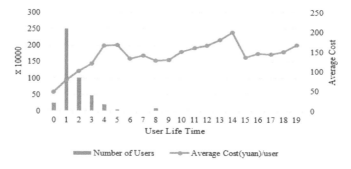

Figure 2. The relationship of user life time and average cost/month.

types of cellphone have the different implicit value. The customers with different terminal brands may indicate the different consumption capacity. According to statistics in several cities, the users with apple devices which are usually more expensive than other cellphone brands cost more on mobile network. So, we can divide the terminal value into several intervals by terminal price.

3.2 Determine user value evaluation index weight

3.2.1 Weight determination algorithm

3.2.1.1 Construct judgment matrix

Suppose comparing the influence of n factors $X = \{x_1, x_2, ..., x_n\}$ on a certain factor T, we can determine the weight of each criterion by comparing each other, this is constructing judgment matrix. Starting from second levels of hierarchy model, for the same layer factors belonging to (or impacting) every upper layer, we can use paired comparison method and 1–9 comparative scale to construct pairwise comparison matrix until the bottom. That is, taking two factors X_i and X_j at a time, and using a_{ij} to indicate the influence ratio of X_i on T and X_j on T. The total comparison result can be represented by matrix $A = (a_{ij})\ n \times n$, and A is the judgment matrix between T and X. If a_{ij} is the influence ratio of X_i on T and X_j on T, and then the influence ratio of X_j on T and X_i on T is $a_{ji} = 1/a_{ij}$ (Ying Xia, 2014). Matrix A is called reciprocal matrix. And if $a_{ij} = a_{ik} \times a_{kj}$, the matrix A is consistency matrix (Fanchun Yan et al. 2003). The assignment criteria of the factors in the judgement matrix is shown in Table 2.

3.2.1.2 Build the new consistency matrixes based on the original judgment matrix A

Based on each row of data in the judgment matrix $A = (a_{ij})_{n \times n}$, we can construct a new group of consistency matrixes:

$$A_i = (a^i_{kl})_{n \times n} = (a_{il}/a_{ik})_{n \times n} \quad i \in \{1, 2, ..., n\} \tag{1}$$

Matrix A_i is built by line i in original matrix A.

$$A_i = \begin{bmatrix}
1 & \cdots & & \cdots & & \cdots & \dfrac{1}{a_{i1}} & \cdots & \cdots \\
\vdots & \ddots & & & & & \vdots & & \vdots \\
& & 1 & \cdots & a^{(i)}_{kl} & \cdots & a_{ik} & & \\
\vdots & \vdots & \vdots & \vdots & \vdots & \vdots & \vdots & \vdots & \vdots \\
& \cdots & \dfrac{1}{a^{(i)}_{kl}} & \cdots & 1 & \cdots & \dfrac{1}{a_{il}} & & \\
\vdots & \vdots & \vdots & \vdots & \vdots & \vdots & \vdots & \vdots & \vdots \\
a_{i1} & \cdots & a_{ik} & \cdots & a_{il} & \cdots & 1 & \cdots & a_{in} \\
\vdots & & & & & & \vdots & \ddots & \vdots \\
\vdots & \cdots & & \cdots & & \cdots & \dfrac{1}{a_{in}} & \cdots & 1
\end{bmatrix} \tag{2}$$

Table 2. Assignment criteria of factors in the matrix.

Comparison between factors	Quantized value
Factor i and factor j are equally important	1
Factor i is a little important than factor j	3
Factor i is more important than factor j	5
Factor i is strongly important than factor j	7
Factor i is extremely important than factor j	9
Median of two adjacent judgment	2, 4, 6, 8

where $a_{kl}^{(i)} = a_{il}/a_{ik}$.

3.2.1.3 Obtain the weight vector of the above consistency matrix by eigenvector method

This can be constructed the N consistency matrixes. Using feature vector method to calculate the weight vector of the corresponding N Group:

$$W_i = (w_{i1}, w_{i2}, ..., w_{in})^T \tag{3}$$

3.2.1.4 Determine the index weight by comprehensive method

Assuming the expert judgment without any preference, each row represents the same and judges the credibility of information. Then all the information can be obtained by the index weights, which are expressed as (4):

$$W = \begin{bmatrix} w_{11} & w_{12} & \cdots & w_{1n} \\ w_{21} & w_{22} & \cdots & w_{2n} \\ \vdots & \vdots & \vdots & \vdots \\ w_{n1} & w_{n2} & \cdots & w_{nn} \end{bmatrix} \times \begin{bmatrix} \frac{1}{n} \\ \frac{1}{n} \\ \vdots \\ \frac{1}{n} \end{bmatrix} \tag{4}$$

3.2.1.5 Compute the combined weights

According to the above steps, we can get the weight vector of a group factor to the upper certain factor. However, the final result we need is the ordering weights that all factors especially the various kinds of projects at the bottom layer for the target layer. The total ordering weights should be calculated by combining the weight of single criterion step by step. The method of weight combination is as follows in Table 3.

3.2.2 Single weight and combined weight calculation

1. Take the factor weights calculation of potential user value as an example

   First, based on the judgment of experts on the influence of user life time, traffic consumption behavior and terminal value to user potential value, we can construct the potential user value judgment matrix $A = \begin{pmatrix} 1 & 0.28 & 2.1667 \\ 3.5714 & 1 & 4.6 \\ 0.4615 & 0.2174 & 1 \end{pmatrix}$.

Table 3. Method of weight combination.

B layer	A layer	A1	A2	...	An	Total ordering weights of B layer
		a1	a2	...	an	
B1		b11	b12	...	b1n	$\sum_{j=1}^{n} b_{1j} \times a_j$
B2		b21	b22	...	b2n	$\sum_{j=1}^{n} b_{2j} \times a_j$
...		...	...	...	...	...
Bn		bn1	bn2	...	bnn	$\sum_{j=1}^{n} b_{nj} \times a_j$

Second, the constructed consistency matrixes by original matrix A are as follows:

$$A_1 = \begin{pmatrix} 1 & 0.28 & 2.1667 \\ 3.5714 & 1 & 7.7381 \\ 0.4615 & 0.1292 & 1 \end{pmatrix}, A_2 = \begin{pmatrix} 1 & 0.28 & 1.288 \\ 3.5714 & 1 & 4.6 \\ 0.7764 & 0.2174 & 1 \end{pmatrix}, A_3 = \begin{pmatrix} 1 & 0.471 & 2.1667 \\ 2.123 & 1 & 4.6 \\ 0.4615 & 0.2174 & 1 \end{pmatrix}.$$

Third, obtain the eigenvectors corresponding to the maximum eigenvalue of the above consistency matrixes by the eigenvector method, and normalize the vector to get the weight:

$$W_1 = (0.1987, 0.7096, 0.0917)^T, W_2 = (0.1870, 0.6678, 0.1452)^T, W_3 = (0.2790, 0.5923, 0.1287)^T.$$

Finally, we get the weights of the indexes "User Life Time", "Traffic Consumption Behavior", and "Terminal Value" in potential value, as $W = (0.2216, 0.6566, 0.1219)$.

2. According to the above computing method, the first layer and the second layer index weights are figured out. Therefore, the whole results are as are as follows in Table 4:

3.3 Compute users' comprehensive value

As discussed in our previous works (Lexi Xu et al. 2014 and Lexi Xu et al. 2015), telecom operators envisage the challenges of utilizing the limited resources to meet different users' diverse service requirements (e.g. hot-spot area). The key is to analyze each user's comprehensive value. User's comprehensive value is a concrete value. The index system paper has constructed two-stage index system, so we should compute the estimated value of each index using these two steps. The first step is to compute the scores of the first-stage indexes using the scores and the weights of the second-stage indexes. The second step is to compute each user's comprehensive value using the scores and the weights of the first-stage indexes. The comprehensive evaluation model is shown in (5):

Table 4. Results of every factors' weights.

Object layer	Criterion layer	Weight	Project layer	Weight
User value	Current value	0.6939	User grade	0.7196
			Package level	0.2804
	Potential value	0.3061	User life time	0.2216
			Traffic consumption behavior	0.6566
			Terminal value	0.1219

Table 5. Ten users' value evaluation result.

Users	Current value	Potential value	Total value
186XXXX8758	97.75701	93.50717	96.45604
186XXXX2418	62.52336	78.85961	67.52425
186XXXX7220	53.83178	46.00453	51.43568
186XXXX6117	89.62617	83.66189	87.80037
186XXXX1339	54.6729	17.39755	43.26208
186XXXX1219	95.51402	76.86655	89.80561
186XXXX5806	98.31776	34.01712	78.63389
186XXXX6748	58.87851	90.08496	68.4315
186XXXX3155	48.50468	27.24821	41.99759
186XXXX1406	66.72897	54.96201	63.12684

$$V_i = \sum_{i=1}^{n} A_i \times W_i \qquad (5)$$

where *A* reflects the first-stage index score, and W indicates the first-stage index weight.

For example, the value result of randomly selected 10 users is shown in Table 5.

4 USER VALUE EVALUATION RESULT ANALYSIS AND APPLICATION

4.1 *User value evaluation result analysis*

When we get the user value, we can analyze the user value by K-means clustering method (Heng Qi, 2014 and Haibin Shi et al. 2014). The overall result of user clustering in a China's city downtown is shown in Table 6. The telecom operator of this city downtown serves more than one million of people. In order to clearly show the cluster result in Figure 3, we randomly select 100 users (considering the scatter plot display effect).

From the Table 6 and Figure 3, we can see that few users have both the high current value and the potential value. Most users have low current value; the proportion is up to 90%, which may be related to the relative short development time of domestic 3G/4G network. Among these low current value users, there are majority (almost 80%) of the users that belongs to the high potential value users. Namely more and more users' mobile Internet traffic behavior is initially formed, but not enough. The result matches the actual situation of the telecom operator in the city basically. The telecom operator should devote to develop the users' traffic consumption and application usage habit by various methods to get more earnings.

Table 6. Overall result of user clustering.

Clustering	Current value	Potential value	Proportion	Quadrant
1	High	high	7%	1
2	Low	high	75%	2
3	Low	low	18%	3

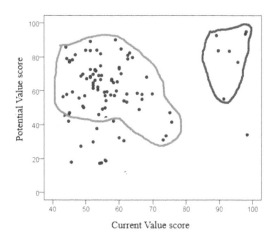

Figure 3. Quadrant distribution map of user clustering results (exemplified).

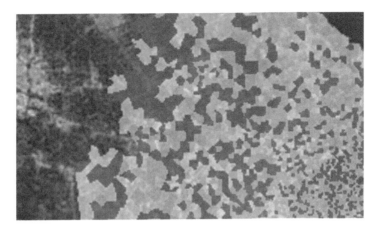

Figure 4.   Users' distribution of high current & high potential value.

The following is the analysis of user clustering based on the first class indicators:

1. The first type of user is the ones who have high current value and high potential value. They are the high-value users to the telecom operators at present. Telecom operators should focus on them in the aspects of network quality and customer service. The network service quality of these users is preferred in the case of limited network resources. That is, the network resources may be allocated at places where more high-value users usually stay, and any network maintenance and optimization related to these users should be of high-priority. At the same time, according to the user's preference, telecom operators can precisely push the application products and other telecom products to them, and also ensure them with good service.

2. The second type of user is the ones who have relatively low current value but high potential value. Although they are not the current-valuable-customers for telecom operators, they may still have great potential in the future. For example, their traffic consumption may gradually increase, which is one of the most important factors to telecom operators. For these group users, telecom operators should input more marketing and service attention. For example, telecom operators can further enhance the traffic consumption development of users and application extension, on the basis of the user's preference in order to improve the user's value. Moreover, the network resources allocation should consider properly the distribution of these group users.

3. The third type of user is the ones who have both low current value and low potential value. They are low-value users to telecom operators now. However, for these group users, based on the user stock management and the precision marketing demand, telecom operators can still take a series of tasks to promote the value of this user group. For instance, telecom operators can take certain marketing strategy for these users and promote suitable traffic packages or mobile application products to them.

4.2   User value evaluation result application

This paper employs the application of network resources assignment as an example. Figure 4 shows the user distribution of high current & high potential value in a China's city downtown. The red areas indicate the hundreds of high-value users per sector. The yellows mean the dozens of high-value users per sector, and the blues represent several users. Figure 4 can assist the local telecom operator to be aware of the corresponding high value area and the levels of value areas (red, yellow, blue in sequence). Then, the local telecom operators can take preference to assign the sufficient network resource to the high value area (red area in Figure 4), which can benefit the income of the telecom operator.

5 CONCLUSION

This paper has proposed a Telecom User Value Evaluation (TUVE) method on the basis of a series of characteristics, which includes the user basic information, the communication consumption, the user business behavior and terminal attribute. The proposed TUVE method evaluates the current value of users. Meanwhile, it also forecasts the potential value based on the user business behavior, user loyalty and user's terminal value with the development of vision. Moreover, we employ the TUVE method to analyze the data in the city of China', then we employ the K-means clustering method to analyze the above result. The result matches the actual situation of the telecom operator in the city basically. Furthermore, based on the tracking and distribution of the users, mobile network resources will be inclined accurately to the regional agglomeration of high-current value users, and then high-potential value users. It can also assist telecom operators to make diverse marketing decision and strategy for different types of users to realize the precisely marketing, such as releasing a pricing package or setting a new market promotion plan. Therefore, the proposed TUVE method is very valuable for telecom operators to assist the network planning, maintenance, optimization, and precise marketing.

REFERENCES

Fanchun Yan, Xiaoming Han, & Weihua Tang. 2003. The extract of expert judging information and the integral method of target scaling in AHP. *Journal of Air Force Engineering University* 4(1): 65–67.

Haibin Shi, Suyang Huang, & Minzheng Huang. 2014. Segmentation of mobile user groups based on traffic usage and mobility patterns. *Proc. IEEE International Conference on Computational Science and Engineering, Chengdu, P.R. China*: 224–230.

Haiming Fu, Lingling Zhang, & Yong Shi. 2007. The application of data mining in mobile subscriber classification. *Proc. International Conference on Fuzzy Systems and Knowledge Discovery, Hainan, P.R. China* 4: 354–358.

Heng Qi. 2014. Telecom customer segmentation model design: a big data perspective. *Master thesis. North China Electric Power University*.

Jiandong Du, & Zhiqiang Zhao. 2011. Dynamic analysis model of mobile Internet customers. *Microcomputer Application* 32(3): 1–6.

Lexi Xu, Xinzhou Cheng, Yue Chen, Kun Chao, Dantong Liu, & Huanlai Xing. 2015. Self-Optimised coordinated traffic shifting scheme for LTE cellular systems. *Proc. EAI International Conference on Self-Organizing Networks, Beijing, P.R. China*: 1–9.

Lexi Xu, Yue Chen, KoK Keong Chai, Dantong Liu, Shaoshi Yang, & John Schormans. 2013. User relay assisted traffic shifting in LTE-Advanced systems. *Proc. IEEE Vehicular Technology Conference, Dresden, Germany*: 1–7.

Lexi Xu, Yuting Luan, Kun Chao, Xinzhou Cheng, Heng Zhang, & John Schormans. 2014. Channel-Aware optimised traffic shifting in LTE-Advanced relay networks. *Proc. IEEE International Symposium on Personal, Indoor and Mobile Radio Communications, Washington, USA*: 1597–1602.

Qiuqing Huang, & Xiongjian Liang. 2005. An analysis of ARPU of telecommunication enterprises. *Journal of Beijing University of Posts and Telecommunications* 7(2): 34–38.

Shu Liu. 2014. The development trend of mobile Internet in 4G era and opportunities and challenges faced by telecom operators. *Guangdong Communication Technology*: 69–72.

Wenbiao Zhang. 2012. Marketing strategy of telecom operators in mobile Internet era. *Electronic World* 17: 6–8.

Xiaofei Tang. 2008. Research on customer behavior for cognitive versus affective loyalty. *China Industry Economics* 3: 101–108.

Ying Xia. 2014. Evaluation and analysis method of mobile Internet user value. *Master thesis. Capital University of Economics and Business*.

Signal and Information Processing, Networking and Computers – Chen & Huang (Eds)
© 2016 Taylor & Francis Group, London, ISBN 978-1-138-02881-4

A novel Big Data based Telecom Operation architecture

Xinzhou Cheng, Lexi Xu, Tao Zhang, Yuwei Jia, Mingqiang Yuan & Kun Chao
China Unicom Network Technology Research Institute, Beijing, P.R. China

ABSTRACT: With the rapid development of big data analysis and cloud computing technology, domestic and foreign telecom operators have started to explore the application of big data in the telecommunication field. How to accelerate the reform of conventional telecom operation architecture, thus realizing the accurate investment and precise operation based on big data has drawn great attention of telecom operators. This paper proposes a novel Big Data based Telecom Operation (BDTO) architecture, containing a series of ideas and algorithms. Then this paper takes the 4G migrated user as the use case, and illustrates the application scenario and the implementation effect of the proposed BDTO architecture.

Keywords: telecom operation; big data; holographic portrait; precise investment

1 INTRODUCTION

1.1 *Opportunity of telecom big data*

Nowadays the telecommunication industry is in the stage of rapid progress (R.A. Becker et al. 2011). The dramatically increased network capacity and transmission rate allow a large number of users to get access to the network with the guaranteed Quality of Service (QoS) (Han Hu et al. 2014). However, the conventional "terminal-channel-cloud" network architecture, which is shown in Figure 1, is difficult to meet these requirements. More specifically, the telecom network is characterized as "channel", which is aimed at delivering multifarious OTT (Over the Top) services from the "cloud" side to the users in the "terminal" side. In this architecture, the cellular networks become the channel to carry the low value-added data.

In the telecom networks, the massive transmitted data contains abundant information, including the network operating status, the details about users and upper layer services. How to make full use of the large volume of telecom big data is key to improve the revenue of telecom operators. Apart from the telecommunication industry, many enterprises in other industries have also realized the value of big data, and have been devoted to deploying their own big data strategy. Compared to these enterprises, telecom big data has dominant advantages, including large user scale, wide geographical scope. Moreover, as for the data quality,

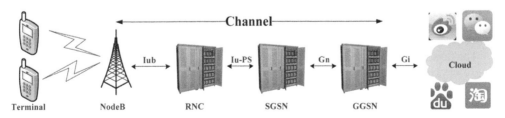

Figure 1. Illustration of the "terminal-channel-cloud" architecture.

it's more comprehensive and precise, which can reach the granularity of anytime, anywhere, anybody and any behavior.

It's generally known that telecom big data brings great opportunity for telecom operators. Telecom engineers are concentrating on developing applications with the assist of big data. This may involve internal applications such as network investment, network optimization etc. In addition, telecom big data can also be used for external applications in order to promote social services. In the era of "Internet+", telecom operators should seize the chance to promote and liquidate big data, thus accelerating the reform of conventional operating architecture and profit model.

1.2 *Challenge of telecom big data*

Telecom big data is characterized as "4-V", which is the abbreviation of "Volume", "Variety", "Velocity" and "Veracity". These four words describe telecom big data from different perspectives, and imply the features of extremely high data volume, distinguished data formats, fast paced data streaming as well as buried data value. How to make full use of the telecom big data to support internal enterprise operation and external multi-dimensional cooperation has become an urgent problem for telecom operators (Han Hu et al. 2014, Karatepe et al. 2014).

To deploy the big data strategy efficiently and effectively, it's really a great challenge to set up a reasonable operation architecture to standardize the implementation of the whole process. This includes data collection, data cleaning, data storage, as well as the ultimate goal of applying telecom data into various fields.

In order to deal with these challenges and effectively utilize telecom big data, this paper presents a novel big data based telecom operation architecture, which contains a series of algorithms. The proposed architecture starts from the user analysis, which is aimed at identifying the characteristics, the resident area, the mainstream services and other preferences of the users. Then based on the full understanding of user behavior, this architecture concentrates on the match degree of the user requirement and the network capability, which will provide reliable reference for network planning, network optimization etc. Moreover, the comprehensive information of users can also benefit market strategies in various industrial fields and social services.

The rest of this paper is organized as follows. Section 2 illustrates the proposed big data based telecom operation architecture, including a series of steps and algorithms. Section 3 takes the 4G migrated user as an example, and presents the implementation effect of the proposed architecture. Section 4 gives the conclusion.

2 BIG DATA BASED TELECOM OPERATION ARCHITECTURE

Telecom operators aim to meet users' diverse QoS requirements. Therefore, the objective of telecom operation is not pursuing the superior Key Performance Indicators (KPIs) of the network. Instead, the objective is to match among three factors, including the user requirement, the service characteristic, and the network capability.

In order to reach this objective, this paper proposes a novel Big Data Based Telecom Operation (BDTO) architecture. The proposed BDTO architecture brings a series of benefits for telecom operators, including the efficient operation strategy adjustment, the effective network evolution, the efficient network planning and optimization etc.

The flow of the proposed BDTO architecture consists of six key steps, including identifying users (1st step), tracking the user-trace (2nd step), capturing the service characteristic (3rd step), the End-to-End perception evaluation (4th step), the network capability correlative adaption (5th step), the implementation and post-evaluation (6th step).

Note that telecom big data is the basis of the proposed BDTO architecture. Telecom operators can obtain telecom big data via a series of methods, including Operation Support System (OSS) collection, S1/Gn/Iu interface collection etc.

2.1 Identify users

The 1st step of the **BDTO** architecture is to identify users, since users are the basis of telecom operation. In this process, the **BDTO** architecture employs telecom big data to draw the 360-degree portrait for each user, in order to be aware of its basic information, its consumption behavior as well as the service behavior.

Figure 2 illustrates the 360-degree portrait for each user, including three factors.

1. The user information includes a series of static attributes, including the age, the gender, the member level, the network type, the user state etc.
2. The user's consumption behavior consists of four sub-factors, including the service character, the consumption character, the trace, and the channel character.
3. The service behavior consists of five sub-factors, including the time preference, the geography preference, the service preference, the terminal preference, and the Application (APP) preference.

On the basis of the 360-degree portrait of each user, the **BDTO** architecture further analyzes the user-group. This user-group analysis aims to search the corresponding group of user (e.g. potential valuable user-group, Iphone terminal user-group, 4G migrated user-group, youth user-group etc.), according to a specific service requirement or a specific character. Typical user-group/clustering analysis algorithms include (Sanqing Hu et al. 2014):

a. K-MEANS: K-MEANS clustering is an algorithms of vector quantization, originally from signal processing, that is popular for cluster analysis in data mining.
b. K-MEDOIDS: The K-MEDOIDS algorithm is a clustering algorithm. The K-MEDOIDS algorithm is partitional (breaking the dataset up into groups) and attempts to minimize the distance between points labeled to be in a cluster and a point designated as the center of that cluster.
c. CLARANS: CLARANS algorithm and Focused-CLARANS algorithm are classical partitioning techniques of clustering, they can be employed in space data base mining.

According to the 360-degree portrait and the user-group analysis, the **BDTO** architecture associates the following four types of factors, including the *'who' relevant factors*, the *'when & where' relevant factors*, the *'what' relevant factors*, and the *'experience' relevant factors*. This process is illustrated in Figure 3.

Overall, the 360-degree portrait and the user-group analysis help telecom operators identify serving users and be aware of the user-group precisely.

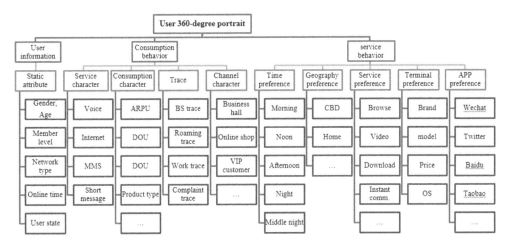

Figure 2. 360-degree portrait for each user.

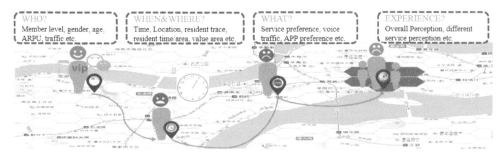

Figure 3. The association of four types of factors.

2.2 Track user-trace

Telecom operation envisages great challenges introduced by the time-varying wireless environment, the user service diversity as well as the user mobility. However, telecom operators should be aware of the user-trace in order to precisely match the user-trace with the user service. Therefore, the 2nd step of the BDTO architecture is to track the user-trace. Specifically, the BDTO architecture correlatively analyses the telecom big data, thus identifying and tracking the user-trace. According to the user-trace, telecom operators can be aware of the user's resident area and the active area as well as the irrelevant area. This can help telecom operators dynamically adjust the network capability and automatically optimize the network.

Note that there are a series of methods to assist telecom operators to obtain user's location, including **GPS/AGPS** based location, WiFi based location, base station based location etc.

Figure 4 illustrates the application of user-trace. Employing the user-trace analysis, the BDTO architecture can obtain the user's active area and the resident area. The proposed architecture further outputs the user-trace map (e.g. resident area map as shown in Figure 4). The BDTO architecture jointly considers the user-trace and service characteristic (e.g. TOP 3 services), and then the BDTO architecture can assist telecom operators to obtain the consumption behavior trace of different user-group, the typical OTT perception. These information benefits the network optimization, the network planning as well as the network construction.

2.3 Capture service characteristic

Due to the user mobility and the service diversity, the service characteristics are time-varying under different time or geography conditions. The typical time-varying service characteristic consists of the service distribution, the service migration as well as the service traffic volume. Therefore, the BDTO architecture captures the service characteristic according to the telecom big data based statistical analysis. In addition, the BDTO architecture also obtains the users' service satisfaction degree under different network configurations.

The mainstream service is identified by three factors: total traffic, user number and visiting times, as shown in Figure 5. Total traffic represents the revenue contribution of a service, and the one consuming more traffic should be guaranteed preferentially. User number is used to evaluate the popularity of a service, high penetration services should be prioritized. The factor of visiting times denotes the activeness of a service. Similarly, highly active services are more valuable.

Figure 6 shows the flowchart of the value area quantization algorithm. In this algorithm, three factors which are service data, terminal data and user data respectively are considered to calculate the value score of each sector.

S1: In this step, the involved data source is collected, parsed and stored for further analysis. This includes the service data, the terminal data and the user data.

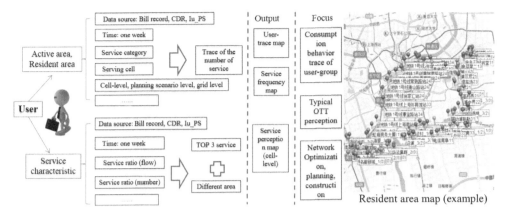

Figure 4.　Diagram of user-trace application.

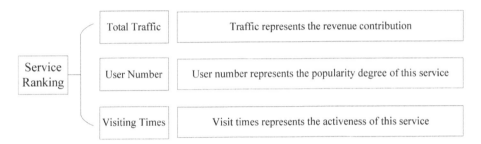

Figure 5.　Factors of service ranking.

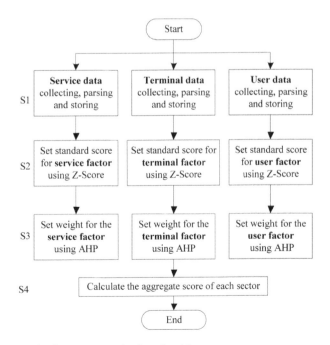

Figure 6.　Flowchart of value area quantization algorithm.

Figure 7.   Presentation of value sectors.

S2: The three factors mentioned above are quantized, and each factor is set as a standard
score using the Z-Score criterion. The following equation is an example of the standard
score calculation for the user factor.

$$U_{user} = \frac{T_{user} - \mu}{\sigma}$$

(1)

where U_{user} is the standard score of the user factor, T_{user} is the actual user number in the
target sector, μ is the mean value of user number in all sectors, and σ is the standard
deviation of user number in all sectors. The calculation of $U_{service}$ and $U_{terminal}$ is almost the
same as U_{user}, and it's omitted here for simplicity.

S3: The AHP method is employed in this step to determine the weight for each factor.

S4: The aggregated score of each sector is calculated.

$$Value = [W_{user}, W_{service}, W_{terminal}] \times [U_{user}, U_{service}, U_{terminal}]^T$$

(2a)

$$= W_{user} \times U_{user} + W_{service} \times U_{service} + W_{terminal} \times U_{terminal}$$

(2b)

where W_{user}, $W_{service}$ and $W_{terminal}$ are the weights of the user factor, the service factor and
the terminal factor respectively. U_{user}, $U_{service}$ and $U_{terminal}$ are the standard scores of the
three factors respectively.

Through the algorithm in Figure 6, the value of each sector is quantized, and it can also be
visually presented on the map, as shown in Figure 7. This can be used as reference for applications such as network planning, network optimization etc.

2.4   End-to-End perception evaluation

Based on the signaling tracking and the service analysis, the BDTO architecture obtains
each user's End-to-End perception under different scenarios, including different time, different areas as well as different service application scenarios. Based on each user's End-to-End

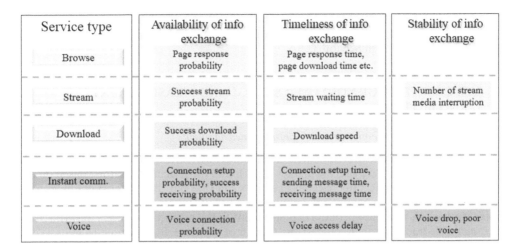

Service type	Availability of info exchange	Timeliness of info exchange	Stability of info exchange
Browse	Page response probability	Page response time, page download time etc.	
Stream	Success stream probability	Stream waiting time	Number of stream media interruption
Download	Success download probability	Download speed	
Instant comm.	Connection setup probability, success receiving probability	Connection setup time, sending message time, receiving message time	
Voice	Voice connection probability	Voice access delay	Voice drop, poor voice

Figure 8. Diagram of End-to-End perception evaluation.

perception, for a specific service under a time period, the BDTO architecture can also obtain the overall End-to-End perception of all users in a specific area (e.g. a specific cell, application scenario, grid, network type etc.).

From the analysis above, each user's End-to-End perception is the basis of overall End-to-End service perception. Furthermore, overall End-to-End service perception benefits telecom operators, especially the network planning and the network optimization.

Figure 8 illustrates the diagram of End-to-End perception evaluation. For each type of service, the End-to-End perception evaluation considers three aspects, including the availability of information exchange, the timeliness of information exchange, and the stability of information exchange.

2.5 Network capability correlative adaption

After above four steps analysis, telecom operators obtain the detailed information of a series factors, including the user, the user-trace, the time, the location, the service, and the End-to-End perception etc. Meanwhile, telecom operators also obtain the correlation and the match results among these factors.

In the 5th step, telecom operators employ the BDTO architecture to obtain the network capability correlative adaption, together with user's requirement and service perception. Under the efficient network capability correlative adaption, telecom operators can provide effective resource investment and optimization scheme. Telecom operators can also maximize the investment efficiency, finally reducing the capital expenditure (CAPEX) and the operational expenditure (OPEX) (R.A. Becker et al. 2011 and Chien-Lung Lee et al. 2014).

2.6 Implementation and post-evaluation

After implementing the network adjustment and the resource investment scheme, the BDTO architecture re-inputs the corresponding factors' data, including the user data and the terminal data as well as the network operation data.

The data re-input process has two objectives. The first is to post-evaluate the performance of the network adjustment and that of the resource investment scheme. The second is to provide updated data for both the future network adjustment and the new round resource investment. In this way, the BDTO architecture can achieve the real-time/normal analysis, thus improving the efficiency and the accuracy of telecom operation.

After introducing the BDTO architecture in Section 2, this paper employs the 4G migrated user as the use case, and illustrates the application scenario and the implementation effect of the BDTO architecture in Section 3.

3 USE CASE OF MIGRATED USER SAFEGUARD MEASUREMENT USING BIG DATA

3.1 General idea of 4G migration safeguarding measurements

The potential 4G users will contribute a lot to the growth of the services traffic in the near future. These 4G users are also the market focus for all the telecom operators. Based on our proposed BDTO architecture, telecom operators can find out the potential 4G users from the existing 2G or 3G users precisely to reduce the cost of sales and improve the sales success rate. Meanwhile, telecom operators can also identify the potential 4G users' resident areas through user-trace and then forecast the growth of traffic if they migrate to 4G services. This can help match the deployment of the network resources with the increasing traffic growth. In this way, telecom operators can achieve the objective of matching the above mentioned three factors, including the user requirement, the service characteristic, and the network capability during the process of 4G migration. The overall migration idea is listed in Figure 9.

3.2 Characteristics of 4G migrated users and the potential 4G user mining

On the basis of each user's 360-degree portrait in China's four different cities, we cluster the 4G migrated user-group. From different points of view, we analyze the telecom big data to find out the characteristics of 4G migrated user group, such as the user level, the user age, the user in-network time, DoU (Data of Usage), ARPU (Average Revenue Per User), MoU (Minutes of Usage), user terminal etc. Based on the extracted characteristics, we can obtain 4G migration user model and dig out the potential 4G users from the existing 2G or 3G users precisely.

We analyze the characteristics of migrated 4G users from different perspectives, including user basic attribute, user consumption and user terminals. The analyzing results are shown in Figure 10.

As shown in Figure 10, we can conclude 4G migrated users' characteristics: Firstly, result (a) shows that almost half of the migrated users are at the age of 20–30 years old. And result (b) reveals the users with long in-network time are more willing to migrate to 4G network comparing with the new users. Secondly, results in (c), (d), (e) show the services consumption of the migrated 4G users. Relatively speaking, the migrated 4G users' services usage and services consumption are at a relatively high level. Thirdly, from the view of the user terminals in (f), we can see that users with the high value terminal are more willing to migrate to 4G network.

Based on the obtained characteristics, we apply our 4G migrated model to dig out the potential 4G users from the existing 2G or 3G users. For different user characteristics, in order to make the user migration policy more effective, we divide 4G potential users into two groups: 4G potential users with 4G terminal, users without 4G terminal. The procedure is shown in Figure 11.

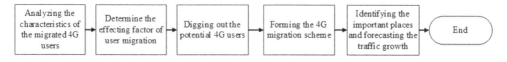

Figure 9.   Flowchart of user migration scheme.

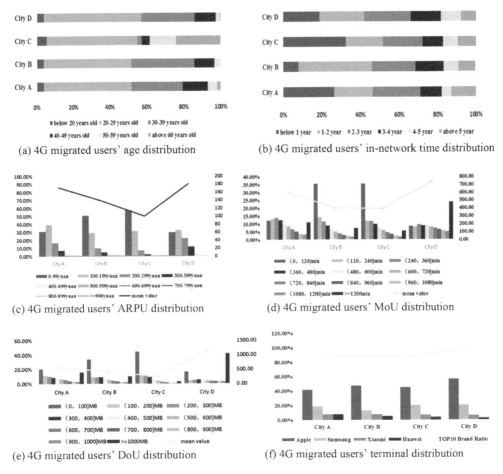

(a) 4G migrated users' age distribution

(b) 4G migrated users' in-network time distribution

(c) 4G migrated users' ARPU distribution

(d) 4G migrated users' MoU distribution

(e) 4G migrated users' DoU distribution

(f) 4G migrated users' terminal distribution

Figure 10.   The characteristics of 4G migrated users.

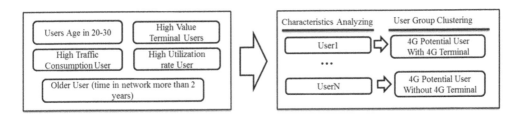

Figure 11.   The procedure of 4G potential user mining.

We take the result of City A as an example. According to the different potential 4G user groups, we propose different migration schemes. For the 4G potential user group with 4G terminal, there are two methods. One method is to open the 4G network license to allow users to camp on 4G network, another method is to encourage these people to purchase the 4G service product. On the contrary, for the 4G potential user group without 4G terminal, one method is also to recommend the 4G service product based on their historical consumption records and another way is to find out the potential 4G mobile phone customers and urge them to adopt 4G terminals. After applying above schemes in City A, we obtain the corresponding users' information under different schemes, as shown in Table 1.

Table 1.   4G migrated user number.

User Origin	4G potential users with 4G terminal	4G potential users without 4G terminal	Total numbers
2G User	21,000	87,000	108,000
3G User	69,000	266,000	335,000

(a) Geographic information of resident area

(b) Enlarged view of downtown resident area

Figure 12.   Geographical distribution of potential 4G users.

3.3   Network capability correlative adaption

3.3.1   Identifying the resident area of potential 4G users

Based on the user-trace and potential 4G user list, we can identify the resident area of potential 4G users. It will provide the location information for the network adjustment, which are resulted from the potential 4G migrated users. Figure 12 presents the geographical distribution of potential 4G users. The result shows that most of the potential 4G users are located in the downtown area.

3.3.2   Service model forecasting of potential 4G users

This sub-section starts from the migrated 4G users' traffic changes before and after network migration. According to the factors of ARPU and DoU, we establish a mapping table of traffic growth value. For different scopes of ARPU and DoU values, we can calculate an average traffic growth rate. By applying this multi-elements linear regression algorithm, we can forecast each potential 4G user's monthly traffic volume after they migrate to the 4G network. The forecasting result is shown in Table 2. We anonymize the user identity and only present a typical example of final result.

3.3.3   Network capability correlative adaption

According to the forecasting result in Table 2 and the resident area in Figure 12, we equally assign the traffic load to the resident area. Firstly, we present the heating map of traffic growth, as shown in Figure 13(a). From Figure 13(a), we can observe the high traffic volume growth location. Secondly, we evaluate each user's End-to-End perception in his resident area and the geographical visualization, as shown in Figure 13(b). We divide the End-to-End perception into four levels, including the high level, the medium level, the normal level and the low level. The overall End-to-End perception evaluation results can determine the direction of network adjustment, network optimization and network planning.

Table 2. 4G migrated user traffic forecasting.

User ID	Traffic volume before migration (MB)	Traffic volume forecasting (MB)
1	126.2001	505.561
2	150.638	419.676
3	221.487	687.939
4	634.178	713.451
–	–	–

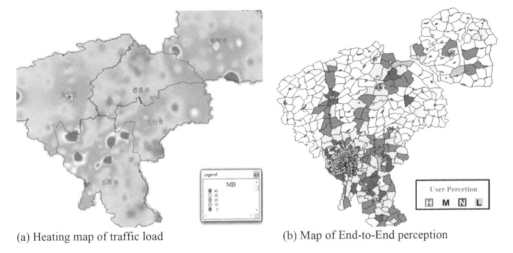

(a) Heating map of traffic load   (b) Map of End-to-End perception

Figure 13. Geographical information of traffic load and End-to-End perception.

Following the above mentioned steps, we realize the great progress in two aspects, compared with the traditional ways. On one hand, we reduce the number of sale-target users by 84%. This can reach the objective of precise marketing. On the other hand, we achieve the purpose of precise investment, and we have saved 12% of the investment of the wireless network construction.

4 CONCLUSIONS

In order to deploy the big data strategy efficiently and effectively, this paper presents a novel big data based telecom operation architecture containing a series of algorithms. It starts from the analysis of users, which is aimed at identifying the characteristics, the resident area, the mainstream services and other preferences of the users. Then based on the full understanding of user behavior, this architecture concentrates on the match degree of the user requirement and the network capability, which will provide reliable reference for network planning, the network optimizing etc. Moreover, the comprehensive information of users also benefits market strategies in various industrial fields as well as a series of social services.

REFERENCES

Becker, R.A., Caceres, R. & Hanson, K. 2011. A tale of one city: using cellular network data for urban planning. *IEEE Pervasive Computing* 10(4): 18–26.

Chien-Lung Lee, Wen-Shu Su, Kai-An Tang, & Wei-I Chao. 2014. Design of handover self-optimization using big data analytics. *Proc. IEEE Asia-Pacific Network Operations and Management Symposium, Hsinchu, China:* 1–5.

Han Hu, Yonggang Wen, Tat-Seng Chua, & Xuelong Li. 2014. Toward scalable systems for big data analytics: a technology tutorial. *IEEE Access* 2: 652–687.

Karatepe, Ilyas Alper, Zeydan, & Engin. 2014. Anomaly detection in cellular network data using big data analytics. *Proc. Verband Deutscher Elektrotechniker European Wireless, Barcelona, Spain:* 1–5.

Sanqing Hu, Ye Ouyang, Yu Dong Yao, Fallah, M.H. & Wenyuan Lu. 2014. A study of LTE network performance based on data analytics and statistical modeling. *Proc. IEEE Wireless and Optical Communication Conference, Newark, USA:* 1–6.

Signal and Information Processing, Networking and Computers – Chen & Huang (Eds)
© 2016 Taylor & Francis Group, London, ISBN 978-1-138-02881-4

Mining characteristics of a 4G cellular network based on big data analysis

G. Li
China Mobile Group Design Institute Ltd. Co., Longjiang, Harbin, China

Q. Fan
Laboratory, China Shipbuilding Industry Corporation, Harbin, China

ABSTRACT: Understanding the characteristics of cellular networks is essential in optimizing network performance and improving user experience. Characteristics of radio Resource Margins (RM) for detailed measurement analysis of spectrum efficiency proposed in Feng Z. et al. (2014), and characterization of the geospatial and temporal dynamics of application usage proposed in Zubair Shafiq M. et al. (2015), are both proved important for cellular network operators handling the explosive growth in the traffic volume in a 4G cellular data network. Therefore, we investigate the two kinds of characteristics using data collected from a cellular network operator in Harbin. First, we test the characteristics of RM include the cells' temporal skewness, diurnal patterns, weekly periodicity and spatial skewness. Second, we cluster cell locations based on their application distributions and study the geospatial and temporal dynamics of application usage across different geographical regions. Finally, the results we obtained have important implications in terms of network design and optimization.

1 INTRODUCTION

Cellular network operators have globally observed an explosive increase in the volume of data traffic in recent years in Feng Z. et al. (2014). New techniques push the development of the network connection speed, smartphones, which make the volume of global cellular data traffic increasing greatly these years, and therefore, cellular network operators suffering. Although 5G will be used as the communication technology, it is no doubt that cellular network operators need to propose creative method for optimizing the network.

Paper Feng Z. et al. (2014) proposed mining the characteristics of spectrum utilization is essential in providing guidelines for resource allocation. As well, they were inspired by the inefficient utilization of radio resources, and then devised an optimization scheme for dynamic radio resources reconfiguration and experimental results proving that it improves radio resources utilization efficiency and traffic load balance significantly. The work is different from others in Willkomm D. et al. (2008), Paul U. et al. (2011), Shafiq M. Z. et al. (2011), Qian F. et al. (2010) in the application of RM.

Several studies have examined cellular network data traffic in Falaki H. et al. (2010), Huang J. et al. (2010), Orstad B.M. et al. (2006), Paul U. et al. (2011), Shafiq M.Z. (2011), Trestian I. et al. (2009), Wittie M.P. et al. (2007), Xu Q. et al. (2011). However, the study of the relationship between application usage and location is just investigated these days. Paper Zubair Shafiq M. et al. (2015) first jointly characterized the geospatial and temporal dynamics of application usage in a 3G cellular data network. The authors argue that significant geospatial and temporal correlations, in terms of traffic volume and application access, provide local optimization opportunities to cellular network operators for handling the explosive growth in the traffic volume.

As a result, our focus in this paper is to provide deep investigations with data collected from a large cellular network in Harbin. Our goal is to show the implications on two important questions: (1) to what extent radio resources in cellular networks are utilized currently and (2) how to utilize them more fairly and efficiently.

The rest of the paper is organized as follows. Background and data are described in Section II. Methods are presented in Section III. Section IV concludes the paper.

2 BACKGROUND AND DATA

As they did in paper Feng Z. et al. (2014), Zubair Shafiq M. et al. (2015), anonymized data sets are used in this paper. The data are generated by thousands of base stations, which serve about 2 million customers. The data, in detail, offer us the information: (1) flow-level information of IP traffic carried in PDP Context tunnels, (2) the following information for each IP flow per minute, (3) the hourly aggregate voice and data traffic of each cell from Jun 1 to Jun 7 in 2012, (4) the daily peak-hour aggregate voice and data traffic for each cell from Jun 1 to Jun 7 in 2012, (5) the carrier configuration and Traffic Channel (TCH) allocation for each cell.

3 METHODS

In this section, we present the two methods proposed in paper Feng Z. et al. (2014) and Zubair Shafiq M. et al. (2015).

3.1 Data processing

Based on the Erlang-B formula, we compute the number of demanded TCHs, which is essential for obtaining each cell's radio resource margins.

According to the test result based on the statistical methods, the call lose rate, denoted as B, is defined as 0.01 in this paper. The traffic load in our data set is the actually accomplished traffic of one cell i in one specific hour t, denoted as X_{it}, so the demanded number of TCHs of that cell in the corresponding hour, denoted as C_{it}, can be calculated by solving the Erlang-B formula [Chen H. & Yao D. D. 2001] below.

$$B = \frac{\left(X_{it}/(1-B)\right)^{C_{it}}/C_{it}!}{\sum_{k=0}^{C_{it}}\left(X_{it}/(1-B)\right)^{k}/k!} \tag{1}$$

For each cell i, its radio resource margins (RM) is denoted as $\{RM_{it}\}$ and computed as

$$RM_{it} = N_i - C_{it} \tag{2}$$

where $\{C_{it}\}$ is the demanded number of TCHs time series, instead of $\{X_{it}\}$, N_i is the total available TCHs. If RM_{it} is positive, Cell i owns spare radio resources at time t.

3.2 Dynamics of radio resource margins

3.2.1 Temporal characteristics
The temporal patterns of radio RM is analyzed firstly by plotting their dynamics in time series. Then the hourly RM of two typical cells and the average hourly RM is showed over the whole network in one week. In the midnight, radio RM exhibits strong diurnal patterns with more RM, and less in the day. The action is accord with the circadian rhythm of human activities. Cell #2 is taken as an example, when people are awakening and active in the day,

398

the cell is short of radio resources, and is sufficient in the midnight when people are taking a rest. The emphasis is Cell#2 is short of radio resources while Cell #1still has a few idle TCHs during the peak hours. In order to explore the long-term periodicity of radio RM, its Autocorrelation Function (ACF) has to be analyzed. ACF is defined as the cross-correlation of the time series itself with different lags, which is estimated as follows.

$$\rho_X(k) = \rho_X(X_n, X_{n+k}) \approx \frac{\frac{1}{N-|k|}\sum_{i=1}^{N-|k|}(X_i - \bar{X})(X_{i+|k|} - \bar{X})}{\frac{1}{N}\sum_{i=1}^{N}(X_i - \bar{X})^2} \qquad (3)$$

where k is the lag of time slots, N is the length of X_n and $\bar{X}$ is the mean of X_n.

The ACF on network-aggregated radio RM is investigated and traffic loaded during daily peak hour, 2012. The two ACF curves cohere with each other very well, which demonstrate the linear relation between RM and traffic load. The fact that a strong negative linear correlation exists between RM and traffic load is consistent with equation (2).We have to point out that the two ACF curves have strong temporal correlation with peaks in every week, and a slow decreasing trend with peaks exist in the rest days. Eight days later, the fact that the curve declines under the empirical threshold 0.3 indicates whether or not correlation exists Correlation Coefficient. In conclusion, the daily peak-hour RM and traffic load are both correlated with their corresponding records in the previous 8 days, which is useful in predict their long-term dynamics.

At last, the temporal skewness of radio RM is investigated, which shows that the low-level hour of radio RM consistent to the peak hour of traffic load on account of the strong negative correlation between them. Statistics suggest that the low-level hour distribution in one day resembles the results of a whole year, which prove the diurnal patterns of RM once more. The temporal diversity of radio RM with low-level hour of RM spanning from 7 AM to 23 PM rather than just focusing in 1 or 2 hours are also be proved. Besides, almost 70% cells are abundant and 30% are short of radio resources in the peak hour of a day. The analyses show that it is quite necessary to recofigure the radio resources over the whole network dynamically, and the radio resources are not utilized efficiently temporally.

3.2.2 Spatial characteristics

We analyze the spatial skewness of radio RM over the whole network using Voronoi cells, to begin with. Each Voronoi cell corresponds to a cell's geographical coverage region, and a negative number indicates that the cell is short of TCHs in peak hour and is more inclined to be blocked. Radio RM is fairly skewed spatially. On one hand, some cells are heavily burdened and starve for radio resources, though most of the cells have fallow resources. On the other hand, the cells in suburban region are more inclined to have idle resources for that cells in urban region are in the contrary situation. We can also see that spatially neighboring cells have similar spectrum utilization patterns from the statistics. Further analysis is performed to quantify to what extent radio RM are skewed spatially. The cells are categorized into 4 classes manually with respect to the average RM ratio in daily peak hour of 2012 using empirical thresholds. For instance, we can look the first interval as the cell lacking in radio resources. We analyze the historical statistics of the four categories. As we can get only 7% of the cells are short of radio resources with each cell in want of 8 TCHs on average during daily peak hour. One fifth of the cells can barely satisfy traffic load while over 70% of the cells can contribute dozens of TCHs to other cells to ease their traffic load. This discovery affirm that radio resources are not utilized efficiently spatially. Clustering coefficient, a metric defined on Cumulative Distribution Function (CDF) curve and ranging from 0 to 1, quantifies to what extent radio resources margins in each category are skewed. The area between the diagonal line and the CDF curve $f(x)$ is denoted as A, the area between $f(x)$ and x-axis is denoted as B, and clustering coefficient, denoted ash, is defined as follows.

$$h = \frac{A}{A+B} = 1 - \frac{B}{A+B} = 1 - 2B = 1 - 2\int_{0}^{1} f(x)dx \qquad (4)$$

Radio RM are distributed over the whole network uniformly with A equals 0, h equals 0. Radio RM are concentrated in only one cell. With B equals 0, h equals 1.

As is shown in the statistics, all of the four categories' clustering coefficients are quite large, which means that the concentration of radio RM in few cells. For instance, 20% of the cells in Light-loaded category account for about 85% of the total radio RM in that category. The spatial skewness of radio RM and the inefficient utilization of radio resources spatially are proved once more if this is the case.

3.3 Aggregate measurement analysis

Temporal analysis, and geospatial analysis are discussed in this section.

3.3.1 Temporal analysis

For the sake of clarity, we investigate the data collected during 24 hours on Jun 7 2012 in terms of aggregate traffic volume and traffic volume for different Apps.

Aggregate traffic would be in terms of byte, packet, flow, and user counts. As reported in prior literature in Paul U. et al. (2011), Shafiq M. Z. (2011), Trestian I. et al. (2009), Wittie M. P. et al. (2007), Xu Q. et al. (2011), we observe a strong diurnal behavior in aggregate traffic. What's more, two peaks are observed. One of both can be different from the study result shown in Trestian I. et al. (2009), Xu Q. et al. (2011). The other is around mid-night, which might reflect users' peculiar activity patterns in the metropolitan area studied in this paper.

Temporal dynamics of traffic is investigated by four Apps in terms of byte, packet, flow, and user counts. Interestingly, strong diurnal characteristics in temporal dynamics are found as above. However, differences across these Apps are shown.

3.3.2 Geospatial analysis

In this section, based on the data we collected, each App's relative popularity across different cells is investigated by calculating the cumulative distribution function of traffic volume of dating, maps, social network, and web browsing Apps with respect to byte, packet, flow, and user counts.

We find people in different cells use different Apps. For instance, all traffic volume of dating App is generated from less than 5 percent of all cells. Furthermore, web browsing is the most ubiquitous application realm. However, even for web browsing, 80 percent of the byte traffic volume is generated from 50 percent of all cells.

4 RESULTS AND CONCLUSION

Until now we have established five major findings: (1) radio resources are not utilized efficiently both temporally and spatially; (2) radio RM and traffic load are negatively and linearly correlated, and both are predictable due to their strong weekly periodicity; (3) traffic volumes of applications exhibit strong diurnal characteristics, (4) the popularity of a given application realm varies across different cell locations, and (5) the traffic volume of a few application realms dominate others overall. These findings suggest the directions for bettering network design and optimization.

REFERENCES

Chen H. & Yao D.D. 2001. Fundamentals of queuing networks: Performance, asymptotics, and optimization. *Springer*.

Correlation Coefficient, http://en.wikipedia.org/wiki/Pearson product moment correlation coefficient.

Falaki H., Mahajan R., Kandula S., Lymberopoulos D., Govindan R. and Estrin D. 2010. Diversity in smartphone usage. In *Proc. 8th Int. Conf. Mobile Syst., Appl, Serv.*, 179–194.

Feng Z., Min J., Yan X., Gao Y., Zhang Q. & Zhang Y. 2014. Characterizing and exploiting temporal-spatial radio resource margins in cellular networks. In *proceedings of IEEE 80th Vehicular Technology Conference.*

Huang J., Xu Q., Tiwana B., Mao Z.M., Zhang M., & Bahl V. 2010. Anatomizing application performance differences on smartphones. In *Proc. 8th Int. Conf. Mobile Syst., Appl., Serv.*, 165–178.

Orstad B.M. and Reizer E. 2006. End-to-end key performance indicators in cellular networks. *Master's thesis. Faculty Eng. Sci., Agder Univ. College*, Norway.

Paul U., Subramanian A.P., Buddhikot M.M. & Das S.R. 2011. Understanding traffic dynamics in cellular data networks. In *Proc. IEEE Conf. Comput. Commun.* 882–890.

Paul U., Subramanian A.P., Buddhikot M.M. & Das S.R. 2011. Understanding traffic dynamics in cellular data networks. In *Proc. of IEEE INFOCOM.*

Qian F., Wang Z.G., Gerber A., Mao Z.M., Sen S. & Spatscheck O. 2010. Characterizing radio resource allocation for 3G networks. In *Proc. of ACM IMC.*

Shafiq M.Z., Ji L., Liu A.X. & Wang J. 2011. Characterizing and modeling internet traffic dynamics of cellular devices. *In Proc. of ACM SIGMETRICS.*

Trestian I., Ranjan S., Kuzmanovic A. & Nucci A. 2009. Measuring serendipity: Connecting people, locations and interests in a mobile 3G network. In *Proc. 9th ACM SIGCOMM Conf. Internet Meas. Conf.*, 267–279.

Willkomm D., Machiraju S., Bolot J. & Wolisz A. 2008. Primary users in cellular networks: A large-scale measurement study. In *Proc. of IEEE DySPAN.*

Wittie M.P., Stone-Gross B., Almeroth K. & Belding E. 2007. MIST: Cellular data network measurement for mobile applications. In *Proc. IEEE 4th Int. Conf. Broadband Commun., Netw. Syst.* 743–751.

Xu Q., Gerber A., Mao Z.M., Pang J. & Venkataraman S. 2011. Identifying diverse usage behaviors of smartphone apps. In *Proc. ACM SIGCOMM Conf. Internet Meas.* 329–344.

Zubair Shafiq M., Ji L., Liu Alex X., Pang J. &Wang J. 2015. Geospatial and temporal dynamics of application usage in cellular data networks. *IEEE transactions on mobile computing*, 2015: 14(7), 1369–1381.

Signal and Information Processing, Networking and Computers – Chen & Huang (Eds)
© 2016 Taylor & Francis Group, London, ISBN 978-1-138-02881-4

Author index

An, S. 203

Cao, L. 35
Cao, L. 289, 299
Cao, X. 249, 261
Chao, K. 249, 289, 299,
  357, 367, 375, 385
Che, H. 99, 195
Chen, D. 27
Chen, J. 59
Chen, N. 69
Chen, W. 281, 367
Chen, X. 27
Chen, Y. 77
Cheng, C. 315, 337
Cheng, D. 131, 185
Cheng, X. 249, 261, 271,
  289, 299, 307, 315,
  329, 337, 347, 357,
  367, 375, 385

de Boer, A. 35

Fan, Q. 397
Feng, H. 175
Fu, C. 157

Gao, J. 249, 289
Ge, H. 43, 209
Guan, J. 315, 347
Guo, J. 175
Guo, X. 203
Guo, Y. 43, 203
Guo, Y. 69, 123

Han, X.N. 115
He, P. 59
Hou, F. 9
Hu, B. 175
Huang, H. 3, 9, 17,
  27, 77, 91, 147, 185

Huang, T. 69
Huang, Y. 43, 209
Huang, Y. 167

Jia, Y. 249, 307, 329,
  367, 385
Jiang, H. 185
Jing, X. 3, 9, 17, 27,
  77, 91
Jing, X.J. 131, 185

Li, B. 109
Li, C. 147
Li, G. 397
Li, J. 209
Li, M. 243
Li, R. 83
Li, S. 43, 203, 209
Li, Y. 3
Li, Y. 131, 185
Liu, W. 147
Liu, X. 329
Lu, D. 139
Lu, M. 157
Luan, Y. 249

Ma, Z. 223
Machu, Q. 99
Mao, C. 35
Mu, M. 281, 347, 375

Qian, C. 139

Ren, P. 357
Rong, B. 83

Shi, F. 35
Shi, J. 115
Song, C. 271, 307, 329
Sun, J. 139
Sun, M. 109

Sun, Q. 35
Sun, S. 3, 17, 27, 69,
  123, 131, 147,
  185, 223
Sun, X. 233
Sun, Z. 35

Tian, T. 109

Wang, B. 35
Wang, J. 215
Wang, L. 17
Wang, P. 375
Wang, S. 249
Wang, X. 27, 131
Wang, X. 185
Wang, Y. 49
Wang, Y. 261, 281, 347
Wang, Z. 195
Wu, D. 375

Xia, X. 35
Xing, Y. 233
Xu, J. 243
Xu, L. 249, 261, 289,
  315, 347, 357,
  375, 385
Xu, S. 83

Yan, H. 167
Yan, Y. 43, 203, 209
Yang, D. 49
Yang, D.C. 215
Yang, H. 83
Yang, T. 175
Ye, X. 131
Ye, Z. 91
Yuan, H. 59
Yuan, J. 243
Yuan, M. 261, 299,
  315, 337, 357, 385

Zhang, H. 271, 307
Zhang, J.Q. 215
Zhang, L. 271
Zhang, M. 233
Zhang, S. 35
Zhang, T. 261, 315,
    347, 367, 385
Zhang, X. 49

Zhang, X. 223
Zhang, X.X. 215
Zhang, Z. 233
Zhao, C. 59, 167
Zhao, C. 109
Zhao, H. 35
Zheng, Y. 167
Zhou, Q. 157

Zhou, S. 315, 337
Zhou, Y. 99
Zhou, Y. 299
Zhou, Z. 69, 233
Zhu, J. 123
Zhu, X.P. 215